Frontiers in Bioactive Compounds

(Vol. 1)

Natural Sources, Physicochemical Characterization and Applications

Edited by:

Constantin Apetrei

Department of Chemistry, Physics and Environment
Faculty of Sciences and Environment
"Dunarea de Jos" University of Galati
47 Domneasca Street
800008 Galati, Romania

Bentham Science Publishers Ltd.
Executive Suite Y - 2
PO Box 7917, Saif Zone
Sharjah, U.A.E.
Email: subscriptions@benthamscience.org

BENTHAM
SCIENCE

CONTENTS

FOREWORD

The field of bioactive compounds is very important for both scientific and applicative point of view. This field is interdisciplinary and in this book the authors propose several themes related to chemical composition, sensors, biosensors, electronic sensory systems, interaction modeling and protection methods applied on bioactive compounds. Therefore, the authors propose an integrated approach regarding bioactive compounds including sources, physicochemical detection, interactions with membranes and the protection of bioactive components against environmental factors.

This integrated approach is not common in books but I believe that is very useful for specialists and researchers working in inter-, multi- and trans-disciplinary fields related to bioactive compounds. The book could be of interest for scientists specialized on chemistry, biochemistry, pharmacy as well as for persons working in food industry, biotechnology, nanotechnology etc. This book will be helpful for a wide range of readers - from university students to researchers.

I think that the authors can be confident that there will be many grateful readers who will have gained a broader perspective of the fields of bioactive compounds as a result of their efforts.

Dr. Eleonora-Mihaela UNGUREANU
Department of Inorganic Chemistry
Physical Chemistry and Electrochemistry
Faculty of Applied Chemistry and Materials Science
Polytechnic University of Bucharest
Romania

PREFACE

This book presents the composition and health benefices of bioactive compounds from natural products, and the principal methods used for separation, detection, quantification and protection of bioactive compounds. A special attention is devoted to membrane model interactions as well as to sensors and biosensors development.

The book is organized in the following way:

Chapter 1 deals with the description of chemical composition of olive oil, highlighting the chemical structure and health benefices of biologic active compounds present in this food.

Chapter 2 describes the composition, chemical structures and benefices of bioactive compounds from red wine. A special attention is devoted to polyphenolic compounds, especially to resveratrol, a compound with multiple benefices and without toxic effects.

Chapter 3 presents the recent advances in the analysis of bioactive compounds based on molecular recognition. Separation techniques (solid phase extraction, chromatography, electrophoresis), and sensing (electrochemical, optical, *etc.*) are revised.

Chapter 4 describes the electrochemical sensors, especially voltammetric sensors, dedicated to the analysis of antioxidants. The principal methods used to improve voltammetric sensors selectivity and sensitivity by modifying the surface using chemical or biological materials are described, also.

Chapter 5 presents the most representative and novel biosensors for the detection of polyphenols, biogenic amines, mycotoxins and allergens in wine, emphasizing the wealth of biosensing strategies and the particularities of wine as a complex analytical matrix.

Chapter 6 provides a review of the most recent works in the field of electronic sensory system for characterization of bioactive compounds, from electronic noses to sensors and biosensors.

Chapter 7 describes the use of Langmuir monolayers as cell membrane models made to interact with bioactive compounds. Trends in the field are discussed as motivation for innovative research to bridge the gap between biology (and medicine) and the physicochemical framework used in cell membrane modeling.

The book closes with Chapter 8 presenting the newest information about the bioactive compounds encapsulation, which ensures the stability of the volatile compounds during thermal processing, the protection of bioactive components against environmental factors, the increase of the solubility of bioactive compounds, and the controlled release of bioactive components.

I appreciatively acknowledge the distinguished specialists who contributed in this book. I consider myself privileged to have had the occasion to work together with colleagues of universities from Spain, Italy, Brazil and Romania. Finally, I want to thank Bentham Science Publishers for the opportunity to edit this book.

Dr. Constantin APETREI
Department of Chemistry, Physics and Environment
Faculty of Sciences and Environment
"Dunarea de Jos" University of Galati
47 Domneasca Street
800008 Galati, Romania
Phone: +40-0236460328
Fax: +40-0236461353
E-mail: apetreic@ugal.ro

List of Contributors

Alina Vasilescu	International Centre of Biodynamics, 1B Intrarea Portocalelor, sector 6, 060101, Bucharest, Romania
Bernardo C. de Gennaro	Dipartimento di Scienze Agro Ambientali e Territoriali, Università di Bari, *via* Orabona 4, Bari, Italy
Bogdan Bucur	Centre of Bioanalysis, National Institute of Research and Development for Biological Sciences, 296, Splaiul Independentei, Bucharest, Romania
Carlo Franchini	Dipartimento di Farmacia - Scienze del Farmaco, Università di Bari, *via* Orabona 4, Bari, Italy
Cecilia Cristea	Analytical Chemistry Department, "Iuliu Haţieganu", University of Medicine and Pharmacy, Cluj-Napoca, Romania
Celia García-Hernández	Department of Inorganic Chemistry, Engineers School, Universidad de Valladolid, Valladolid, Spain
Constantin Apetrei	Department of Chemistry, Physics, and Environment, "Dunarea de Jos", University of Galati, 47 Domneasca Street, Galati, Romania
Cristian Dima	Faculty of Food Science and Engineering, "Dunarea de Jos", University of Galati, Romania
Cristina García-Cabezón	Department of Materials Science, Engineers School, Universidad de Valladolid, Valladolid, Spain
Cristina Medina-Plaza	Department of Inorganic Chemistry, Engineers School, Universidad de Valladolid, Valladolid, Spain
Daniel Matatagui	Centro de Ciencias Aplicadas y Desarrollo Tecnológico - UNAM, México D.F
David Paniagua-Albillos	Department of Inorganic Chemistry, Engineers School, Universidad de Valladolid, Valladolid, Spain
Ede Bodoki	Analytical Chemistry Department, "Iuliu Haţieganu", University of Medicine and Pharmacy, Cluj-Napoca, Romania
Felippe J. Pavinatto	São Carlos Institute of Physics, University of São Paulo, São Carlos, São Paulo, Brasil
Filomena Corbo	Dipartimento di Farmacia - Scienze del Farmaco, Università di Bari, *via* Orabona 4, Bari, Italy
Ioana Silvia Hosu	National Institute for R&D in Chemistry and Petrochemistry, Splaiul. Independentei 202, Bucharest, Romania Institute of Electronics, Microelectronics and Nanotechnology (IEMN), UMR-CNRS 8520, Université Lille 1, Avenue Poincaré—BP 60069, Villeneuve d'Ascq, France

Jesús Lozano Escuela de Ingenierías Industriales, Universidad de Extremadura, Badajoz, Spain

José Antonio de Saja-Sáez Department of Condensed Matter Physics, Faculty of Sciences, Universidad de Valladolid, Valladolid, Spain

José Pedro Santos GRIDSEN, Instituto de Tecnologías Físicas y de la Información (ITEFI-CSIC), Madrid, Spain

Luciano Caseli Institute of Environmental, Chemical, and Pharmaceutical Sciences, Federal University of São Paulo, Diadema, São Paulo, Brasil

Ma Carmen Horrillo GRIDSEN, Instituto de Tecnologías Físicas y de la Información (ITEFI-CSIC), Madrid, Spain

Maria Lisa Clodoveo Dipartimento di Scienze Agro Ambientali e Territoriali, Università di Bari, *via* Orabona 4, Bari, Italy

María Luz Rodríguez-Méndez Department of Inorganic Chemistry, Engineers School, Universidad de Valladolid, Valladolid, Spain

Osvaldo N. Oliveira Jr São Carlos Institute of Physics, University of São Paulo, São Carlos, São Paulo, Brasil

Pasquale Crupi Dipartimento di Scienze Agro Ambientali e Territoriali, Università di Bari, *via* Orabona 4, Bari, Italy

Radu Oprean Analytical Chemistry Department, "Iuliu Hațieganu", University of Medicine and Pharmacy, Cluj-Napoca, Romania

Robert Săndulescu Analytical Chemistry Department, "Iuliu Hațieganu", University of Medicine and Pharmacy, Cluj-Napoca, Romania

Silvia Rodríguez-Sánchez Department of Inorganic Chemistry, Engineers School, Universidad de Valladolid, Valladolid, Spain

Stefan Dima Department of Chemistry, Physics and Environment, Faculty of Sciences and Environment, "Dunarea de Jos", University of Galati, Romania

Thatyane M. Nobre São Carlos Institute of Physics, University of São Paulo, São Carlos, São Paulo, Brasil

Tiziana Dipalmo Dipartimento di Scienze Agro Ambientali e Territoriali, Università di Bari, *via* Orabona 4, Bari, Italy

Frontiers in Bioactive Compounds

(Vol. 1)

Natural Sources, Physicochemical Characterization and Applications

CHAPTER 1

Extra Virgin Olive Oils: Bioactive Compounds and Health Benefits

Maria Lisa Clodoveo[1,*], Tiziana Dipalmo[1], Pasquale Crupi[1], Bernardo C. de Gennaro[1], Carlo Franchini[2], Filomena Corbo[2] and **Constantin Apetrei[3]**

[1] *Dipartimento di Scienze Agro Ambientali e Territoriali, Università di Bari, via Orabona 4, 70126 Bari-Italy*

[2] *Dipartimento di Farmacia, Scienze del Farmaco- Università di Bari, via Orabona 4, 70126 Bari-Italy*

[3] *Department of Chemistry, Physics, and Environment, "Dunarea de Jos" University of Galati, 47 Domneasca Street, 800008 Galati-Romania*

Abstract: Due to the unique extraction process, extra virgin olive oils conserve numerous substances, which are often excluded from other vegetable oils during the extraction and refining steps. Previous works showed that fatty acids profile of lipids are important for the dietary. In the last five decades many researchers demonstrated that monounsaturated fatty acids are involved in prevention of diseases such as the metabolic syndrome and cardiovascular ones. Moreover, vitamin E is active against weakened immunity, aging, environmental toxic substances, diabetic syndrome symptoms, enhancing the health status from different points of view, such as cardiovascular and neoplastic ones. Hydroxytyrosol has many effects; it is an anti-inflammatory compound, with antibacterial, antioxidant and cardioprotective properties. However, not all extra virgin olive oils are equal. Inside the market category "extra virgin" there is a wide number of products ranging from the highest quality to the standard quality ones. In 2012 the European Commission approved three health claim applicable to the extra virgin olive oil but they are still scarcely used by the producers.

Keywords: Bioactive compounds, Extra virgin olive oils, Health claim.

* **Corresponding author Maria Lisa Clodoveo:** Department of Agricultural and Environmental Science, University of Bari, Italy; E-mail: marialisa.clodoveo@uniba.it

Constantin Apetrei (Ed.)

INTRODUCTION

The Mediterranean Diet is inscribed in 2010 on the "Representative List of the Intangible Cultural Heritage of Humanity". This is a lifestyle, which has its roots in antiquity. It is a legacy of very ancient traditions having been part of the life of those countries bordering the Mediterranean Sea. According to many authors, in fact, people who follow the principles of this diet get various health benefits, preventing heart attacks and strokes. Actually this regimen embraces natural and healthy food in balance with each other, such as cereals and their by-products, vegetables, fresh and dried fruit, wine, meats and low-fat cheese, and of course the "Prince" of all: Extra Virgin Olive Oil (EVOO). As reported by the European Regulation, EVOOs are "the oils obtained from the fruit of the olive tree (*Olea europaea* L.) (Fig. **1**) solely by mechanical or other physical means under conditions, particularly thermal conditions, that do not lead to alterations in the oil, and which have not undergone any treatment other than washing, decantation, centrifugation and filtration".

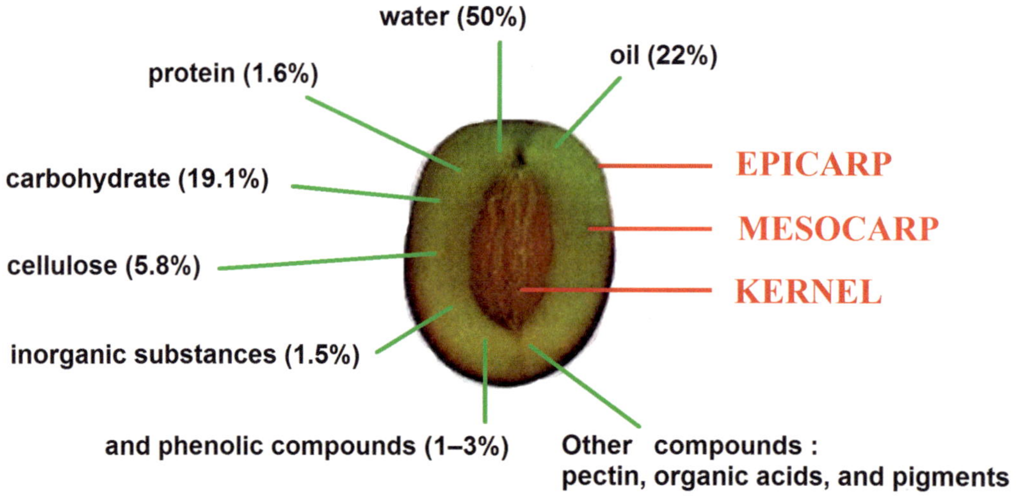

Fig. (1). The fruit of olive component (the epicarp or epidermis, the mesocarp or flesh and the endocarp or kernel, composed by a woody shell enclosing a seed) and average composition.

The Olive Fruit

The olive fruit, an oval-shaped drupe, is composed by epicarp, mesocarp and endocarp; the epicarp changes the color from green to black during the maturation stage, the mesocarp represents about 87% of the whole fruit and the hard endocarp or stone contains the seed or kernel. It represents about the 23% of fruit mass.

The seed contains 2–4 g oil /100 g. Olive fruit weight may range from 2–12 g, although some varieties may weigh as much as 20 g [1]. The chemical composition of resulting olive oil depends on parameters including (i) cultivar, (ii) agronomic procedures, (iii) geographical area, and the (iv) maturity index. Variety, territorial areas and agronomic practices have a heavy effect on the fruit development. The extraction methods and storage conditions can modify oil composition [2].

The Olive Oil

Due to the unique extraction process, extra virgin olive oils conserve numerous substances, which are often excluded from other vegetable oils during the extraction and refining steps. The "unsaponifiable fraction" is composed by more than 200 minor components and represents about 2% of the total mass [3]. This fractions is composed by chemical substances that show a high value from the nutritional and healthy point of view. Moreover, they contribute to enhance the shelf life and the sensory properties of olive oil, useful to identify adulteration. This fraction includes:

- tocopherols,
- sterols,
- pigments (carotenoids and chlorophylls),
- phenols,
- hydrocarbons,
- aromatic and aliphatic alcohols,
- triterpene acids,
- waxes and minor constituents.

The composition of "unsaponifiable fraction" depends from many factors, such as: (i) the variety, (ii) the maturity index, (iii) the crop year (alternate bearing: "on year" and "off year"), (iv) olive harvesting methods, (v) the storage time of fruits,

(vi) the oil extraction methods, and (vii) the storage conditions of the oil.

The "saponifiable fraction" of EVOO, which represents about 98 percent of the oil, is composed of saturated and unsaturated fatty acids, which can be arranged as triglycerides, diglycerides and monoglycerides. Depending on the market classification of the product, a variable percentage of free fatty acids can be also found.

Fatty Acids

EVOO contains mainly monounsaturated fatty acids, among them the oleic acid (Fig. **2**) represents the major percentage (55 and 83% of the total fat). Moreover, linoleic acid is present with a variable concentration ranging from 3.5% to 21%. (*Commission of Codex Alimentarius. Revised Norm for olive oil. CI 1993/15-FO, 1993*). Several agronomical factors such as (i) variety, (ii) agronomic practices, (iii) geographical origin (the latitude), and the (iv) maturity index, strongly affect fatty acid composition of EVOO. In fact, olive fruit modifies its chemical composition due to activation and inhibition of different enzymatic activities during the maturation stage. During this period, the content of oleic acid (C18:1) is not modified; on the contrary, the saturated fatty acids, mainly represented by palmitic acid (C16:0), decrease, while, poly-unsaturated acids such as linoleic acid (C18:2) rises. The MUFA/PUFA ratio, responsible for oxidative stability of virgin olive oils, as well as the natural antioxidant content, decreases during the ripening [4].

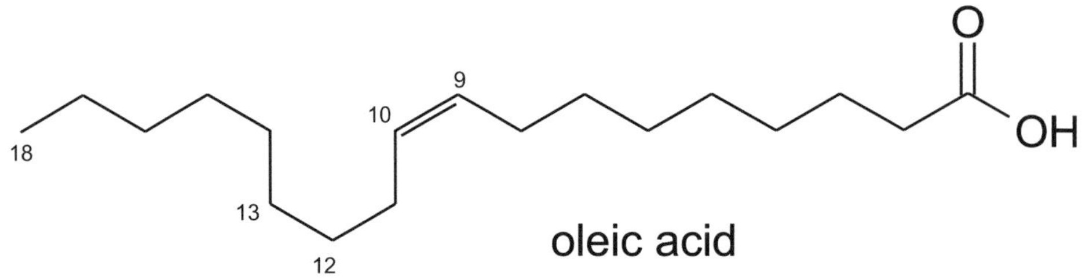

Fig. (2). Structural formula of oleic acid.

Previous works showed that fatty acids profile of lipids are important for the dietary [5]. Over 50 years of research has evidenced the efficacy of dietary monounsaturated fatty acids for reduction of traditional risk factors defining metabolic syndrome and cardiovascular disease.

Recent researches on human nutrition ascribes to the MUFA rich diets beneficial effects on chronic disease. In fact, dietary guidelines recommend an increment of dietary MUFA intake, accompanied by a reduction of saturated fatty acids intake (SFA), because is well demonstrated that MUFA intake can promote an improvement of blood lipid profiles, blood pressure, insulin sensitivity and glucose levels. Moreover, recent data demonstrated their key role into determine the body composition preventing the risk of obesity, by means preferential oxidation reactions. Epidemiological studies based on clinical trial underline the cardioprotective activity of the MUFA into the dietary [5]. Moreover, other healthy effects are demonstrated against cancers of the breast [6], of colorectum [7], of ovary [8], and of prostate [9].

Tocopherols

Tocopherols and tocotrienols, (Fig. **3**) also known as vitamin E, are a set of antioxidant molecules, non polar and fat soluble, characterized by a chromanol ring and a hydrophobic side chain. Tocopherols are identifiable by a phytyl chain, while tocotrienols by an isoprenyl one. Tocopherols (α-, β-, γ-, and δ-tocopherol) and the corresponding tocotrienols vary also by the number and positions of methyl substituents on the phenolic part of the chromanol ring [10]. However, due to its capacity to be incorporated into lipoproteins, α-tocopherol is the only tocol that is able to satisfy the vitamin E need [11]. The α-tocopherol in membranes and lipoproteins acts as a radical-chain-breaking antioxidant. Vitamin E deficiency produces a decrement of antioxidant defense above all in those cells damaged by the stock of free radical. Free radicals developed during normal aerobic metabolism can be the cause of cellular damage [12].

Basic research and observational studies suggest vitamin E protects against exercise fatiguc, weakened immunity, aging, diabetic damage, environmental toxins [13], and reduces the risk of cardiovascular disease [14], cancer of prostate [15] and bladder [16].

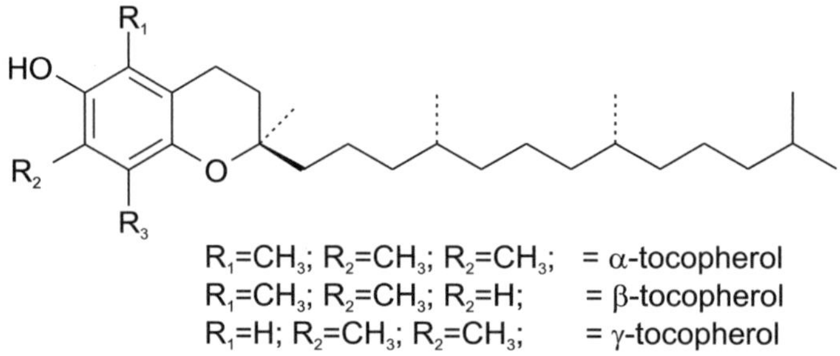

Fig. (3). Structural formulas of tocopherols.

Sterols

Sterols are isoprenoid compounds formed by a sterol nucleus and an alkyl chain; most of them show an unsaturated bond in position C-5 in the nucleus. If completely saturated they are named stanols [17]. Their nutritional interest are due to their capacity to ameliorate the serum cholesterol profile in humans both inhibiting the absorption of dietary cholesterol and regulating the reabsorption of cholesterol excreted into the enterohepatic cycle [18]. Sterols are biologically active against cancer. The content of selected sterols is widely accepted as one of the most important markers for the detection of adulterated olive oils, as reported into the European Regulation. Sterols determine the genuineness of olive oil by their concentration and composition. The binding limits for the content of the sterols are set forth in the Commission Regulation (EEC) No. 2568/91 (1991), which can be considered as being the essential standard to assess the quality of all types of olive oils.

As reported in the Fig. (**4**), the various sterol structures vary in the R tail and in the alchyl chain. Beta-sitosterol, campesterol, and stigmasterol are structurally similar to cholesterol: they do not have methyl groups at the 4th carbon atom (4-desmethyl sterols). Sterols are present in a free (80%) and esterified form and account for 30–60% of the non-saponifiable fraction of EVOO which is 0.4–0.8% of virgin olive oil. Sterol components with the highest representation are β-sitosterol (more than 80%), campesterol (2.9–4%), and Δ^5–avenasterol (7%). Cholesterol is also found in olive oil at a small amount of about 1–3 mg/kg [19].

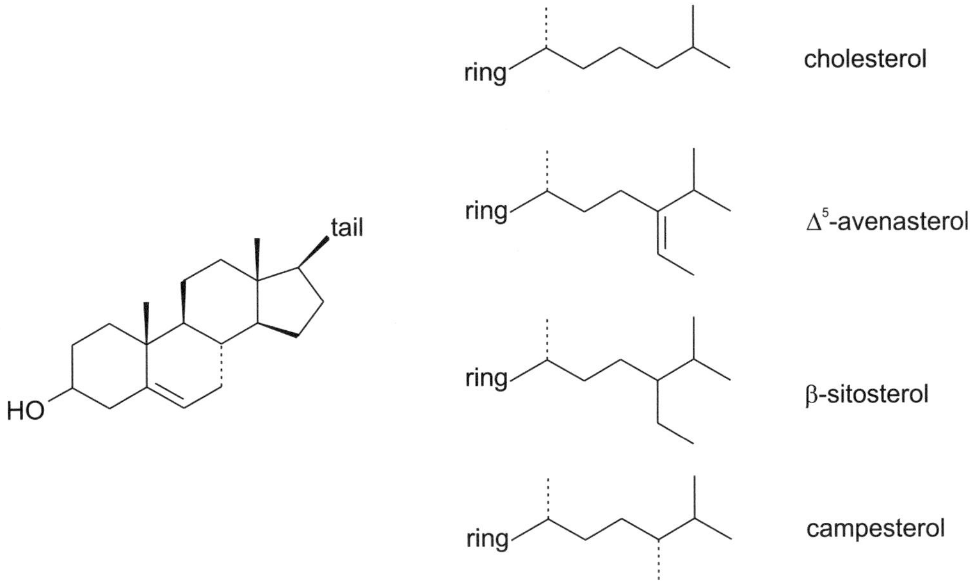

Fig. (4). Structural formulas of sterols.

Carotenoids and Chlorophylls

Olive oil color is due solely to the solubilisation of pigments, carotenoids and chlorophylls, present in the original fruit. The EVOO can vary from a green-to a yellowish color, although the green hue may vary depending on (i) the variety and (ii) the maturity index of fruits used [19]. As the harvesting time of the fruits increases, pigment content decreases. During storage of olive fruits, the presence of lipoxygenase has been detected, as well as a considerable decrease in chlorophylls and a small decrease in carotenoids. During the extraction process, the chlorophyllic fraction is destroyed in the greater part, and although the carotenoid fraction is also affected, its concentration increases in the oil with respect to that in the fresh fruit [19]. Because of their antioxidant properties in the dark and pro-oxidant effect in the light, carotenoids and chlorophylls assume a crucial role during the storage of EVOO. Chlorophylls are mainly present as pheophytins; pheophytin α (Fig. **5**) is the most abundant, pheophytin β and chlorophyll β are present only in trace while chlorophyll α has not been revealed.

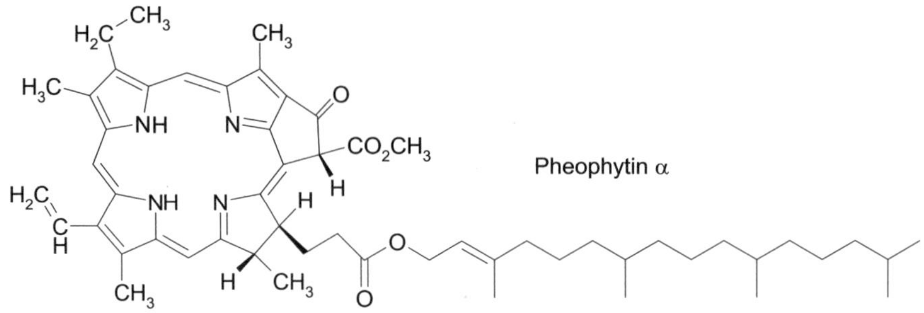

Fig. (5). Structural formula of pheophytin α.

β-carotene (Fig. **6**) and lutein are the main carotenoids identified in olive oil; however, many other xanthophylls are also present, such as antheraxanthin, β-cryptoxanthin, luteoxanthin, mutatoxanthin, neoxanthin, and violaxanthin. Chlorophylls are photosensitizers because, if exposed to light, favour the conversion of ground state triplet oxygen ($^3O_2$) to highly reactive excited state singlet oxygen ($^1O_2^*$), thus initiating the olive oil oxidation.

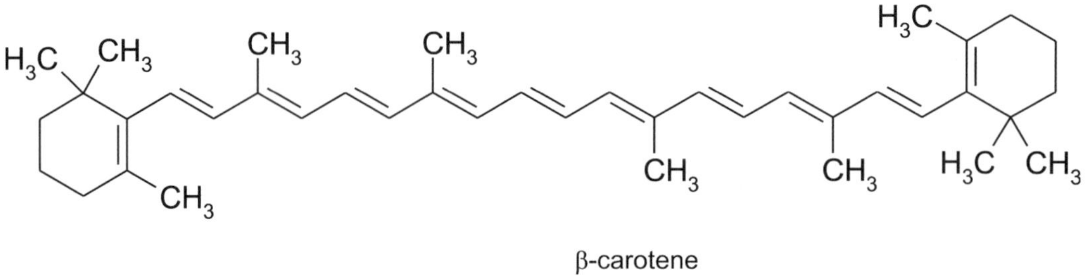

Fig. (6). Structural formula of β-carotene.

The photoxidation by $^1O_2^*$ proceeds more rapidly (~1000 to 1500 times) than the common oxidation mediated by $^3O_2$. Conversely, β-carotene and the other carotenoids improve the oxidative stability of oil against photoxidation, quenching $^1O_2^*$.

Lutein has an antioxidant effect preventing skin aging and cancer risk; its presence makes VOO as an highly skin protector foodstuff [1]. Dietary all *trans* α-Carotene, β-carotene, and β-cryptoxanthin are provitamin A carotenoids and constitute the major source of our vitamin A needs, indeed they can be metabolized in the liver to retinol (vitamin A). Vitamin A is fundamental for the promotion of general growth, maintenance of visual function, regulation of differentiation of epithelial tissues, and embryonic development. Low intake of food rich in carotenoids and chlorophylls is associated with risk of cancer [20].

Hydrocarbons

The composition of hydrocarbon fractions of EVOO oils is influenced by (i) the variety, (ii) the geographical origin, and (iii) the growing conditions of the olive trees. EVOO contains n-alkanes from C14 to C35 together with low amounts of other hydrocarbons, such as n-alkenes, sesquiterpenes (α-farnesene), terpenes (kaurene), and low-molecular-mass mono- and polycyclic aromatics (such as benzene, styrene, and naphthalene).

In particular, squalene, a triterpenoid hydrocarbon formed by six isoprene units, represents ~80−90% of the EVOO hydrocarbon fraction and its concentration ranges from 200 to 7500 mg/kg of olive oil. Squalene acts as a biochemical precursor of cholesterol and other steroids [1]. All plants and animals including humans are capable of producing squalene. In humans, squalene is synthesized in the liver and the skin, transported in the blood by very low density lipoproteins (VLDL) and low density lipoproteins (LDL), and secreted in large quantities by the sebaceous glands [21].

Squalene (Fig. 7) acts as a potent scavenger of $^{1}O_2^{*}$, inhibiting the lipoperoxidation induced by UV radiations [22]; moreover, it showed immune-stimulating, antineoplastic and anticancer (against cancer of colon, breast, and prostate) properties, thus it seems to be responsible of several health effects of olive oil [23].

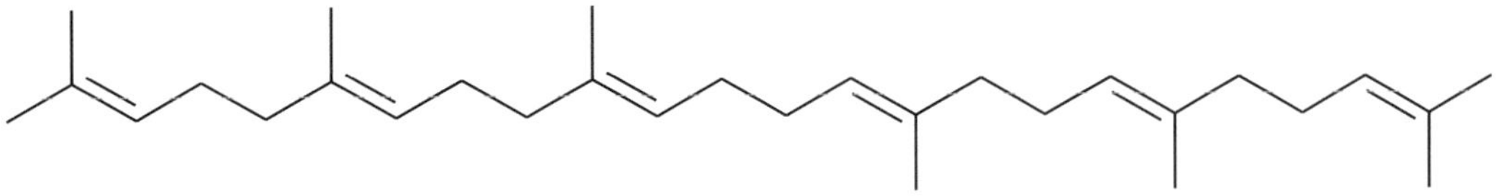

squalene

Fig. (7). Structural formula of squalene.

Triterpene dialcohols and acids

Erythrodiol and uvaol (Fig. **8**) are the principal triterpene dialcohols found in virgin olive oil; the former is present both in the free and esterified form with level ranging between 20 and 70 mg/kg, the latter plays a protective role in the oxidation of lipoproteins of low density *in vitro*, and has a protective effect on induced hepatic injuries. Also pentacyclic triterpenic acids, such as oleanolic acid, are present in VOO; oleanolic acid and its isomer, ursolic acid, have been studied for many pharmacological properties [24]. Numerous researchers confirm that pentacyclic triterpenes have antibacterial, antifungal, anticariogenic, antiallergic, anti-inflammatory, hepatoprotector, gastroprotector, hypolipidemic, antiathero-sclerotic, and antidiabetic effects [25 - 29]. They can also interfere in several phases of the development of different types of neoplasias [30].

Oleanolic acid also has beneficial effects on multiple sclerosis [31].

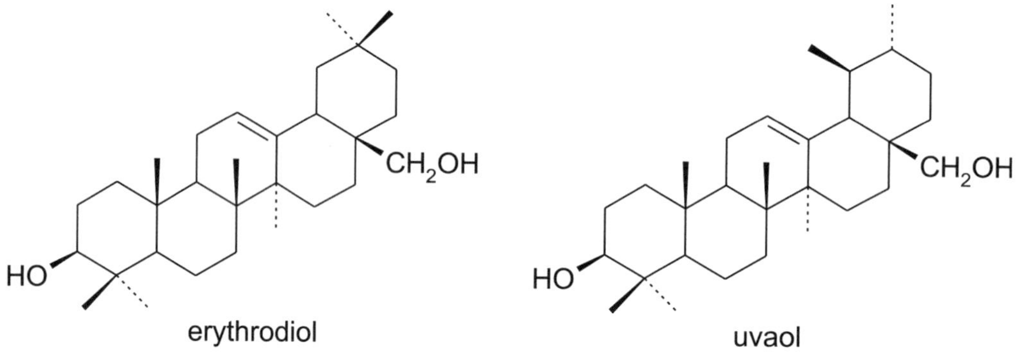

Fig. (8). Structural formulas of erythrodiol and uvaol.

Phenols

Polyphenols (Fig. **9**) are compounds possessing several phenol moieties. Plant phenols are generally involved in defense against UV radiation or aggression by pathogens, parasites, and predators. Into EVOO, phenols, representing ~18-37% of the unsaponifiable fraction, are responsible for its bitterness and astringency, due to their interaction with the glycoprotein of saliva.

Fig. (9). Polyphenols classification.

Polyphenols can be divided into flavonoids and nonflavonoids. Flavonoids consist of six subgroups, namely flavonols, flavan-3-ols (monomeric and polymeric structures), flavones, isoflavones, flavanones, and anthocyanidins; while, phenolic acids, stilbenes, and lignans belong to non-flavonoid group. Polyphenols of olive oil show autoprotective, sensory, and nutritional-therapeutic effects. The total content of phenolic compounds in EVOO generally ranges between 40 and 1000 mg/kg [1], with various concentrations of individual phenolic compounds being present. These phenolic compounds in EVOO originate from those originally present in the olive fruits used for oil extraction but are substantially modified depending on the conditions used for oil extraction and subsequent storage. Only a small fraction (approximately 2%) of the phenolic compounds present in the olive fruit is transferred to the extracted oil, the rest (98%) is either retained in the pulp or washed off with the wastewater [1]. Yet, the phenolic composition of the olive fruit at harvest is probably the most important variable involved in the phenolic composition of the final EVOO.

Olive oil phenols can be divided into five categories: phenolic alcohols, phenolic acids, flavonoids, secoiridoids, and lignans.

Phenolic alcohols

The phenethyl alcohols, hydroxytyrosol and tyrosol, and their precursors, such as oleuropein, (Fig. **10**) are the most abundant phenolic compounds in EVOOs. Indeed, hydroxytyrosol (Fig. **11**) and tyrosol generally derived from the hydrolysis of the secoiridoid aglycones of oleuropein and ligstroside, respectively. Hydroxytyrosol (3,4-dihydroxyphenylethanol) contributes to the bitter taste of olive oil and it has been correlated to many health benefits; it shows several biological functions, especially due to its strong antioxidant and radical-scavenging properties.

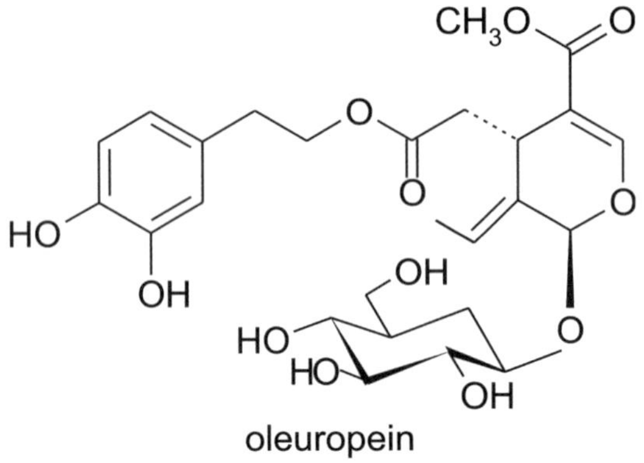

Fig. (10). Structural formula of oleuropein.

Hydroxytyrosol is more active than antioxidant vitamins and synthetic antioxidants and exerts its antioxidant activity by changing itself into a catechol quinone [32]. The European Regulation on health claims in food products, in the case of olive oil (Reg. 1924/2006; Reg. 432/2012), states that the health effect can be claimed only if the oil contains more than 5 mg of hydroxytyrosol and its derivatives (*e.g.*, oleuropein complex and tyrosol) in 20 g of olive oil [33].

Phenolic acids

Phenolic acids can be distinguished in hydroxybenzoic acids (such as vanillic acid, syringic acid, and gallic acid) and hydroxycinnamic acid (such as caffeic acid, ferulic acid, and sinapic acid) characterized by a basic structure containing C6–C1 or C6–C3 skeleton, respectively. The main hydroxycinnamic acid derivative reported in olive fruit is verbascoside (Fig. **12**).

Flavonoids

The flavonoids, with a chemical structure of C6–C3–C6 atoms, include the glycosides of flavonol (luteolin-7-glucoside and rutin) (Fig. **13**), and anthocyanins, such as the glucosides of cyanidin and delphinidin.

Secoiridoids have been exclusively detected in plants of the family of *Olearaceae*. Oleuropein, (Fig. **10**) demethyloleuropein, and ligstroside are the main secoiridoids present in virgin olive oil, and appear as secondary glycosidic compounds similar to coumarins; they are characterised in the molecular structure by the presence of elenolic acid in its glucosidic or aglyconic form [34].

Oleuropein is the most concentrated secoiridoid revealed in unripe olive fruit and it is recognized as the mainly responsible for the bitterness of the olive fruit; during ripening of olives, the content of oleuropein diminishes up to undetectable concentration in mature fruit, because it is completely degraded to hydroxytyrosol, tyrosol, and verbascoside. Ligstroside (deacetoxy-ligstroside aglycon) is another secoiridoid typically found, which contributes to the pungent odor of extra virgin olive oil [1].

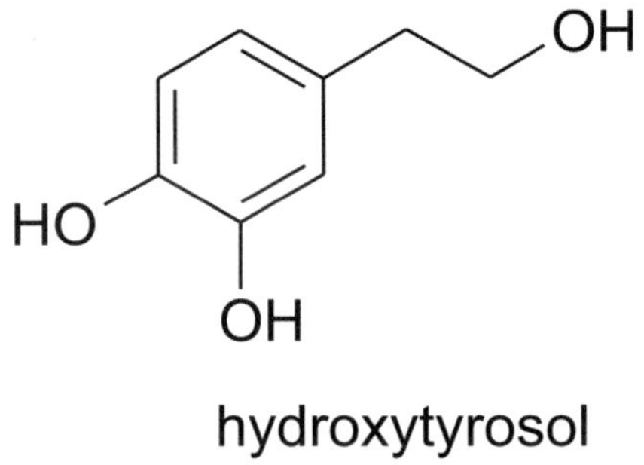

Fig. (11). Structural formula of hydroxytyrosol.

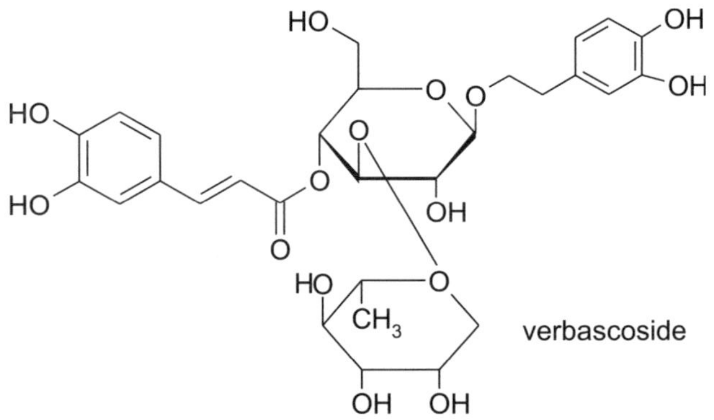

Fig. (12). Structural formula of verbascoside.

Oleocanthal (decarboxy methyl ligstroside aglycone) (Fig. **14**) is a common secoiridoid present in olive oil, known for its organoleptic characteristic ("pungency", from which its name derives: *oleo-* for olive,*-canth-* for sting, and

-*al* for aldehyde) and bioactivity. Indeed, oleocanthal is homologous with the non-steroidal anti-inflammatory drug, *ibuprofen*, used for relieving pain, helping with fever, and reducing inflammation. Moreover, it showed positive effects against many chronic inflammatory states including cancer, neurodegenerative, and joint degenerative diseases; then, it is efficient in decreasing the markers of arthritis and disrupting processes vital to the formation of Alzheimer's disease. Furthermore, oleocanthal reduces proliferation, migration, and also promotes apoptosis in cancer cells as well as prevents tumour induced cell transformation [35].

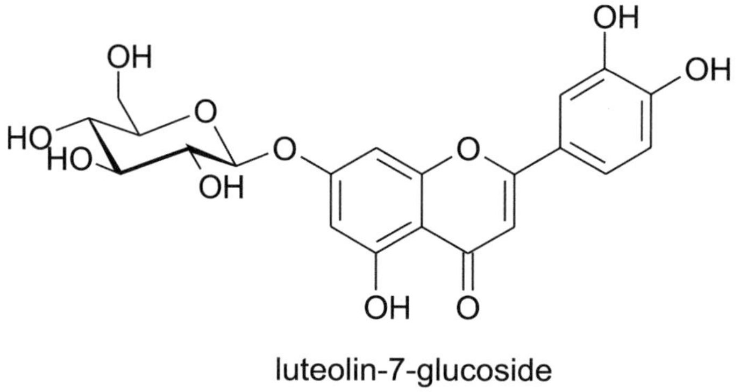

Fig. (13). Structural formula of luteolin-7-glucoside.

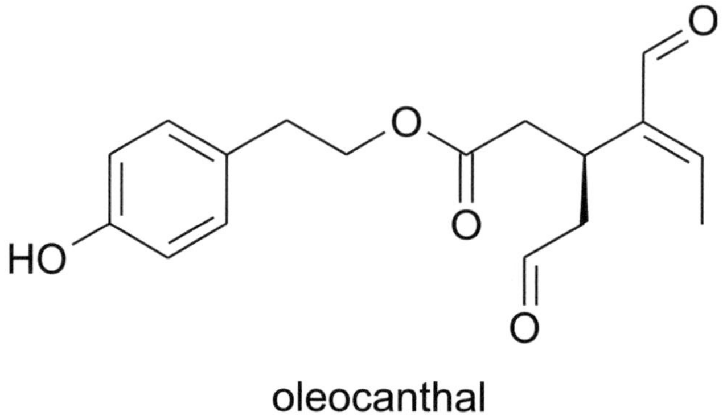

Fig. (14). Structural formula of oleocanthal.

Lignans

Lignans (such as 1-acetoxypinoresinol and pinoresinol) are present in EVOOs up to a concentration of 100 mg/kg. However, cultivar, environmental conditions, and oil production technologies are all factors which can determine a considerable variation in lignans concentrations. Several biological and pharmacological effects can be attributed to lignans, in particular the two aforementioned compounds may contribute to the reported beneficial effects which are attributed to polyphenols on human health of a diet rich in olive oil [36].

Is All Olive Oil Created Equal?

No wonder Homer deemed olive oil liquid gold. But not all olive oil is created equal, including those bearing the words Extra Virgin on their label. Numerous factors come into play in order to produce a superior extra virgin olive oil in terms of taste, nutrition, and healthy qualities. Table **1** shows the main health effects of EVOO.

The system of classification of the EU legislation on olive oils quality grades (Table **2**) is not more enough to explain the great quality variability inside the EVOO category, creating difficulties for the consumer, mainly when the product is exported and purchased by who lives in countries where no habit of consuming EVOO is. However, also in the countries traditionally consuming olive oil, due to this system of classification, the consumers cannot perceive the exact value of the products and can't distinguish between a superior quality EVOO and a second choice ones. Following the European classification "Virgin" is an adjective that indicates the oil obtained from the fruit of the olive tree (*Olea europaea* L.) exclusively by mechanical or other physical means under conditions, especially thermal conditions, that do not cause any alteration in the oil, and which have not experienced any treatment other than washing, decantation, centrifugation and filtration. When the product, on the basis of chemical and sensory evaluation, is classified as no suitable to human consumption, it has to undergone to a refining process with charcoal or other chemical or physical treatments. Oil refining deals with the correction of undesirable defects in the free acidity, color, odor and taste of oils with a perspective to improving and/or homogenizing their properties.

Where olive-pomace oil is concerned, the objective is to make the oil suitable for consumption in accordance with the IOC trade standard. Virgin and refined oils differ little in fatty acid composition, while phenolic compounds and other antioxidants are almost completely destroyed by high temperature treatments during the refining process.

Inside the category named "extra virgin" there is a wide range of products ranging from the highest to the standard quality. The highest quality EVOOs correspond to the products characterized by the lowest values of the hydrolytic and oxidative parameters, by the highest value of sensory evaluation and richness in healthy molecules. The standard quality products correspond to the products that fall within the limits of the law, but are poor in terms of the healthy properties, the amount of antioxidant molecules, parameters not considered by the law. Both the products are labelled as "extra virgin" because the range considered for the legal parameters is too large and do not allow a clear differentiation, having elapsed more than 25 years of their first establishment, and being the subsequent modifications not enough to inform the consumers about the different healthy quality of the olive oils.

Table 1. EVOO and health.

EVOO health effects	
PREGNANCY AND CHILDHOOD	It has been showed that the post-natal development, in terms of height, weight, behavior, and psychomotor reflexes, is better in babies from mothers who consumed olive oil when pregnant. Vitamin E is very important for the infant because of, for instance, the favorable effect it has on the hepato-biliary system; during labor, it transfers from mother's blood to the breast glands and so, during breast feeding, the mother continues to supply it to the baby. Of course, it is essential to maintain the levels of this vitamin during breast feeding by consuming food such as EVOO which is particularly rich of it. Moreover, olive oil also provides enough essential fatty acids for the development of the new-born child; its ratio of linoleic acid to linolenic acid (essential fatty acids) is similar to that of breast milk.
CHOLESTEROL	Total blood cholesterol as well as LDL-cholesterol and triglycerides can be decreased by consuming EVOO, which, furthermore, does not alter the levels of HDL-cholesterol (and may even raise them), playing a protective role in preventing the formation of fatty patches.
CARDIOVASCULAR DISEASES	It has been observed that olive oil rich diets can prevent the formation of blood clots and platelet aggregation, avoiding the excessive blood coagulation, thus contributing to the low incidence of heart failure.

(Table 1) contd.....

EVOO health effects	
BLOOD PRESSURE	The regular consumption of olive oil into the diet has been proved to lower the blood pressure, by decreasing both systolic (maximum) and diastolic (minimum) blood pressure. There is late proof that the daily olive oil intake can diminish the dose of drugs expected to control blood pressure in hypertensive patients, maybe due to a reduction in nitric acid caused by polyphenols.
DIABETES	An olive oil rich diet is not just a natural remedies in the treatment of diabetes, but it might also counteract or postpone the onset of the disease by preventing insulin resistance and its possible negative implications, increasing HDL-cholesterol, decreasing triglycerides, and ensuring better blood sugar level control and lower blood pressure.
OBESITY	It has been shown that an olive oil rich diet, stimulating to eat more vegetables, leads to greater and longer-lasting weight loss than a low-fat diet.
SKIN	Having a lipid profile very similar to that of human skin, the olive oil can reduce the cell damage caused by free radicals (generated by external factors, such as sun's rays), supporting the natural mechanisms that neutralize their action, and delay the ageing process of the skin. Moreover, olive oil has a large proportion of antioxidants such as polyphenols and vitamins (A, D, E and K) which are the main source of protection against the free radicals that produce cell oxidation, and makes it a good aid in specific therapies to treat skin disorders such as acne, psoriasis and seborrheic eczemas.
IMMUNE SYSTEM	It has been reported that olive oil intake reinforces the immune system against external attacks from microorganisms, bacteria or viruses. Late research has concluded that the unsaturated fats in the make-up of olive oil are great partners in bringing down critical immunological parameters, for example, the expansion of lymphocytes prompted by particular mitogens of both B-and T-cells. These fatty acids play an important role in various immune functions. They are included in regulating inflammatory processes and they may be effective in the treatment of some autoimmune diseases.
CANCER	A recent study has revealed that the type of fat seems to be more correlated to cancer incidence than the quantity of fat; indeed, olive oil, rich in unsaturated fatty acids (mainly oleic acid), provides protection against colon-rectal, prostate, and breast cancer. It has been observed that this fatty acid brings down the generation of prostaglandins derived from arachidonic acid, which thus has huge impact in the creation and development of tumors. Even though, it cannot be excluded the positive effect of other constituents of olive oil, such as polyphenols and squalene, for instance against the incidence of melanomas.

(Table 1) contd.....

EVOO health effects	
AGEING	Olive oil contains a lot of antioxidants (vitamin E, polyphenols) which act mainly as free radicals scavengers, which are the compounds responsible of chronic diseases and ageing, and contribute in extending life expectancy, as shown in several epidemiological studies. Many ageing-related diseases are influenced by diet, in particular osteoporosis and deteriorated cognitive function; olive oil appears to have a favorable effect on bone calcification and mineralization and may prevent memory loss and age-related cognitive decline in healthy elderly people, due to the large amount of monounsaturated fats.

In this condition, the olive oil market is characterized by an asymmetric information [37]. When the seller knows more about the product being traded than the buyer, this condition causes economic imbalances that result in an adverse selection of sellers. Usually, consumers are unable to distinguish quality differences and, thus, incline toward cheaper EVOOs, giving preference to large bottlers that sell low-cost and low quality products.

Table 2. EU legislation on olive oils quality grades following the Commission Delegated Regulation (EU) 2015/1830 of 8 July 2015.

Extra virgin olive oil: virgin olive oil with free acidity (expressed as oleic acid) ≤ 0.8 g/100 g, and other typical characteristics corresponding to the standards fixed by Annex 1 of Regulation (EU) 2015/1830 for this category. With regard to the sensory evaluation: median of defects = 0; median of fruity aroma > 0.
Virgin olive oil: virgin olive oil with free acidity (expressed as oleic acid) ≤ 2 g/100 g, and other typical characteristics corresponding to the standards fixed by Annex 1 of Regulation (EU) 2015/1830 for this category. With regard to the sensory evaluation: median of defects > 0 but ≤ 3.5; median of fruity aroma > 0.
Lampante virgin olive oil: virgin olive oil with free acidity (expressed as oleic acid) > 3.3 g/100 g, and other typical characteristics corresponding to the standards fixed by Annex 1 of Regulation (EU) 2015/1830 for this category. With regard to the sensory evaluation: median of defects > 3.5. Inedible as such, it is adapted for refining or technical use.
Refined olive oil: obtained from virgin olive oils by refining methods without altering the initial glyceridic structure. It has a free acidity (expressed as oleic acid) ≤ 0.3 g/100 g and other typical characteristics corresponding to the standards fixed by Annex 1 of Regulation (EU) 2015/1830 for this category. This product may only be sold direct to the customer if allowed in the nation of retail deal.
Olive oil composed of refined and virgin olive oils: oil consisting of a blend of refined olive oil and virgin olive oils which results edible. It has a free acidity (expressed as oleic acid) ≤ 1 g/100 g and other typical characteristics corresponding to the standards fixed by Annex 1 of Regulation (EU) 2015/1830 for this category. The country of retail sale may require a more specific designation.

(Table 2) contd.....

Crude pomace olive oil is an oil whose characteristics correspond to the standards fixed by Annex 1 of Regulation (EU) 2015/1830 for this category. It is adapted for refining or technical use. Refined pomace olive oil is the oil obtained from crude olive pomace oil by refining methods without altering the initial glyceridic structure. It has a free acidity (expressed as oleic acid) ≤ 0.3 g/100 g and other typical characteristics corresponding to the standards fixed by Annex 1 of Regulation (EU) 2015/1830 for this category. This product may only be sold direct to the customer if allowed in the nation of retail deal. **Olive pomace oil**: Oil consisting of a blend of refined olive-pomace oil and virgin olive oils oil. It has a free acidity (expressed as oleic acid) ≤ 1 g/100 g and other typical characteristics corresponding to the standards fixed by Annex 1 of Regulation (EU) 2015/1830 for this category.

The correct communication of the healthy properties of EVOOs is a strategy to distinguish the highest quality extra virgin olive oils; it can be easily made through the use of health claim authorized in 2012 by the European Commission (Reg. UE 432/2012) regarding the polyphenols of olive oil, but still scarcely used by the producers. Taking in mind that the concentration of total polyphenols in EVOO can range from the minimum of 45.3 mg/kg to the maximum of 931.2 mg/kg [38]. and that the higher healthy quality is linked with the higher concentration of polyphenols, this health claim represent a powerful tool to improve the market value of the high quality products and it is able to improve the actual olive oil trade market regulation which cause the phenomenon of "information asymmetry" that penalize producers of best quality olive oil and consumers.

An informed consumer is a better consumer. The health claims are a tool to segment the commercial trade category named "Extra" because only the best products are healthy. Table **3** shows the list of the four health claims that can be used for EVOO labelling and communication.

The health claims are effectively a tool to define the trade category named "Extra virgin" because only the best products show the healthy properties due to the high presence of minor compounds. High quality extra virgin olive oil can be obtained only from healthy and properly ripened olive fruits.

The fine sensory features and the remarkable nutritional and biological properties of EVOOs are mainly due to the presence of its minor components (*i.e.* volatile and phenolic compounds). It is known that the presence and amount of these minor components in virgin olive oil depend on agronomic practices and technological variables: (i) the variety, (ii) the maturity index, (iii) the extraction process. Therefore, the use of suitable agronomic and processing conditions clearly offers the chance to modify the final contents and profiles of phenolic in the EVOO and, thus, to modulate the characteristics of the final product permitting the olive oil mill industry to obtain the desired kind of virgin olive oil. It must be a shared responsibility between growers and producers, all of which must achieve the correct prerequisites of a specially appointed procedure control and quality confirmation techniques.

The preservation of phenols starts inside the olive orchard. Olive phenolic content depends both quantitatively and qualitatively on its genetic makeup. However, environmental factors and agronomic practices, such as fertilization and irrigation, have also been shown to affect the phenolic composition of virgin olive oil. Phenol content significantly decreased in virgin olive oil as nitrogen added by fertilization increased in fruit [39]. Phenol level shows an inverse relationship with the amount of water applied to the olive trees due to their involvement in defense against oxidative stress [40 - 45]. Early harvest results in oils with higher polyphenol values [2, 46 - 48]. Regard to the choice of the harvesting method and its influence on phenol content of EVOO, it is related to its effect on preservation of fruit integrity. Damage and bruises of olive fruits during the harvest, results in a reduction of phenol content of the final product, because pressure within the olive pile during storage may cause secretion of fluid from the olives favoring the contact of the phenolic substances with the oxidative enzymes polyphenoloxidase (PPO) and peroxidase (POD). In order to preserve the oil phenol content during the whole extraction process, without compromise the quality of fruits obtained applying the best cultivation practices, fruit storage before oil processing is not encouraged.

Table 3. List of permitted health claims for OO.

Substance	Claim	Conditions of use of the claim
Olive oil polyphenols	Olive oil polyphenols contribute to the protection of blood lipids from oxidative stress	The claim may be used only for olive oil which contains at least 5mg of hydroxytyrosol and its derivatives (*e.g.* oleuropein complex and tyrosol) per 20g of olive oil. In order to bear the claim information shall be given to the consumer that the beneficial effect is obtained with a daily intake of 20 g of olive oil.

(Table 3) contd.....

Substance	Claim	Conditions of use of the claim
Oleic acid	Replacing saturated fats in the diet with unsaturated fats contributes to the maintenance of normal blood cholesterol levels. Oleic acid is an unsaturated fat.	The claim may be used only for food which is **high in unsaturated fatty acids**, as referred to in the claim high unsaturated fat as listed in the Annex to Regulation (EC) No 1924/2006 and subsequent amendments A claim that a food is high in unsaturated fat, may only be made where at least 70% of the fatty acids present in the product derive from unsaturated fat under the condition that unsaturated fat provides more than 20% of energy of the product.'.
Vitamin E	Vitamin E contributes to the protection of cells from oxidative stress	The claim may be used only for food which is at least a **source of vitamin E** as referred to in the claim source of vitamin e as listed in the Annex to Regulation (EC) No 1924/2006 and subsequent amendments.
Monounsaturated and/or polyunsaturated fatty acids	'Replacing saturated fats with unsaturated fats in the diet has been shown to lower/reduce blood cholesterol. High cholesterol is a risk factor in the development of coronary heart disease	The claim may be used only for food which is **high in unsaturated fatty acids**, as referred to in the claim high unsaturated fat as listed in the Annex to Regulation (EC) No 1924/2006 and subsequent amendments. The claim may only be used on fats and oils

Table 4. SWOT analysis used to evaluate the need for health claims for the EVOO producers.

Strengths	Weaknesses
- Strong ability to add value to products - Use of already authorized health claims - Strong ability to differentiate products - Strong ability to resolve the gap of the of "asymmetry information" - To communicate and to influence the eating culture - Strong credibility of the claim deriving from the evaluation of EFSA panel of expert	- Lack of skilled employees with specialized knowledge (*e.g.* about legal requirements) - Focus on just authorized health claims - A tool to predict the permanence of the claim validity during the shelf life of the product lacks - Knowledge about health claims on EVOO and their impact on purchase attitude and consumption behavior of consumers is still limited. - Regard to the claim of polyphenols, an official method of analysis lacks - Regard to the claim of polyphenols, the analytical description by EFSA of the molecules to be quantified is unclear. - Regard to the claim of polyphenols, it is difficult to understand for the average consumer.

(Table 4) contd.....

Strengths	Weaknesses
Opportunities	**Threats**
- Trade inside and outside the EU - Increasing support of EFSA and European Commission - Growing consumer demand for healthy food - Increased segmentation of markets into target groups with specific needs (*i.e.* elderly people; infants and children, pregnant women; sportsmen)	- Increased power of EVOO bottling firms - Increased competition from emerging producing countries - Aging and declining European population affecting employment and markets

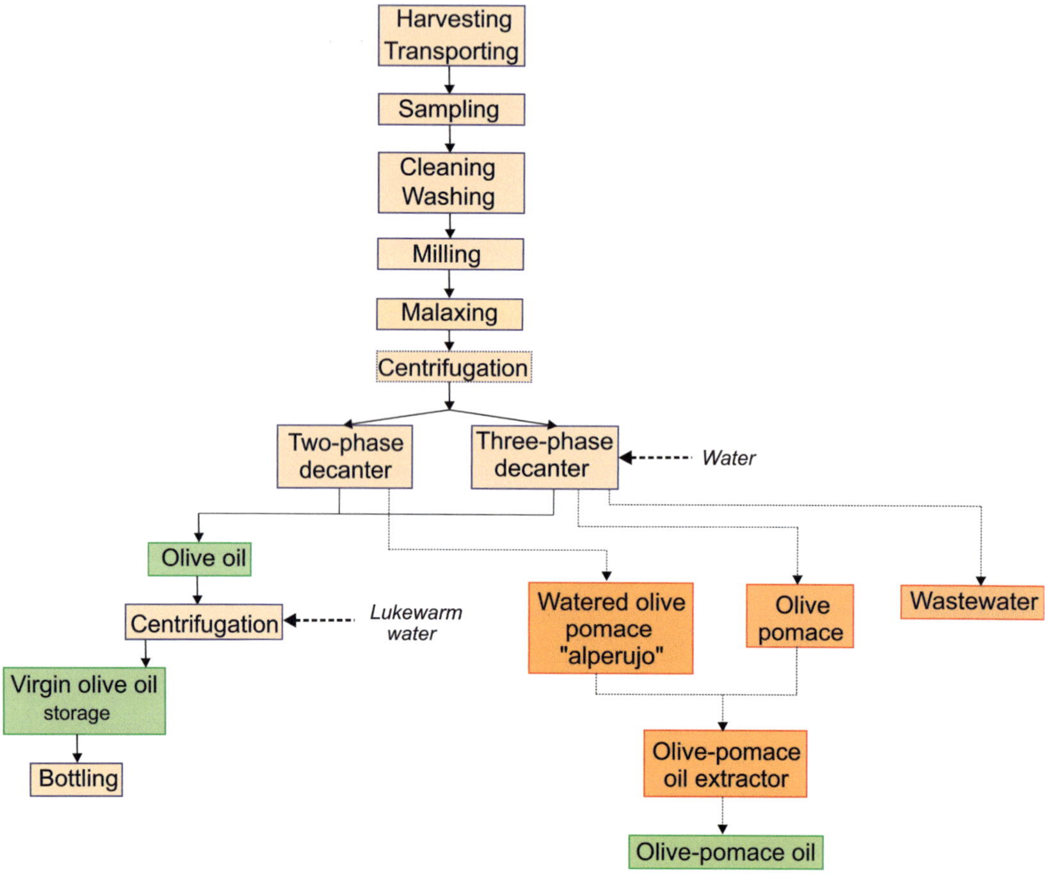

Fig. (15). Olive oil extraction process.

The first step to extract VOO from fruits consists in crushing the drupes (Fig. **15**). The type of crusher chosen is critical to determine the quantity of phenols into the

EVOO because of the possible physical, chemical, and biochemical variations [49]. Crushing is normally carried out using four kinds of machines: the traditional stone mill, the hammer or disk crusher, and, finally, the innovative de-stoner [49]. The employment of stone mill, due to the exposition of the olive paste to the atmospheric oxygen, causes the oxidation of polyphenols, and simultaneously a decreasing of the bitter and pungent taste of the product [49]. The hammer crusher, cutting deeply the skin, extracts more phenols, so the resultant oil has longer shelf life than the oil obtained by stone mill. The employment of the disk-crusher avoids the overheating of the olive paste minimizing the risk of oxidation, however, the oil can result stronger in spicy taste but lesser bitter than the oil obtained by the hammer-crusher. The destoner, also called "depitting" machine, crushes only the pulp tissues excluding the seed rich in oxidoreductases. The resulting EVOO has a higher phenol content than those obtained by other crushing system [50].

Malaxation is one of the most important critical points of the olive oil mechanical extraction process [51]. During malaxation of olive paste a complex bioprocess occurs modifying deeply the phenol content of the final product [52 - 55]. The endogenous enzymatic activities can be modulated by controlling three main process parameters: temperature, time, and atmosphere composition inside the malaxer [51]. In general, increasing temperature from 25 to 35°C can reduce the enzymatic oxidative reaction causing an increment in both total phenolics and the pungent phenolic oleocanthal. Polyphenols are much more affected by the temperature rather than the time of malaxation [56]. The increment of malaxing time favors the activity of β-glucosidases which produce the aglycon molecules of phenols increasing their affinity to the oily phase. The oxidative enzymatic activities could be regulated through a strategic control of the atmosphere composition of the head space of the tank [57]. The presence of inert gas inside the head-space of the malaxer produces a twofold effect: it reduces the activity of the oxidase enzymes preserving the polyphenolic substances. However, it is sufficient a hermetic closure of the malaxer to reduce the concentration of oxygen and a reduction of the activity of oxidoreductases. The separation of oil from solid and liquid phases of olive paste is usually performed by using the centrifugation system [58]. There are two type of decanter: a two- or a three-phase centrifugal decanter. The two-phase process requires no dilution or only a little dilution of the olive paste. As consequence, the oils obtained after extraction by the two-phase centrifugal system exhibited a higher content of polyphenols because these substances are more soluble in the water than in the oily phase. VOO should be stored after the extraction process in stainless steel and maintained at a constant temperature of between 10-18 °C before bottling [51]. During the storage, the concentration of all polyphenols tend to decrease, except for lignans, which seem to be the most stable to the action of oxygen, light, and heat. During filtration a loss of a significant part of polyphenols occurs favoring a reduction of oxidative stability. Packaging is an important factor influencing olive oil shelf life, and phenol concentration. A good packaging solution have to guarantee a barrier against pro-oxidant factors such as light, oxygen, and heat. The best solution is a small size green glass bottle.

CONCLUDING REMARKS

The training of olive producers and stone millers about the strategy to obtain high quality EVOOs could help the increment of availability of healthy product and the benefits that can be acquired in the long run due to the potential reduction in the risk of diseases for the population related to oxidative stress, convertible into a potential reduction of future health costs.

CONFLICT OF INTEREST

The authors confirm that they have no conflict of interest to declare for this publication.

ACKNOWLEDGEMENTS

Declared none.

REFERENCES

[1] Ghanbari R, Anwar F, Alkharfy KM, Gilani A-H, Saari N. Valuable nutrients and functional bioactives in different parts of olive (*Olea europaea L.*)-a review. Int J Mol Sci [Internet] 2012 Jan [cited 2015 Oct 19]; 13(3): 3291-40. Available from: http://www.scopus.com/inward/record.url?eid=2-s2.0-84858966810&partnerID=tZOtx3y1

[2] Baccouri O, Guerfel M, Baccouri B, *et al.* Chemical composition and oxidative stability of Tunisian

monovarietal virgin olive oils with regard to fruit ripening. Food Chem 2008; 109(4): 743-54. [http://dx.doi.org/10.1016/j.foodchem.2008.01.034] [PMID: 26049987]

[3] Bulotta S, Celano M, Lepore SM, Montalcini T, Pujia A, Russo D. Beneficial effects of the olive oil phenolic components oleuropein and hydroxytyrosol: focus on protection against cardiovascular and metabolic diseases. J Transl Med 2014; 12: 219. [Internet]. [http://dx.doi.org/10.1186/s12967-014-0219-9] [PMID: 25086598]

[4] Beltrán G, Del Rio C, Sánchez S, Martínez L. Influence of harvest date and crop yield on the fatty acid composition of virgin olive oils from cv. Picual. J Agric Food Chem 2004; 52(11): 3434-40. [Internet]. [http://dx.doi.org/10.1021/jf049894n] [PMID: 15161211]

[5] Gillingham LG, Harris-Janz S, Jones PJ. Dietary monounsaturated fatty acids are protective against metabolic syndrome and cardiovascular disease risk factors. Lipids 2011; 46(3): 209-28. [http://dx.doi.org/10.1007/s11745-010-3524-y] [PMID: 21308420]

[6] Alegre MM, Knowles MH. Robison R a., O'Neill KL. Mechanics behind breast cancer prevention - focus on obesity, exercise and dietary fat. Asian Pac J Cancer Prev 2015; 2013(14): 2207-12.

[7] Vargas AJ, Thompson PA. Diet and nutrient factors in colorectal cancer risk Nutr Clin Pract 2012; 27(5): 613-23. [http://dx.doi.org/10.1177/0884533612454885]

[8] Blank MM, Wentzensen N, Hollenbeck a, Park Y. Dietary fat intake and risk of ovarian cancer in the NIH-AARP Diet and Health Study. Br J Cancer [Internet] Nature Publishing Group 2012; 106(3): 596-602. Available from: http://www.pubmedcentral.nih.gov/articlerender.fcgi?art-id=3273348&tool=pmcentrez&rendertype=abstract

[9] Pelser C, Mondul AM, Hollenbeck AR, Park Y. Dietary fat, fatty acids, and risk of prostate cancer in the NIH-AARP diet and health study. Cancer Epidemiol Biomarkers Prev 2013; 22(4): 697-707. [Internet]. [http://dx.doi.org/10.1158/1055-9965.EPI-12-1196-T] [PMID: 23549401]

[10] Schwartz H, Ollilainen V, Piironen V, Lampi AM. Tocopherol, tocotrienol and plant sterol contents of vegetable oils and industrial fats. J Food Compos Anal 2008; 21(2): 152-61. [http://dx.doi.org/10.1016/j.jfca.2007.07.012]

[11] Institute of medicine, 2000. Vitamin E. In: Dietary reference intakes for vitamin C, vitamin E, selenium, and carotinoids. pp. 186–283. National academy press, Washington, DC. Accessible at www.nap.edu.

[12] Burton GW. Vitamin E: molecular and biological function. Proc Nutr Soc 1994; 53(2): 251-62. [http://dx.doi.org/10.1079/PNS19940030] [PMID: 7972139]

[13] Winter R. Vitamin E: Your protection against exercise fatigue, weakened immunity, heart disease, cancer, aging, diabetic damage, environmental toxins. Three Rivers Press 2013.

[14] Sesso HD, Buring JE, Christen WG, Kurth T, Belanger C, MacFadyen J, *et al.* Vitamins E and C in the prevention of cardiovascular disease in men: the Physicians' Health Study II randomized controlled trial. JAMA 2008; 300(18): 2123-33. Available from: http://www.scopus.com/inward/record.url?eid=2-s2.0-56149111693&partnerID=tZOtx3y1

[15] Klein EA, Thompson IM, Tangen CM, Crowley JJ, Lucia MS, Goodman PJ, *et al.* Vitamin E and the

risk of prostate cancer: the Selenium and Vitamin E Cancer Prevention Trial (SELECT). JAMA 2011; 306(14): 1549-56. Available from: http://www.scopus.com/inward/record.url?eid=2-s2.0-80053954605&partnerID=tZOtx3y1

[16] Lotan Y, Goodman PJ, Youssef RF, Svatek RS, Shariat SF, Tangen CM, *et al.* Evaluation of vitamin E and selenium supplementation for the prevention of bladder cancer in SWOG coordinated SELECT. J Urol 2012; 187(6): 2005-10. Available from: http://www.scopus.com/inward/record.url?eid=2-s2.0-84861094601&partnerID=tZOtx3y1

[17] Moreau RA, Whitaker BD, Hicks KB. Phytosterols, phytostanols, and their conjugates in foods: structural diversity, quantitative analysis, and health-promoting uses. Prog Lipid Res 2002; 41(6): 457-500. [Internet]. [http://dx.doi.org/10.1016/S0163-7827(02)00006-1] [PMID: 12169300]

[18] Harrabi S, Herchi W, Kallel H, Mayer P, Boukhchina S. Liquid chromatographic–mass spectrometric analysis of glycerophospholipids in corn oil. Food Chem Elsevier Ltd 2009; 114(2): 712-6. Available from: http://linkinghub.elsevier.com/retrieve/pii/S0308814608011874

[19] Giuffrè AM, Louadj L. Influence of crop season and cultivar on sterol composition of monovarietal olive oils in reggio calabria (Italy). Czech J Food Sci 2013; 31(Patel): 256-63.

[20] Minguez-Mosquera MI, Gandul-Rojas B, Garrido-Fernandez J, Gallardo-Guerrero L. Pigments present in virgin olive oil. J Am Oil Chem Soc Springer-Verlag 1990; 67(3): 192-6. Available from: http://www.scopus.com/inward/ record.url?eid=2-s2.0-0025402294&partnerID=tZOtx3y1

[21] Bendich A, Olson JA. Biological actions of carotenoids. FASEB J 1989; 3(8): 1927-32. [PMID: 2656356]

[22] Gabás-Rivera C, Barranquero C, Martínez-Beamonte R, Navarro MA, Surra JC, Osada J. Dietary squalene increases high density lipoprotein-cholesterol and paraoxonase 1 and decreases oxidative stress in mice. PLoS One. Public Library of Science 2014; 9(8): e104224. Available from: http://www.scopus.com/inward/record.url?eid=2-s2.0-84905906211& partnerID=tZOtx3y1

[23] Kostyuk V, Potapovich A, Stancato A, *et al.* Photo-oxidation products of skin surface squalene mediate metabolic and inflammatory responses to solar UV in human keratinocytes. PLoS One 2012; 7(8): e44472. [http://dx.doi.org/10.1371/journal.pone.0044472] [PMID: 22952984]

[24] Caramia G, Gori A, Valli E, Cerretani L. Virgin olive oil in preventive medicine: From legend to epigenetics. Eur J Lipid Sci Technol 2012; 114(4): 375-88. [Internet]. [http://dx.doi.org/10.1002/ejlt.201100164]

[25] Guinda A, Rada M, Delgado T, Gutiérrez-Adánez P, Castellano JM. Pentacyclic triterpenoids from olive fruit and leaf. J Agric Food Chem 2010; 58(17): 9685-91. [http://dx.doi.org/10.1021/jf102039t] [PMID: 20712364]

[26] Tian L ting, Ma L, Du N sheng. Survey of pharmacology of oleanolic acid [Internet]. Zhongguo zhong yao za zhi = Zhongguo zhongyao zazhi = China journal of Chinese materia medica 2002; 901: 884-6. Available from: http://www.scopus.com/inward/record.url?eid=2-s2.0-0142090553&partner-ID=tZOtx3y1

[27] Dzubak P, Hajduch M, Vydra D, *et al.* Pharmacological activities of natural triterpenoids and their

therapeutic implications. Nat Prod Rep 2006; 23(3): 394-411.
[http://dx.doi.org/10.1039/b515312n] [PMID: 16741586]

[28] Liu J. Oleanolic acid and ursolic acid: research perspectives. J Ethnopharmacol 2005; 100(1-2): 92-4. [http://dx.doi.org/10.1016/j.jep.2005.05.024] [PMID: 15994040]

[29] Juan ME, Wenzel U, Ruiz-Gutierrez V, Daniel H, Planas JM. Olive fruit extracts inhibit proliferation and induce apoptosis in HT-29 human colon cancer cells. J Nutr 2006; 136(10): 2553-7. [PMID: 16988125]

[30] Kuo R-Y, Qian K, Morris-Natschke SL, Lee K-H. Plant-derived triterpenoids and analogues as antitumor and anti-HIV agents. Nat Prod Rep 2009; 26(10): 1321-44. [http://dx.doi.org/10.1039/b810774m] [PMID: 19779642]

[31] Yamai H, Sawada N, Yoshida T, *et al.* Triterpenes augment the inhibitory effects of anticancer drugs on growth of human esophageal carcinoma cells *in vitro* and suppress experimental metastasis *in vivo*. Int J Cancer 2009; 125(4): 952-60. [http://dx.doi.org/10.1002/ijc.24433] [PMID: 19462449]

[32] Martín R, Hernández M, Córdova C, Nieto ML. Natural triterpenes modulate immune-inflammatory markers of experimental autoimmune encephalomyelitis: therapeutic implications for multiple sclerosis. Br J Pharmacol 2012; 166(5): 1708-23. Available from: http://www.scopus.com/inward/record.url?eid=2-s2.0-84862729245&partnerID=tZOtx3y1 [http://dx.doi.org/10.1111/j.1476-5381.2012.01869.x]

[33] Rietjens SJ, Bast A, Haenen GR. New insights into controversies on the antioxidant potential of the olive oil antioxidant hydroxytyrosol. J Agric Food Chem 2007; 55(18): 7609-14. [http://dx.doi.org/10.1021/jf0706934] [PMID: 17683140]

[34] Tsimidou MZ, Boskou D. The health claim on "olive oil polyphenols" and the need for meaningful terminology and effective analytical protocols. Eur J Lipid Sci Technol Wiley-VCH Verlag; 2015; 117(8): 1091-4. Available from: http://www.scopus.com/inward/ record.url?eid=2-s2.--84938751635&partnerID=tZOtx3y1

[35] Bianco A, Uccella N. Biophenolic components of olives. Food Res Int 2000; 33: 475-85. [http://dx.doi.org/10.1016/S0963-9969(00)00072-7]

[36] Parkinson L, Keast R. Oleocanthal, a phenolic derived from virgin olive oil: a review of the beneficial effects on inflammatory disease. 2014; 15(7): 12323-34. [http://dx.doi.org/10.3390/ijms150712323]

[37] Brenes M, Hidalgo FJ, García A, Rios JJ, García P, Zamora R, *et al.* Pinoresinol and 1-acetoxypinoresinol, two new phenolic compounds identified in olive oil. JAOCS, J Am Oil Chem Soc. Soc 2000; 77(7): 715-20. Available from: http://www.scopus.com/ inward/record.url?eid=2-s2.--0034226996&partnerID=tZOtx3y1

[38] Fernández-Escobar R, Beltrán G. Sánchez-Zamora M a., García-Novelo J, Aguilera MP, Uceda M. Olive oil quality decreases with nitrogen over-fertilization. HortScience 2006; 41(1): 215-9.

[39] Morales-Sillero A, Jiménez R, Fernández JE. Troncoso a., Beltrán G. Influence of fertigation in "Manzanilla de Sevilla" olive oil quality. HortScience 2007; 42(5): 1157-62.

[40] Berenguer MJ, Vossen PM, Grattan SR, Connell JH, Polito VS. Tree irrigation levels for optimum

chemical and sensory properties of olive oil. HortScience 2006; 41(2): 427-32.

[41] Gómez-Rico A, Salvador MD, Moriana A, Pérez D, Olmedilla N, Ribas F, *et al.* Influence of different irrigation strategies in a traditional Cornicabra cv. olive orchard on virgin olive oil composition and quality. Food Chem 2007; 100(2): 568-78. [http://dx.doi.org/10.1016/j.foodchem.2005.09.075]

[42] Grattan SR, Berenguer MJ, Connell JH, Polito VS, Vossen PM. Olive oil production as influenced by different quantities of applied water. Agric Water Manage 2006; 85(1-2): 133-40. [http://dx.doi.org/10.1016/j.agwat.2006.04.001]

[43] Jose Motilva M, Jess Tovar M, Paz Romero M, Alegre S, Girona J. Influence of regulated deficit irrigation strategies applied to olive trees (Arbequina cultivar) on oil yield and oil composition during the fruit ripening period. J Sci Food Agric 2000; 80(14): 2037-43. [http://dx.doi.org/10.1002/1097-0010(200011)80:14<2037::AID-JSFA733>3.0.CO;2-0]

[44] Romero MP, Tovar MJ, Girona J, Motilva MJ. Changes in the HPLC phenolic profile of virgin olive oil from young trees (*Olea europaea* L. Cv. Arbequina) grown under different deficit irrigation strategies. J Agric Food Chem 2002; 50(19): 5349-54. [Internet]. [http://dx.doi.org/10.1021/jf020357h] [PMID: 12207473]

[45] Tovar MJ, Motilva MJ, Romero MP. Changes in the phenolic composition of virgin olive oil from young trees (*Olea europaea* L. cv. Arbequina) grown under linear irrigation strategies. J Agric Food Chem 2001; 49(11): 5502-8. [http://dx.doi.org/10.1021/jf0102416] [PMID: 11714351]

[46] Beltrán G, Aguilera MP, Del Rio C, Sanchez S, Martinez L. Influence of fruit ripening process on the natural antioxidant content of Hojiblanca virgin olive oils. Food Chem 2005; 89(2): 207-15. [http://dx.doi.org/10.1016/j.foodchem.2004.02.027]

[47] Dag A, Kerem Z, Yogev N, Zipori I, Lavee S, Ben-David E. Influence of time of harvest and maturity index on olive oil yield and quality. Sci Hortic (Amsterdam) 2011; 127(3): 358-66. Available from: http://www.scopus.com/inward/ record.url?eid=2-s2.0-78650225364&partnerID=tZOtx3y1 [http://dx.doi.org/10.1016/j.scienta.2010.11.008]

[48] Salvador MD, Aranda F, Fregapane G. Influence of fruit ripening on "Cornicabra" virgin olive oil quality: A study of four successive crop seasons. Food Chem 2001; 73(1): 45-53. [http://dx.doi.org/10.1016/S0308-8146(00)00276-4]

[49] Dugo G, Pellicano TM, La Pera L, Lo Turco V, Tamborrino A, Clodoveo ML. Determination of inorganic anions in commercial seed oils andin virgin olive oils produced from de-stoned olives andtraditional extraction methods, using suppressedion exchange chromatography (IEC). Food Chem 2007; 102: 599-605. [http://dx.doi.org/10.1016/j.foodchem.2006.05.039]

[50] Clodoveo ML, Hbaieb RH, Kotti F, Mugnozza GS, Gargouri M. Mechanical strategies to increase nutritional and sensory quality of virgin olive oil by modulating the endogenous enzyme activities 2014; 13(2): 135-54. [http://dx.doi.org/10.1111/1541-4337.12054]

[51] Clodoveo ML, Dipalmo T, Schiano C, La Notte D, Pati S. What's now, what's new and what's next in virgin olive oil elaboration systems? A perspective on current knowledge and future trends. J Agric

Eng 2014; 45(2): 49. Available from: http://www.scopus.com/inward/ record.url?eid=2-s2.--84907527486&partnerID=tZOtx3y1

[52] Clodoveo ML. Malaxation: Influence on virgin olive oil quality. Past, present and future – An overview. Trends Food Sci Technol 2012; 25(1): 13-23. [http://dx.doi.org/10.1016/j.tifs.2011.11.004]

[53] Jimenez Marquez A, Hermoso Fernandez M, Uceda Ojeda M. Extraction of virgin olive oil by two-phase continuous system. Influence of different variables of the process on certain parameters related to oil quality. Grasas Aceites 1995; 46(4–5): 299-303.

[54] Ranalli A, Contento S, Schiavone C, Simone N. Malaxing temperature affects volatile and phenol composition as well as other analytical features of virgin olive oil 2001; 103(4): 228-38. [http://dx.doi.org/10.1002/1438-9312(200104)103:4<228::AID-EJLT228>3.0.CO;2-7]

[55] Di Giovacchino L, Costantini N, Ferrante ML, Serraiocco A. Influence of malaxation time of olive paste on oil extraction yields and chemical and organoleptic characteristics of virgin olive oil obtained by a centrifugal decanter at water saving. Grasas Aceites 2002; 53(2): 179-86. [http://dx.doi.org/10.3989/gya.2002.v53.i2.302]

[56] Amirante R, Cini E, Montel GL, Pasqualone A. Influence of mixing and extraction parameters on virgin olive oil quality Influencia del batido y de los parámetros de extracción en la calidad del aceite de oliva virgen. Grasas Aceites 2001; 52(3-4): 198-201. [http://dx.doi.org/10.3989/gya.2001.v52.i3-4.357]

[57] Gómez-Rico A, Inarejos-García AM, Salvador MD, Fregapane G. Effect of Malaxation Conditions on Phenol and Volatile Profiles in Olive Paste and the Corresponding Virgin Olive Oils (*Olea europaea* L. Cv. Cornicabra). J Agric Food Chem 2009; 57(9): 3587-95. Available from: http://www.scopus.com/inward/record.url?eid=2-s2.0-66149117090&partnerID=tZOtx3y1

[58] Clodoveo ML, Hachicha Hbaieb R. Beyond the traditional virgin olive oil extraction systems: Searching innovative and sustainable plant engineering solutions. Food Res Int Elsevier Ltd 2013; 54(2): 1926-33. [http://dx.doi.org/10.1016/j.foodres.2013.06.014]

[59] Amirante P, Clodoveo ML, Leone A, Tamborrino A, Patel VB. Influence of Different Centrifugal Extraction Systems on Antioxidant Content and Stability of Virgin Olive Oil. Olives and Olive Oil in Health and Disease Prevention. Elsevier Inc 2010; 85-93. [http://dx.doi.org/10.1016/B978-0--2-374420-3.00010-3]

CHAPTER 2

Wine: Biologic Active Compounds and Health Benefits

Constantin Apetrei*

Department of Chemistry, Physics and Environment, Faculty of Sciences and Environment, "Dunarea de Jos" University of Galati, Romania

Abstract: Wine contains several hundreds of compounds, most of them with impact on quality attributes and health benefices. Among these compounds, phenolic compounds play a significant role in sensory properties of wines as well as in antioxidant properties. Phenolic compounds are highly efficient compounds displaying antioxidant properties and preventing human cardiovascular diseases, also. Furthermore, the polyphenolic compounds are capable of quenching free radicals, thereby inhibiting the oxidation process of low density lipids (LDL). This effect about LDL could reduce the risk of atherosclerosis, cardiovascular diseases, heart attack, cancer, DNA degradation *etc*. Moderate drinking of red wine is beneficial against coronary disease and several forms of cancer. The phenolic compounds thought to be responsible are catechins and tannins. Catechins are powerful antioxidants, preventing cellular damage. Resveratrol and quercetin may improve the immune system, prevent towards heart diseases, inhibit cancer formation, and even prolong life. The amino acids are essential nutrients and have various benefits on human body such as muscle maintenance, tissue repair, and immunity. Amino acids yield biogenic amines that have allergenic or toxic effects on humans. Beneficial nitrogen compounds such as melatonin and serotonin are also found in red wines. Among vitamins, only vitamins B_1, B_2 and B_{12} are present in wines.

Keywords: Antioxidant, Catechin, Free radical, Polyphenols, Resveratrol, Wine.

* **Corresponding author Constantin Apetrei:** Department of Chemistry, Physics and Environment, Faculty of Sciences and Environment, "Dunarea de Jos" University of Galati, 47 Domneasca Street, 800008 Galati, Romania; E-mail: apetreic@ugal.ro

INTRODUCTION

A definition of "wine" is "the drink resulting from the fermentation by the yeast-cells, and also in certain cases by the cells of lactic bacteria, of the juice from the crushing or maceration of grape-cells" [1].

Wine is a natural multicomponent alcoholic beverage, which includes various bioactive compounds. The chemical composition is related to different factors, among them could be highlighted quality of grape, winemaking process, storage and ageing of wine, transport and storage of bottles *etc* [2 - 5].

Wine compounds include water, alcohols, acids, sugars, phenolic compounds, nitrogen containing compounds, vitamins and different aroma compounds [2 - 5]. The phenolic compounds from grape and oak barrels are a huge group of biologic active compounds with vital functions in winemaking [2 - 8]. Sensory characteristics of wine, for example the taste (sweet, sour, salty, bitter and *umami*), astringency, and colour are related to several compounds, the phenolic compounds being the most important [9 - 12]. Excellent antioxidant properties of wine is often correlated with principal health benefices [13 - 16]. A reasonable drinking of red wine is recommended and it is part of Mediterranean diet [17, 18].

PHENOLIC COMPOUNDS

Phenolic compounds are secondary metabolites present in the vegetable kingdom [19 - 22]. Phenolics have a similar structure involving a benzene ring substituted with one or more hydroxyl groups [23, 24].

Phenolics are produced from one amino acid phenylalanine. The phenylpropanoid biosynthetic pathway is recognized as the principal one. In plants, it have been identified more than 4,000 different phenolic compounds. Their function in plants is related to numerous functions: defence from UV radiation, defence against some pathogens, pigmentation, attraction of pollinators, and spreading of seeds [25 - 27].

White wines contains basically phenolic compounds from the flesh of the grapes, for instance gallic acid, hydroxynammic acid esters, catechin, epicatechin, gallocatechin gallate, procyanidin and catechin gallate. The phenolics amounts in white wines varies and usually they are in the range of few mg/L [28, 29]. In addition, red wines contains some special classes of phenolic compounds such as: flavonol-3 group (catechin), flavane (3,4)-diol group (leucocyanidin), flavonol-3 (quercetin), anthocyanins and tannins, in addition to compounds present in white wines [30 - 32].

Chemical structures of phenolic compounds comprise from aromatic based compounds with low molecular mass to polymeric, high molecular mass compounds such as tannins. There are two groups of phenolic compounds: flavonoids (having a common C6-C3-C6 skeleton) and non-flavonoids [33 - 35].

Classification and principal phenolic compounds found in grape and wine is summarized in Table **1**.

Table 1. Classification of phenolic compounds

Phenolic compounds	
Non- Flavonoids	Flavonoids
Phenolic acids - Hydroxybenzoic acids - Hydroxycinnamic acids Stilbenes	Flavonols Flavononols and flavones Flavanols - Catechins - Condensed Tannins - Procyanidins - Prodelphinidins Anthocyanins

Phenolic acids from wines include hydroxybenzoic acids and hydroxycinnamic acids [36 - 38]. Hydroxybenzoic acids have as principal characteristics a C6-C1 skeleton. The most frequent hydroxybenzoic acids (Fig. **1**), from wine are gallic acid, gentisic acid, p-hydroxybenzoic acid, protocatechuic acid, syringic acid, salicylic acid, and vanillic acid [39, 40].

Gallic acid is the hydroxybenzoic acid present in greater amounts in wine. It yields from the grape and from the hydrolysis of tannins [41, 42].

Hydroxycinnamic acids have as identity characteristics a C6-C3 skeleton. The different compounds present in wine are caffeic acid, caftaric acid, p-coumaric acid, p-coutaric acid, ferulic acid, fertaric acid and sinapic acid [43, 44].

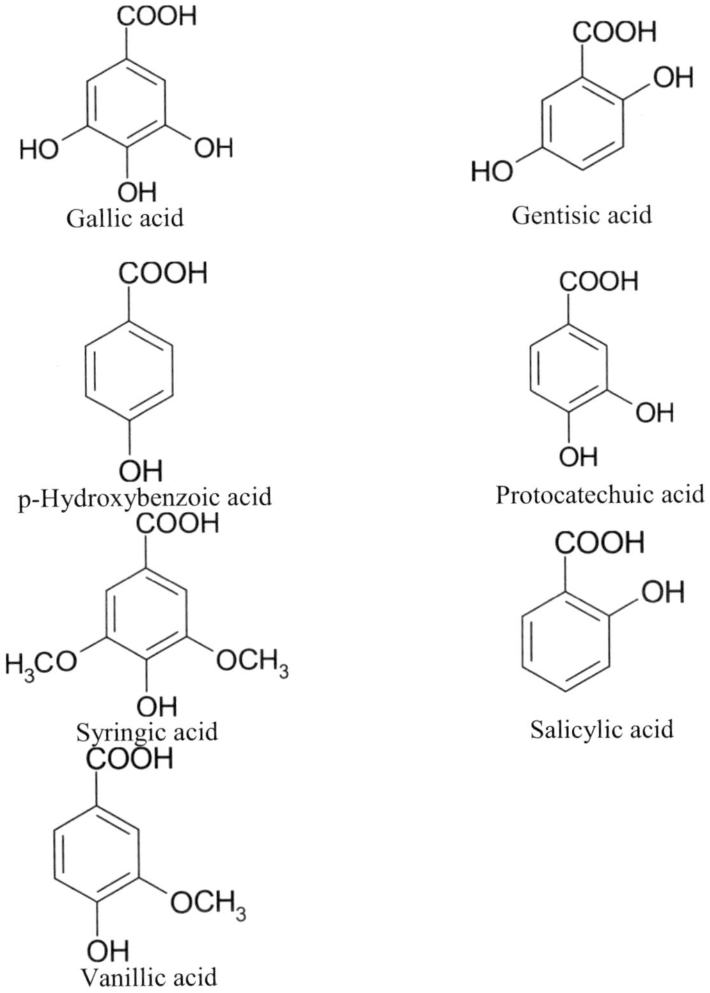

Fig. (1). Molecular structures of hydroxybenzoic acids.

From the class of hydroxycinnamic acids, caftaric acid (Fig. **2**), is the hydroxybenzoic acid presents in greater amounts in wine (more than 50% from total hydroxycinnamic acids) [45].

Stilbene derivatives (Fig. **3**), have two benzene rings in the basic structure, usually bonded by an ethylene chain. The principal stilbenes found in wine are trans-resveratrol, cis-resveratrol, trans-piceid, cis-piceid, trans-piceatannol, trans-pterostilbene, trans-δ-viniferin, trans- ε-viniferin, α-viniferin and hopeaphenol [46, 47].

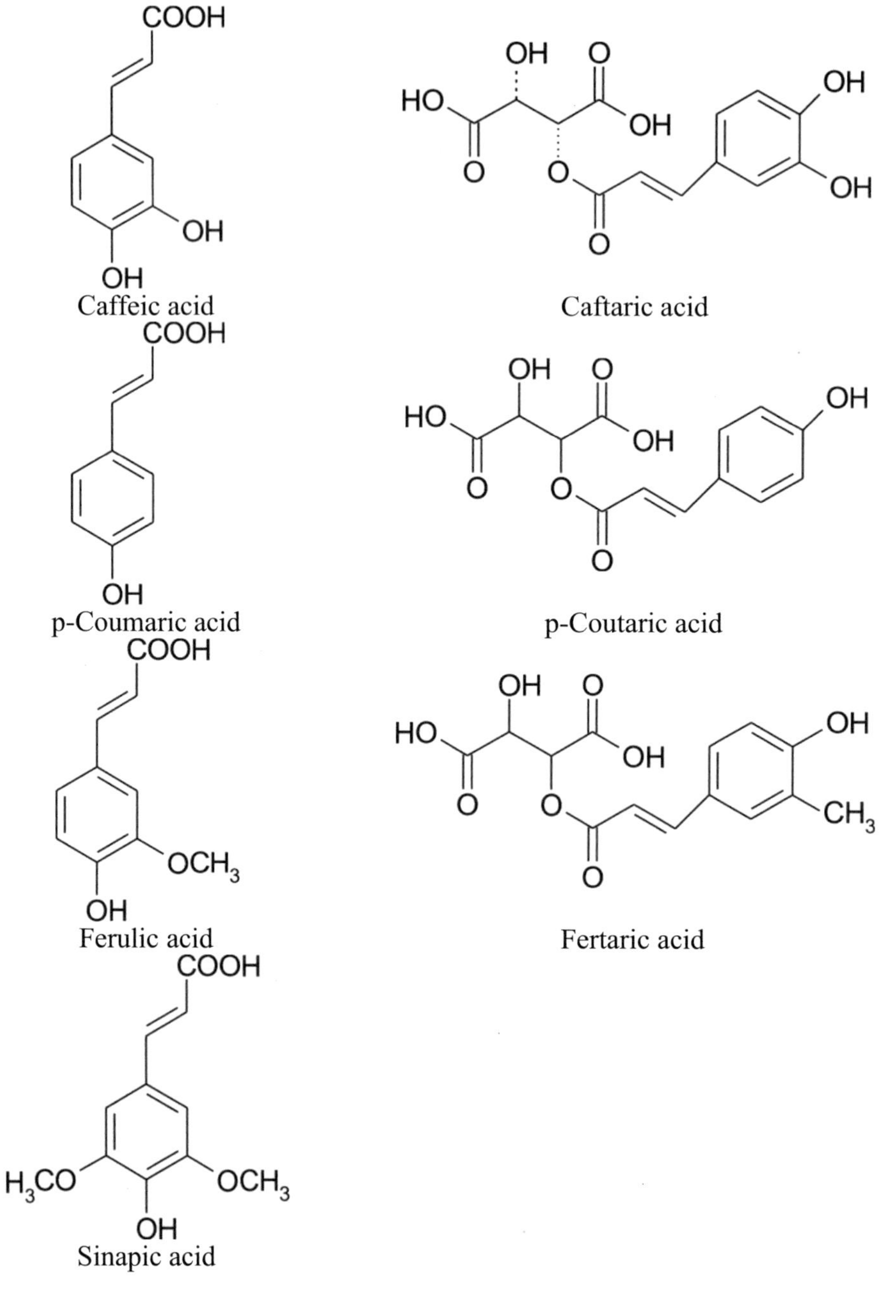

Fig. (2). Molecular structures of hydroxycinnamic acids.

The most relevant and studied stilbenic compound is trans-resveratrol, formed in grapevines as reaction of plant to fungal disease produced by *Botrytis cinerea* [48].

Flavanols, flavonols, dihydroflavonols and anthocyanins take part from the flavonoid family. These compounds are characterized by a C6-C3-C6 skeleton formed from two phenolic rings linked together by a heterocyclic pyran ring [49, 50].

Several classes can be distinguished on the basis of the oxidation stage of the pyran ring.

Flavanols (flavan 3-ols) found in grape and wine are (+)-catechin, (–)-epicatechin, (–)-epicatechin 3-gallate, (+)-gallocatechin, (–)-epigallocatechin, (+)-gallocatechin 3-gallate and (–)-epigallocatechin 3-gallate [51, 52].

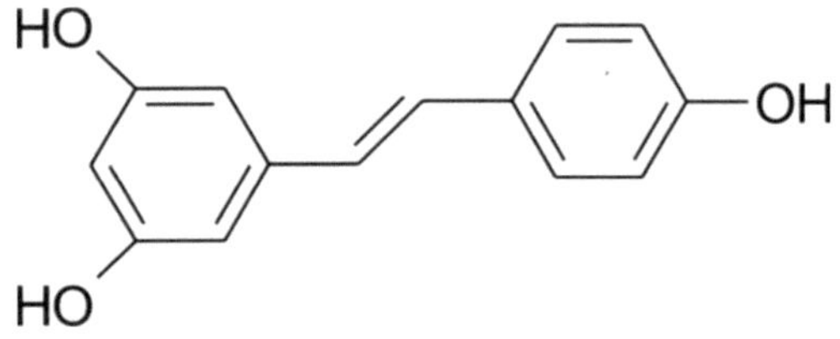

trans-Resveratrol

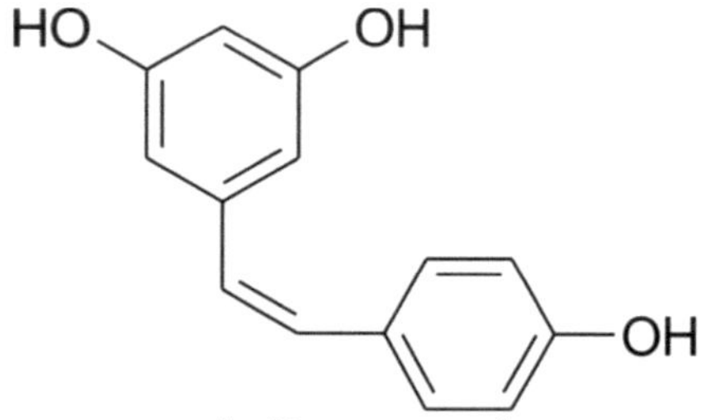

cis-Resveratrol

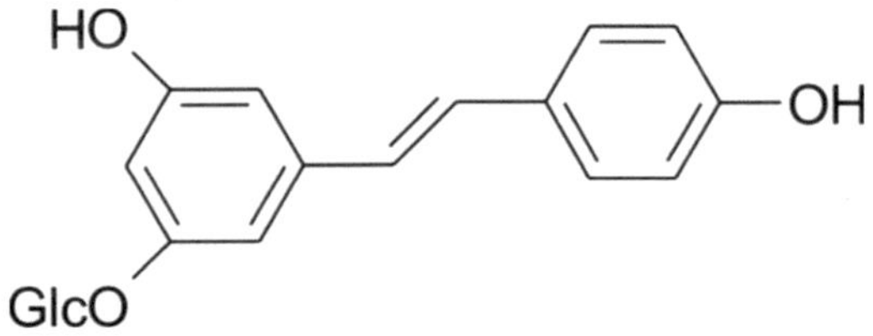

trans-Piceid

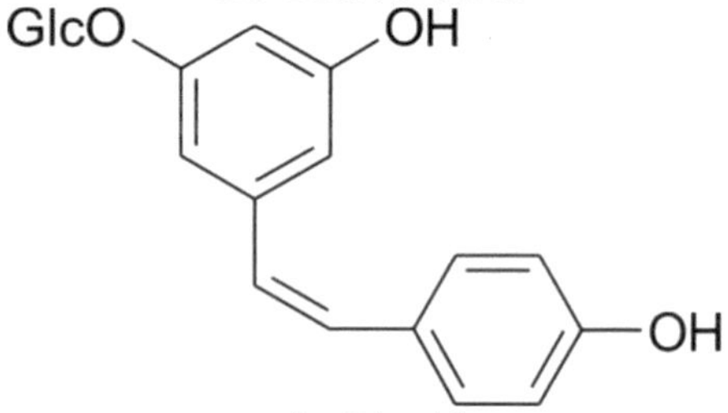

cis-Piceid

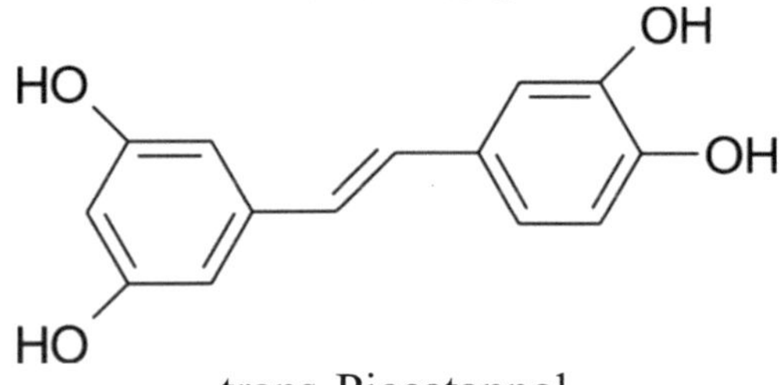

trans-Piceatannol

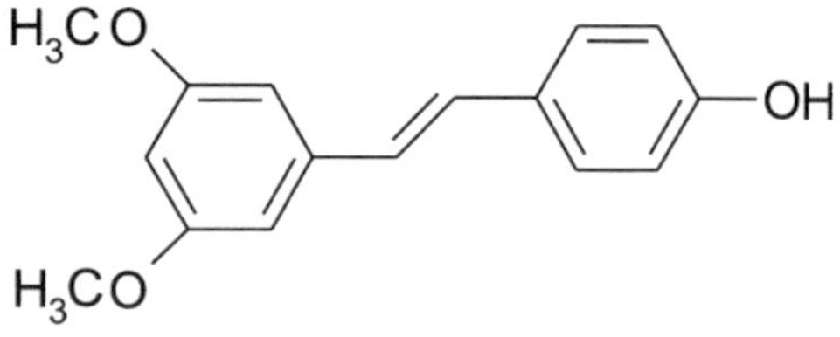

trans-Pterostilbene

Fig. 3 contd.....

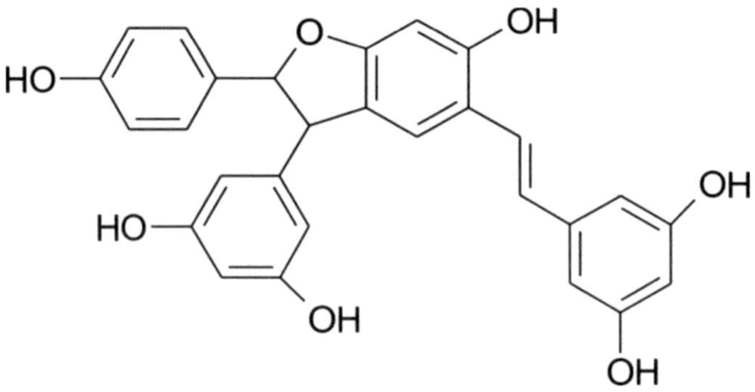

trans-δ-Viniferin

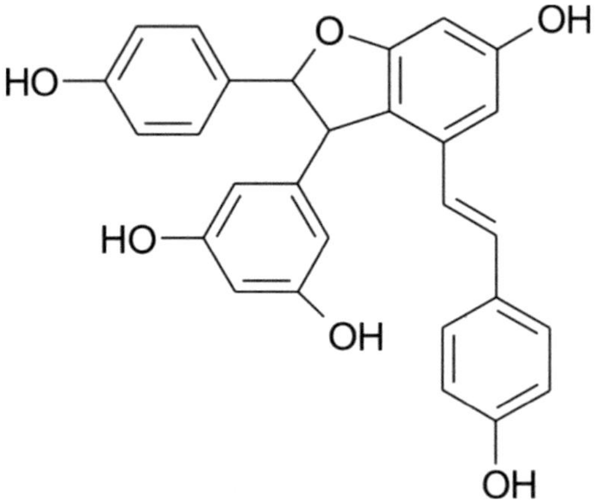

trans- ε-Viniferin

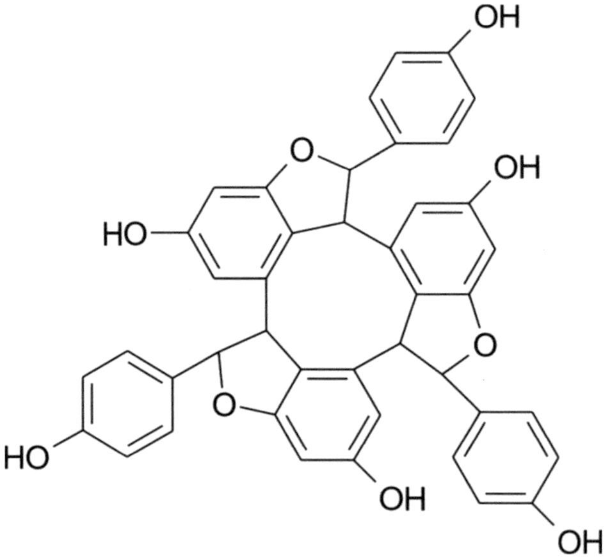

α-Viniferin

Fig. 3 contd.....

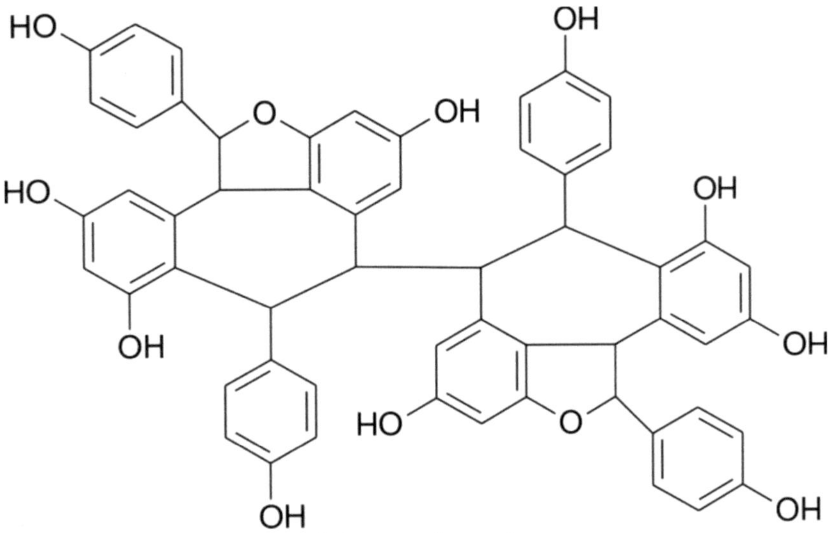

Hopeaphenol

Fig. (3). Molecular structures of stilbenes from wine.

Flavanols molecules (Fig. **4**), have a natural tendency towards condensation forming stable compounds of two, three, four or more molecules bound together. Dimers, trimers, tetramers and pentamers are known as oligomeric proanthocyanidins. The clusters larger than pentamers are known as tannins or polymeric proanthocyanidins [53 - 55].

The term "tannin" refers to the ability of specific compounds to interact with proteins, which are mostly involved in the astringency perception [56]. The term "proanthocyanidin" is related to the capacity to release anthocyanidins when heated under acidic conditions [57]. In Fig. (**5**) is presented the hypothetical tannin made up from condensation of four subunits: catechin, epicathechin, epigallocatechin, epicatechin gallate.

Tannins (principally condensed tannins) are found in red wine. These compounds found in wine can be originates from several sources and the tactile properties vary in function of the source: grape skins, stems, and grape seeds. Wine tannins extraction depends on the winemaking process performed [58].

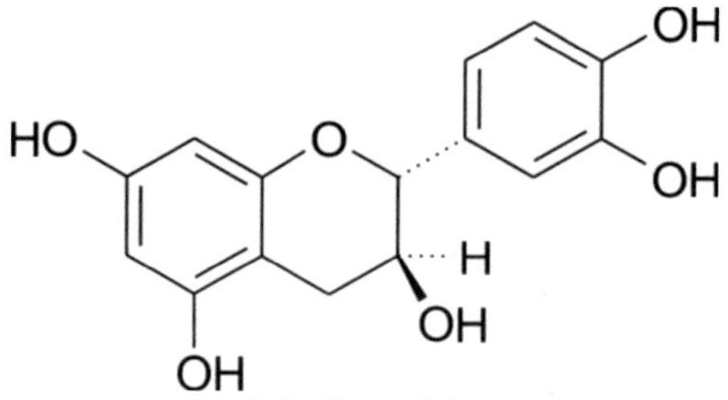

(+)-Catechin

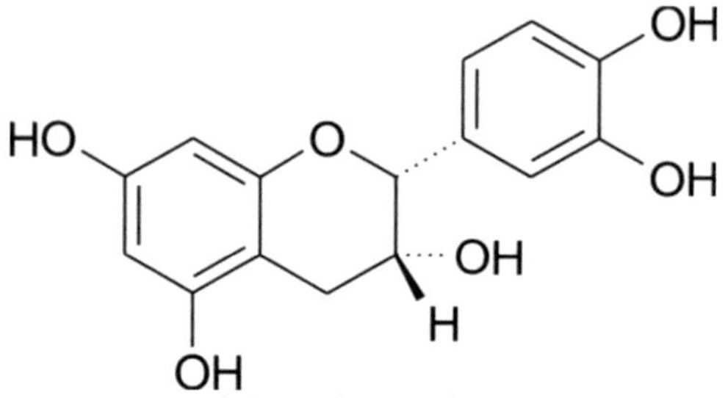

(–)-Epicatechin

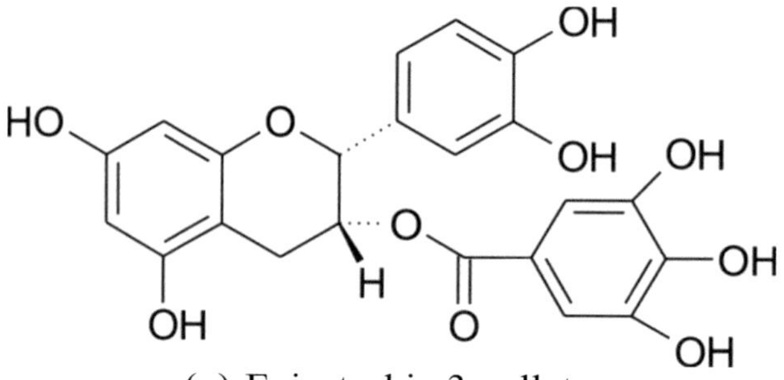

(–)-Epicatechin 3-gallate

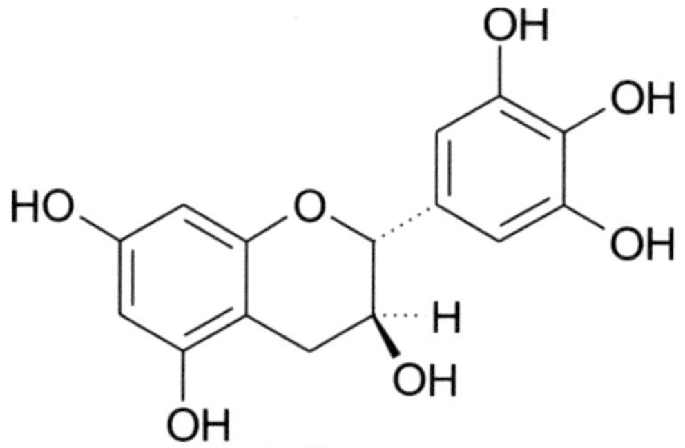

(+)-Gallocatechin

Fig. 4 contd.....

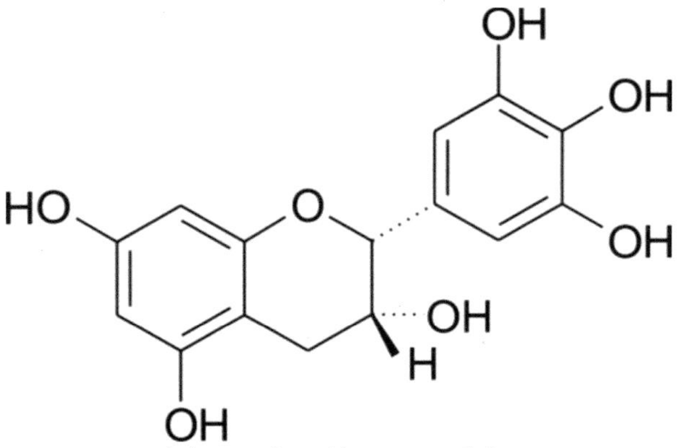

(–)-Epigallocatechin

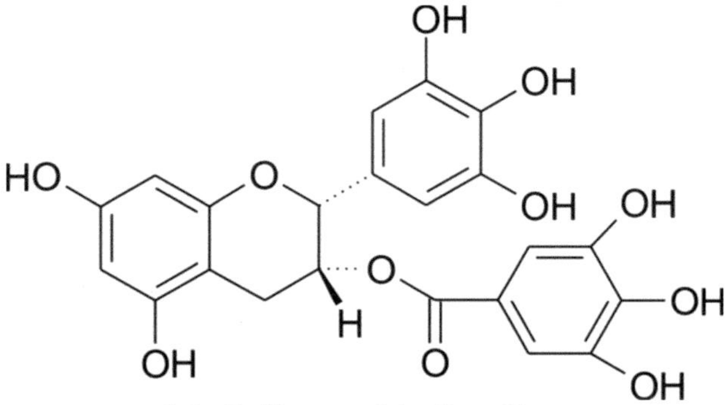

(+)-Gallocatechin 3-gallate

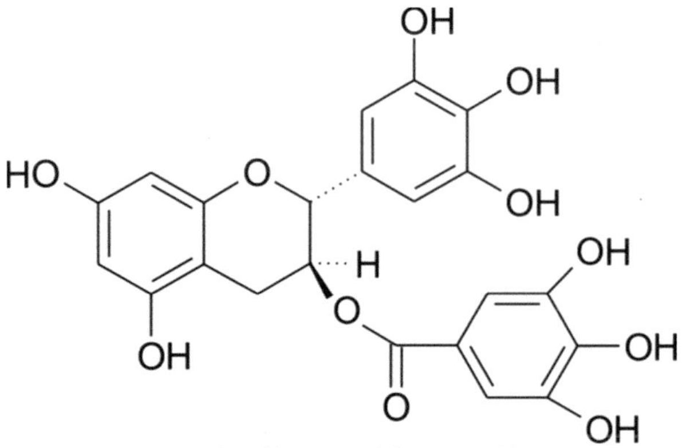

(–)-Epigallocatechin 3-gallate

Fig. (4). Molecular structures of flavanols.

An important amount of tannin originates from oak barrels, mostly new ones, where oak barrels are employed in ageing process of wine. The tannins are transformed during the winemaking and ageing processes. The length of the chain and the phenolic subunits changes [59]. These changes are important in the sensorial properties of wines, colour, taste and aroma. The complex interaction between tannins and proteins are also very important in stability and equilibrium among wine components [60].

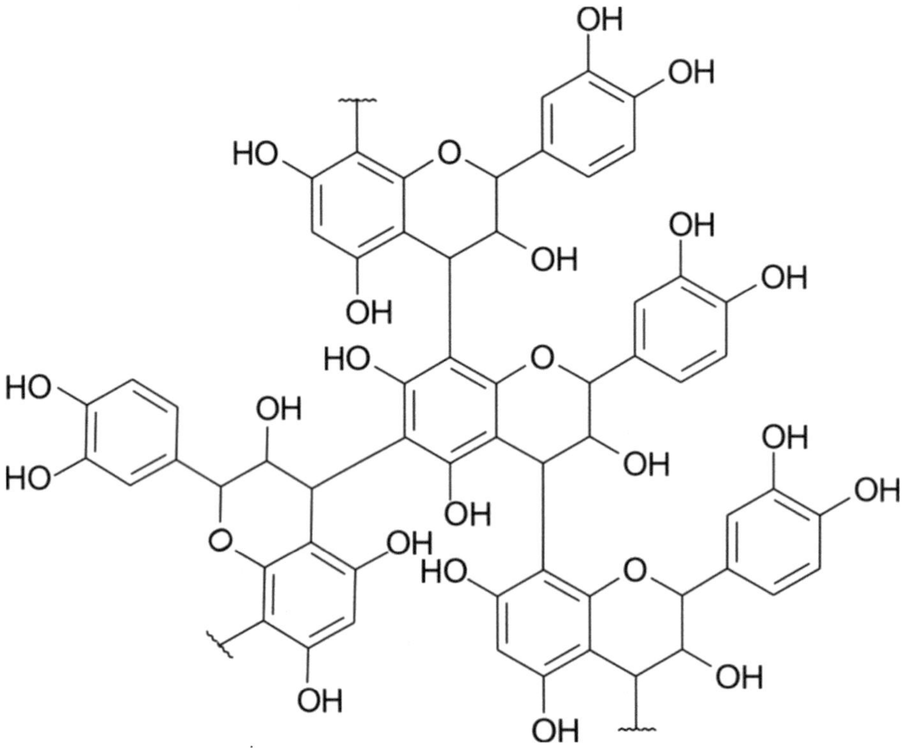

Fig. (5). Molecular structure of tannin made up from condensation of four subunits.

Flavonols (Fig. **6**), are present in skins and pulp of the grape and leaves of the grapevine. Different flavonols and dihydroflavonols (flavanonols) were detected and identified in wine. These are kaempferol, quercetin, myricetin and isorhamnetin [61, 62].

The most important flavonols in grape and wine are the glycosides of quercetin [61, 62].

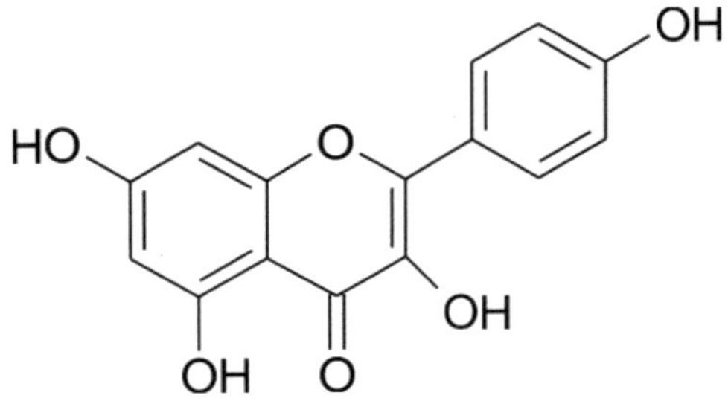

Kaempferol

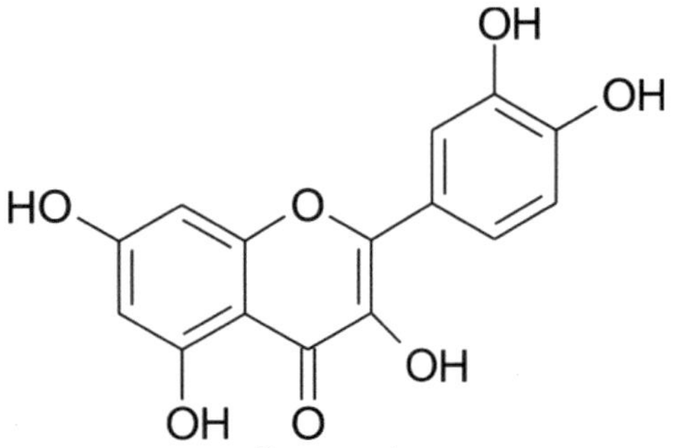

Quercetin

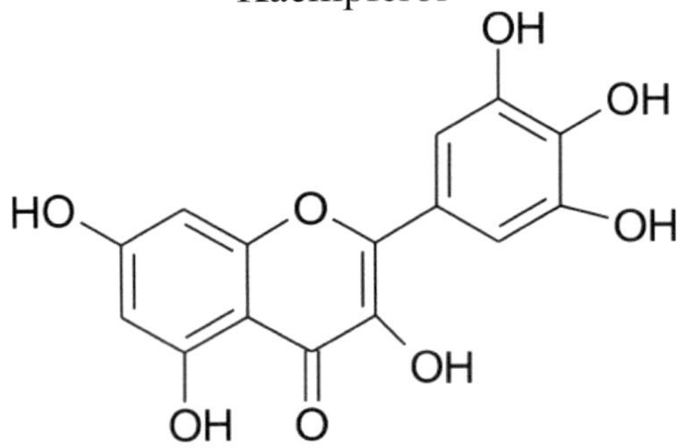

Myricetin

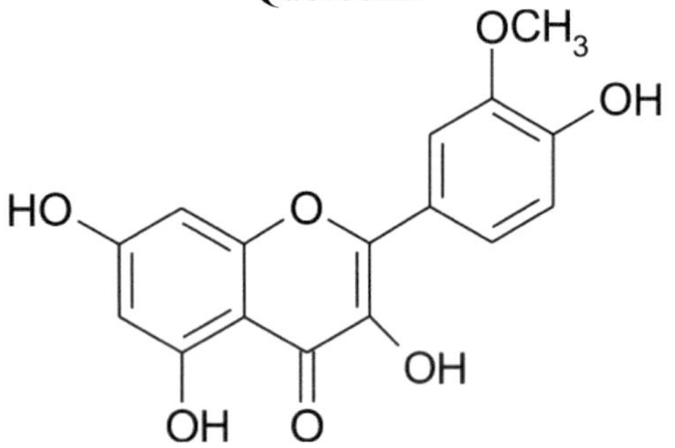

Isorhamnetin

Fig. (6). Molecular structure of flavonols.

Anthocyanins are characteristic to red varieties of grape and contained specially in berry skins. In teinturier varieties anthocyanins are also present in pulp, which is coloured [63, 64]. Their structure is characterized by the flavylium cation, including one benzene ring bonded by an oxygenated heterocycle in the cationic form [65]. Anthocyanins from wine are glucosylated derivatives of five aglycones (called anthocyanidins): cyanidin, peonidin, petunidin, delphinidin and malvidin (Fig. 7) [66, 67].

Wood has a significant influence on the flavour profile and organoleptic characteristics of a wine. By means of physico-chemical interaction between the wood and the wine, numerous substances are extracted from the wood to the wine. Some of these substances interact with other compounds from the wine forming different flavours to the original ones [68, 69].

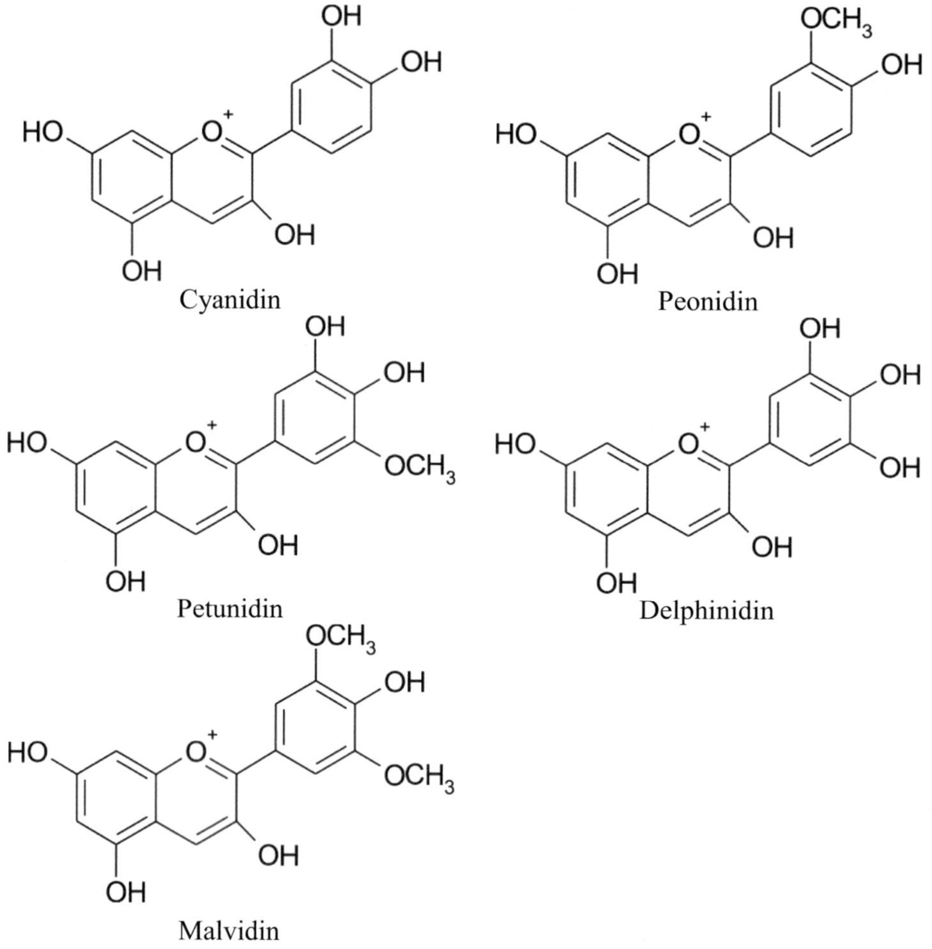

Fig. (7). Molecular structure of anthocyanidins from wine.

As a consequence of lignin decomposition in oak wood, several volatile phenols, such as guaiacol or syringynol derivatives (Fig. **8**), are formed and eliberated into wine [70]. The guaiacol derivatives have superior aroma and taste thresholds than the syringynol derivatives and contribute to wine flavour considerably more [71, 72].

Vanillin is the most representative compound of this group. Amounts of vanillin raise considerably with toasting of oak wood [73, 74]. Eugenol, the major aroma compound characteristic of cloves, is also found in oak wood [75, 76].

Phenolics compounds have antioxidant activity delaying oxidation processes [77]. The antioxidants inhibit oxidation processes playing the role of chain inhibitors by donating hydrogen atom to alkylperoxyl radicals and forming stable derivatives [78].

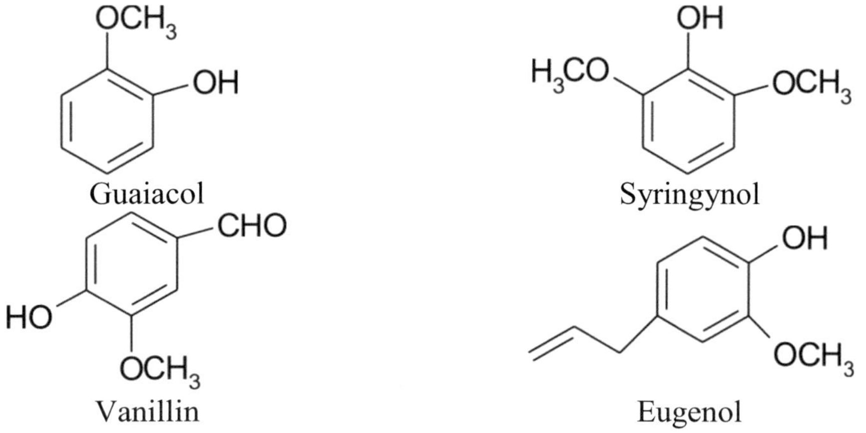

Fig. (8). Molecular structure of volatile phenols from oak wood.

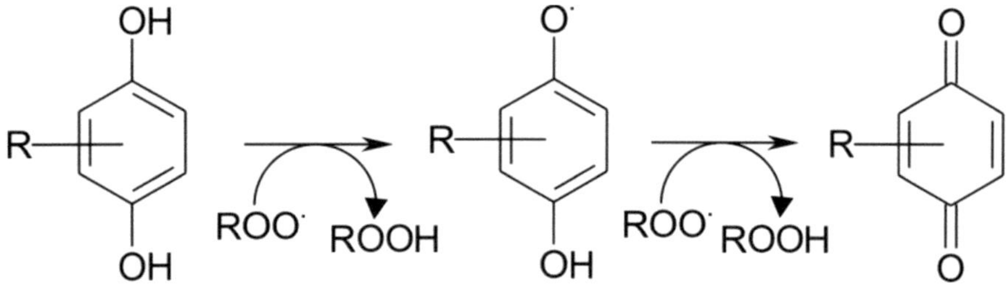

Fig. (9). Mechanism of scavenging free radicals of para-dipolyphenolic compounds.

The mechanism of scavenging free radicals was demonstrated by using DPPH (1,1-diphenyl-2-picrylhydrazyl) method (Fig. **9**), confirming that phenolic compounds act as hydrogen donors [79, 80].

HEALTH BENEFITS OF POLYPHENOLS

The interests for particular phytochemicals and nutraceuticals increased in the last period of time because of their medicinal and pharmaceutical importance [81].

Grape and wines have various in health promoting compounds. Among these compounds are the bioactive metabolites such as phenolic and polyphenolic compounds [82]. Grape and wine phenolics are important for researchers and the industry because of their bio-active properties that could be valorified by the development of commercial products [83]. The health benefit results from their high content in flavonoids and other compounds from the polyphenol category, which are responsible for the colour, taste and aroma of wines [84, 85].

Quercetin and resveratrol are the main responsible for heart protection effects [86]. These biologic active phenolic compounds can protect LDL from oxidative processes, that is related to arterial damage [87]. Furthermore, these compounds decrease blood clotting due to their antiaggregant properties [88].

The resveratrol is a bioactive compound with great significance in medicine and pharmacy. Numerous scientific papers demonstrated that trans-resveratrol is a powerful antioxidant and diminishes the risks of heart disease and block the development of cancer cells [89].

The phenolic compounds (+)-catechin, (-)-epicatechin, both as monomeric as well as dimers, trimers and tetramers have antimutagenic and antiviral effects [90]. The health benefits of procyanidins and catechins have been demonstrated. For instance, the grape extract could be used as antioxidant supplement in the functional foods [91].

Quercetin, kaempferol, myricetin, isorhamnetin and their glycosides have significant antioxidant properties [92].

Some of these natural antioxidants have also antimicrobial properties and could be used for food preservation [93].

Resveratrol, catechins and epicatechins are ones of the main components in red wine. These phenolic compounds have a high effectiveness to diminish the activity of cyclooxygenases COX-1 and COX-2 [94]. Furthermore, resveratrol, catechins and epicatechins prevent the differentiation of fibroblasts into myofibroblasts as result of the collagen formation [95].

Resveratrol has antimicrobial effects against numerous bacteria such as *Escherichia coli* and *Staphylococcus aureus* [96].

Resveratrol have a great importance in protection of humans against benzopyrene, the toxic compound from cigarette smoke that provoke lung cancer inhibiting the cell receptor. As consequence benzopyrene is not able to bind tothe membrane of the cells [97]. It is also available in the case of other carcinogenic polycyclic aromatic hydrocarbons.

Resveratrol inhibits the formation of endothelin-1 and influence the activity of hearth cells [98, 99]. Furthermore, resveratrol inhibits angiotensin II, which is a potent vasoconstricting hormone [100].

Tannins have antidiarrheal, haemostatic, and antihemorrhoidal properties. Furthermore, tannins have antiinflammatory effects and can be employed in gastritis, esophagitis, and enteritis. Tannins are also beneficial in protecting the kidneys. Tannins were used for diarrhea, dysentery, and hemorrhaging. Tannins could provoke retrogradation of tumours that are until that time present in tissue [101]. Tannins have anti-viral, antibacterial, and antiparasitic effects [102 - 104].

NITROGEN FRACTION OF WINES

Amino acids, peptides, proteins and ammonium ion are the main components of the nitrogen fraction of wines. Their presence in wine could be related to grape composition, to metabolites of yeasts, and to degradation or hydrolysis of grape proteins [105].

In wines were found different amino acids (Fig. **10**), the contents varying in function of grape variety, geographic origin and vintage, viticulture and winemaking techniques, *etc* [106].

The amino acids are used as nitrogen source for growth for yeasts and lactic bacteria having a nutritional role [107]. Amino acids have an important function in organoleptic properties of wine [108].

Several amino acids are present in enough amounts in wine to play an fundamental role in developing the taste and flavour profile of high quality wines [108].

The amount of total amino acids in red wine is from 300 to 1300 $mg \times L^{-1}$ and include the amino acids proline, alanine, glycine, glutamic acid, glutamine, arginine, and γ-aminobutyric acid [109].

Glutamic acid is an amino acid present in wine grapes. Glutamic acid is the main precursor of glutamate taste compounds. The contribution of glutamic acid and its salts (glutamates) to the global taste of wine is recognized [110].

Alanine, proline, glycine and arginine contribute to sweet taste in different foods and are present in significant amounts in wine grapes and wines. The amino acid proline interact with flavonol compounds to contribute bitter taste [111].

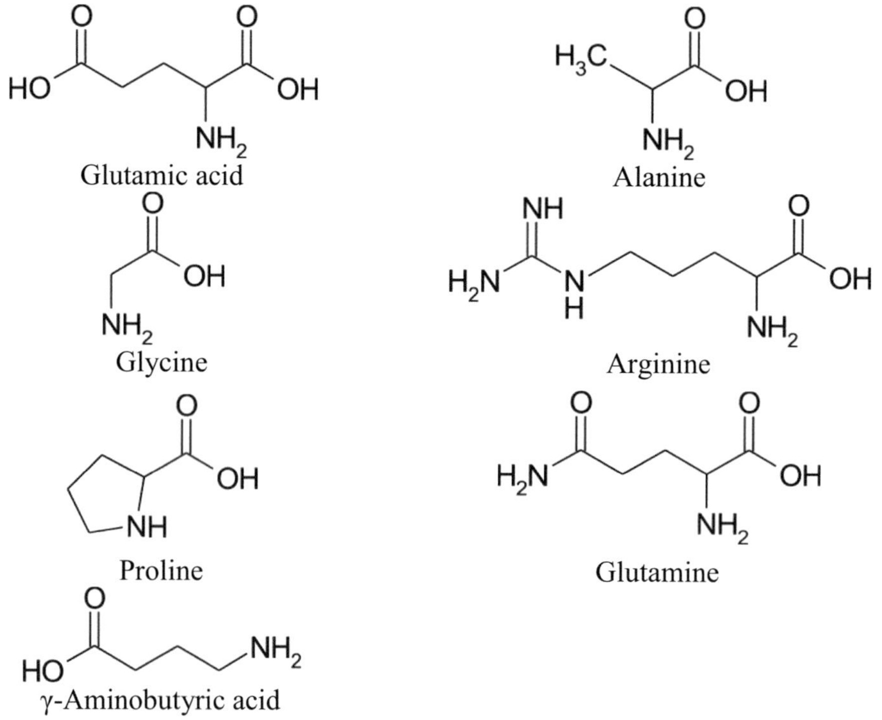

Fig. (10). Molecular structure of amino acids from wine.

Red wine contains several amino acids that are essential nutrients and have a lot of benefits in human body. Amino acids are critical nutrients in muscle maintenance, tissue repair, immunity and control of all human body's processes [112]. Additionally, from amino acids are formed biogenic amines, which are undesired in wine being allergenic or toxic. Biogenic amines appears in all food containing proteins or free amino acids because of microbial or biochemical activity [113].

Biogenic amines (Fig. **11**), result by enzymatic decarboxylation of amino acids during the metabolism of bacteria or yeasts [114]. Firstly, yeasts are active during alcoholic fermentation. Second, lactic acid bacteria are active during malolactic fermentation. Wines contain only a few milligrams per liter of each of the most important amines, such as histamine, tyramine, putrescine, tryptamine and phenethylamine [115].

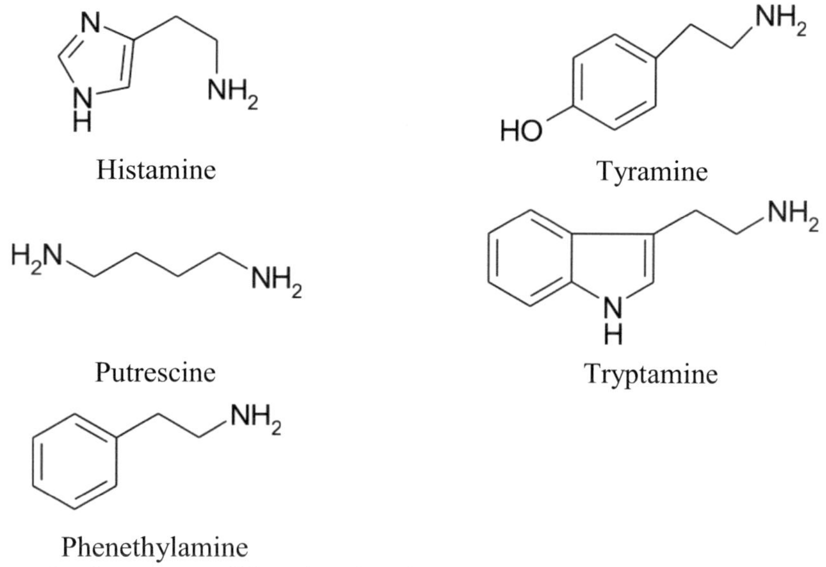

Fig. (11). Molecular structure of biogenic amines from wines.

Decarboxylase enzymes are present in living organisms but the enzyme activity is influenced by several physico-chemical parameters such as temperature, pH, sulfur dioxide content, *etc* [116].

Biogenic amine contents vary in large limits. The winemaking techniques have the major influence in the formation of these compounds in wines [117].

Quantification of biogenic amines contents is very important in wines due to biological and toxicological aspects [117]. In general, biogenic amines content could be a indicator of wine quality since their occurrence is usually associated with inappropriate sanitary conditions during vinification [118].

It is well known that the consumption of huge amounts of histamine and tyramine have toxic effects on humans. Biogenic amines amounts in wine are quite low comparing with other fermented food products rich in proteins. Presence of other compounds in wines, such as ethanol and acetaldehyde could enhance their toxicological or allergenic effects [119].

The intoxication symptoms associated with histamine are headaches, heart palpitations, rushes, nausea, low blood pressure and digestive problems. The intoxication symptoms associated with tyramine are related to migraine and hypertension [120]. Amount limits of biogenic amines have not been regulated by the Organisation Internationale de la Vigne et du Vin (OIV). However, some countries have established the maximum limits for histamine amount in wine (2 mg×L^{-1} in Germany, 10 mg×L^{-1} in Switzerland) [121].

A very important nitrogen based compound is glutathione (Fig. **12**). Glutathione (γ-L- glutamyl-L-cysteinylglycine) is a tripeptide of L-glutamate, L-cysteine and glycine. Its properties are to sulfhydryl group of the cysteine residue. Sulfhydryl group confers to glutathione particular oxido-reduction and nucleophilic properties. Glutathione can be present in the cell in the reduced or oxidized state (glutathione disulfide) (Fig. **13**) [122].

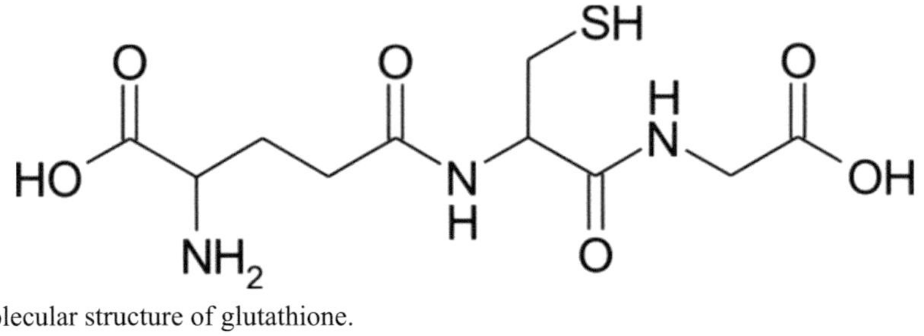

Fig. (12). Molecular structure of glutathione.

Glutathione is a powerful antioxidant that is naturally present in grapes. Glutathione in one antioxidant commonly used in oenology besides SO_2 and ascorbic acid. The reduction potential is five and three times higher than that of SO_2 and ascorbic acid respectively. Therefore, glutathione is a much more powerful antioxidant [123].

Browning is a spontaneous oxidative process taking place during the vinification of wine influencing negatively the appearance, aroma and flavor of wines. *Ortho*-diphenols are the principal compounds responsible for the oxidative browning in wine. These compounds are oxidized by the oxygen from the air, processes biocatalyzed by polyphenoloxidases (*i.e.* laccase, tyrosinase). Wine must be protect against oxidation. From the class of antioxidants, glutathione is one of the most effective against browning. Regarding the mechanism of action against browning, glutathione acts directly on quinones formed by oxidation [124]. The mechanism is presented in the Fig. (**14**).

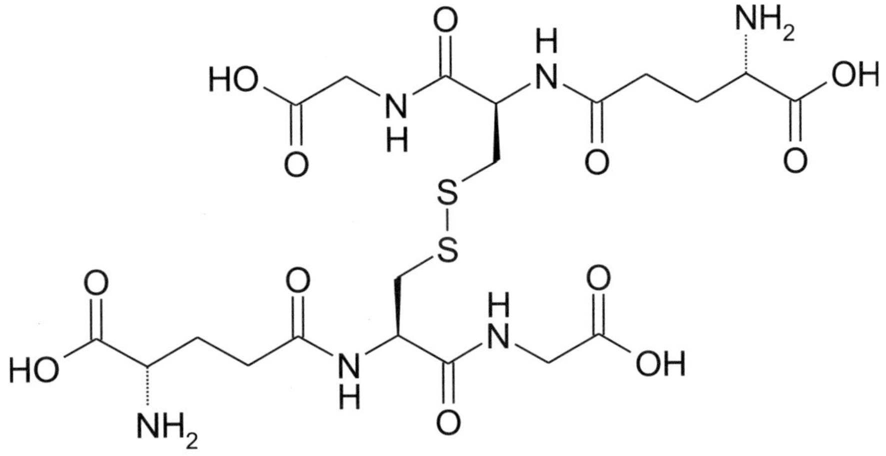

Fig. (13). Molecular structure of glutathione disulfide.

The product of reaction is a thioether that is not a substrate for further oxidation by polyphenoloxidase even it has a *ortho*-dihydroxyphenol structure.

Glutathione protects different compounds from wine against oxidation such as esters, terpenes, volatile thiols, *etc.* [124].

Glutathione content in must and wine is highly variable. Numerous factors can influence the glutathione concentration in must and wine. These are exposure to tyrosinase and laccase activity, oxygen content, grape skin maceration and pressing [125].

Glutathione is found in every cell of the body. Glutathione stimulates the action of ingested antioxidants from whole foods and allows the appropriate utilization of phytonutrients essential for health and longevity [126].

Glutathione is a powerful regenerator of immune cells and plays a major role in optimizing the immune system. Glutathione is crucially important in the synthesis and the maintenance of DNA, and in the biosynthesis of protein and amino acids in the body. Furthermore, glutathione is very important in the removal of chemical toxins and carcinogens. Additionally, glutathione is indispensable in protecting cells from oxidization and attack of free radicals by activating intracellular enzymes for the removal of all toxic species at the cellular level [127, 128].

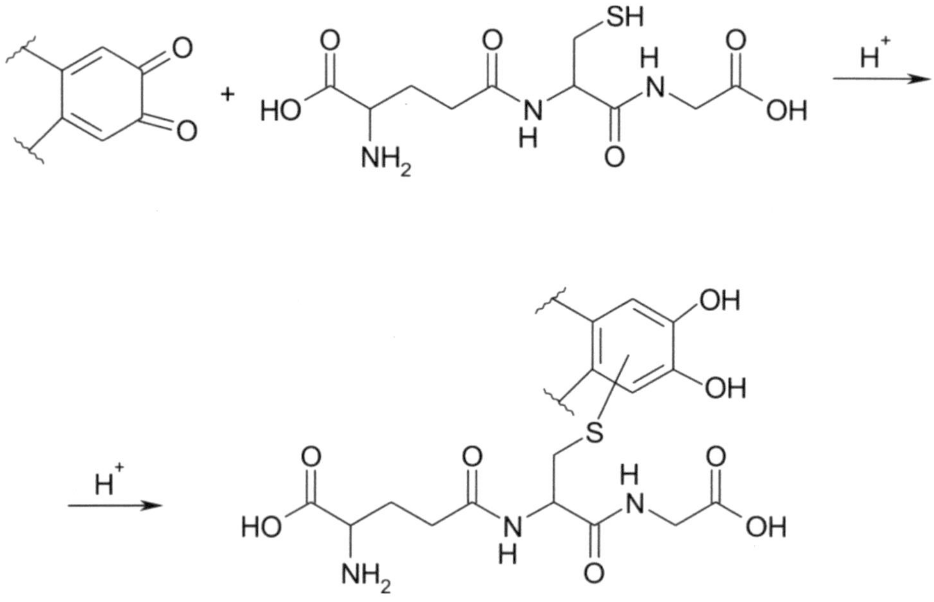

Fig. (14). Reaction of glutathione with ortho-quinone derivatives.

Other nitrogen compounds found in wines are melatonin and serotonin (Figs. **15**, **16**) [129, 130]. Among all alcoholic beverages, red wines have the highest concentration of melatonin. The melatonin found in red wines provide the same benefits as the hormone synthesized in the human body. Melatonin or N-acetyl-5-methoxytriptamine is the pineal hormone. It is of considerable interest due to its modulating function of a series of physiologic processes. Melatonin is usually employed in therapy for disturbances of sleep and jetlag, epilepsy, Parkinsonism, Alzheimer's disease, movement disorders, *etc* [131].

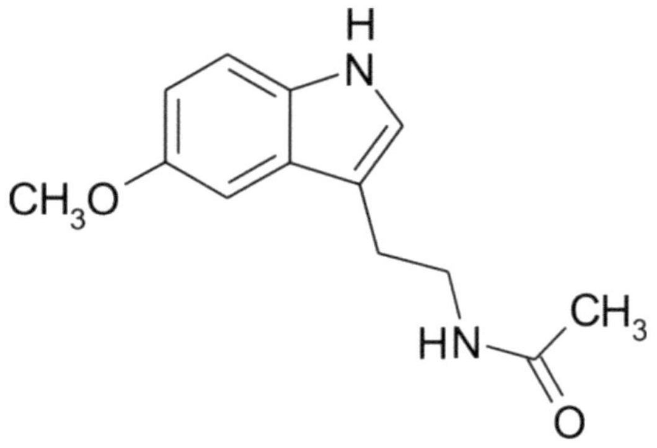

Fig. (15). Molecular structure of melatonin.

Serotonin is a neurotransmitter that is created mostly in the gut and is responsible for mood changes.

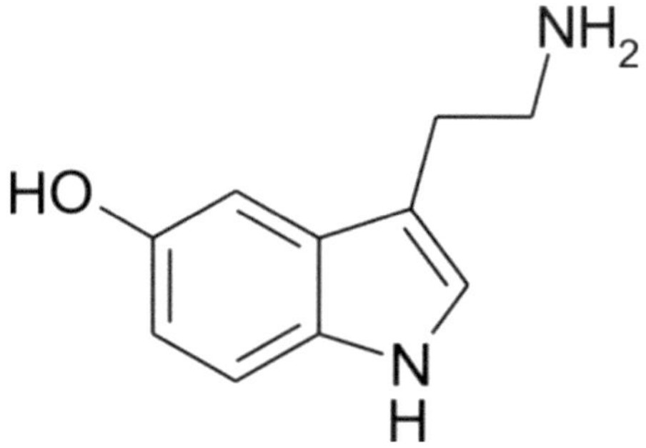

Fig. (16). Molecular structure of serotonin.

Serotonin transmit signals across a synapse from one neuron to another targeted neuron. The effects of serotonin are related with central nervous system. The functions of serotonin are very different and is involved in control of appetite, memory and learning, cardiovascular function, temperature regulation, endocrine regulation, sleep and depression [132].

VITAMINS IN WINES

Wine contains small quantities of hydrosoluble vitamins, notably vitamin B_1 (thiamine), vitamin B_2 (riboflavin) and vitamin B_{12} (cyanocobalamin) [133].

Thiamin or vitamin B_1 is synthesized by plants, bacteria, algae and some fungi. The molecule comprises a piridiminic ring coupled with a thiazolic ring [134]. The chemical structure of vitamin B_1 is shown in the Fig. (**17**).

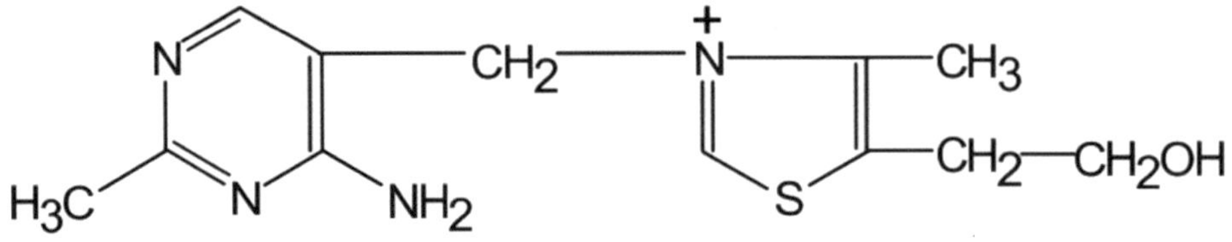

Fig. (17). Molecular structure of vitamin B_1.

Thiamine is absorbed from the intestine by active transport. In the case of large doses the diffusion process become important. Maximum capacity of absorption is 10-15 mg/day. The intestinal mucosa convert enzymatically the thiamine to thiamine pyrophosphate (Fig. **18**).

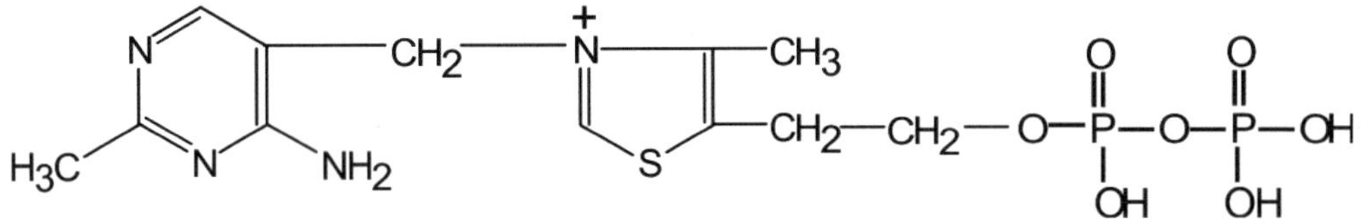

Fig. (18). Molecular structure of thiamine pyrophosphate.

Low doses, corresponding to minimum food intake, is entirely metabolized; higher doses fill the deposits and the excess is eliminated unchanged by urine.

Thiamine pyrophosphate or carboxylase is the active form of vitamin B_1. It is a coenzyme and intervene in many biochemical reactions important of carbohydrate metabolism: oxidative decarboxylation of pyruvate and alpha-ketoglutarate, *etc.*

Vitamin deficiency can occur in a low intake associated with increased consumption of carbohydrates or increased metabolic needs - during pregnancy, breastfeeding, in hyperthyroidism, in malabsorption, *etc.* Clinical syndromes characteristic is *beri-beri* disease. It is known in the East as due to the consumption of large quantities of husked rice; in the West, one of the common causes is chronic alcoholism, that cause deficiency of thiamine. Thiamine deficiency clinically manifested by muscle atrophy and degenerative peripheral poly neuritis (dry *beri-beri*), edema and cardio-megaliths (wet *beri-beri*), and encephalopathy phenomena (Wernicke syndrome, Korsakoff syndrome) [135 - 137].

Vitamin B_2

Vitamin B_2 or riboflavin (Fig. **19**), is a derivative of dimethyl isoalloxazine (flavinic cycle), containing ribitol [138].

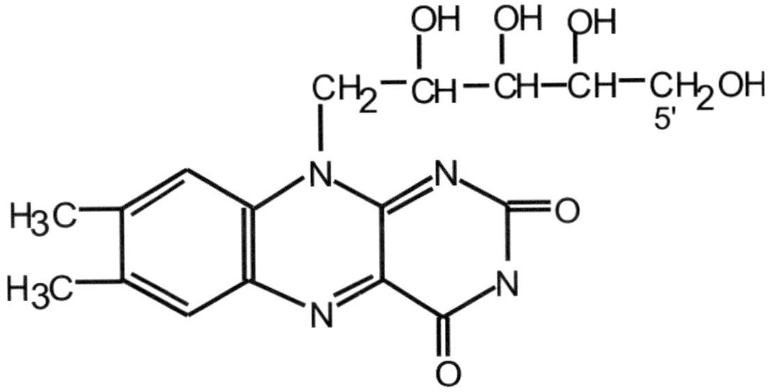

Fig. (19). Molecular structure of vitamin B_2.

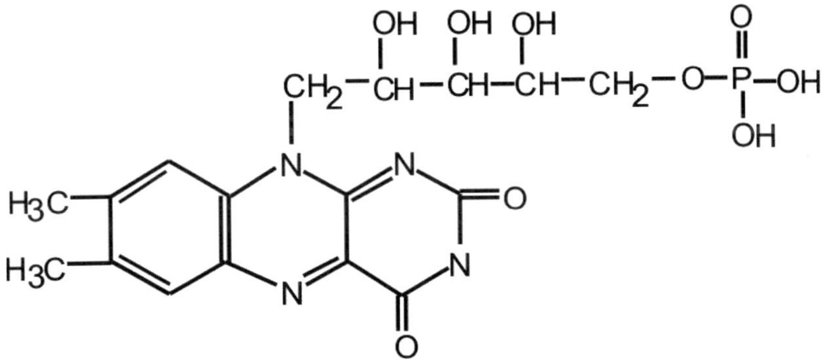

Fig. (20). Molecular structure of flavin mononucleotide.

Vitamin B_2 molecule has the oxido-reductive properties. Containing two hydrogens in the positions 1 and 10, riboflavin is reduced to colourless leuco derivative, and by oxidation, the riboflavin drops back to the yellow. Therefore, the formation of semiquinone forms of riboflavin prove that it is important in electrons transport [139].

The ribitol part of riboflavin may be esterified at C-5' with phosphoric acid resulting a mononucleotide: flavin mononucleotide (FMN) (Fig. **20**) [140].

Flavin mononucleotide can be combined with 5'-adenylic acid forming a dinucleotide: flavin adenine dinucleotide (FAD) (Fig. **21**) [141].

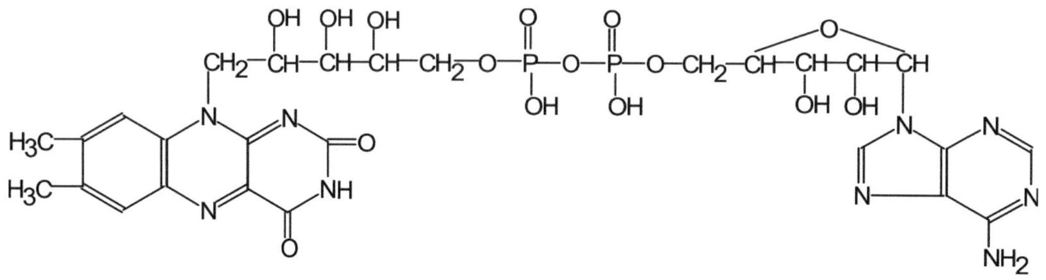

Fig. (21). Molecular structure of flavin dinucleotide.

Riboflavin has an important role in biochemical processes. Absorbed in the intestine, it is phosphorylated to form the two coenzymatic forms: flavin mononucleotide (FMN) and flavin dinucleotide (FAD). Flavoproteins including FMN or FAD as coezymatic component are redox enzymes and assure hydrogen transportation in the processes of cellular respiration. A number of flavoproteins have role of catalysts in the cycle of citric acid, to β-oxidation of fatty acids, to oxidative phosphorylation, as well as to reactions for the synthesis of purine bases of nucleotides, therefore, indirectly, of the protein. Thus, riboflavin is involved in the use of feed nutrients and, providing a sufficient intensity of cellular respiration, contributes to the formation of macroergic compounds, which are used in the process of biosynthesis, particularly for protein synthesis [142, 143].

Cyanocobalamin (vitamin B_{12})

Cyanocobalamin structure (Fig. **22**), is characterized by a great complexity. Vitamin B_{12} molecule contains four pyrrole rings reduced, of which three are related to the coordinating the cobalt ion and the fourth by a covalent bond; the cobalt ion is bound by a cyano radical (-C≡N) and by a nitrogen atom of the 5,6-

dimethylbenzimidazole; 5,6-dimethylbenzimidazole is linked to the tetrapyrrole ring by ribose - phosphate and aminopropanol [144].

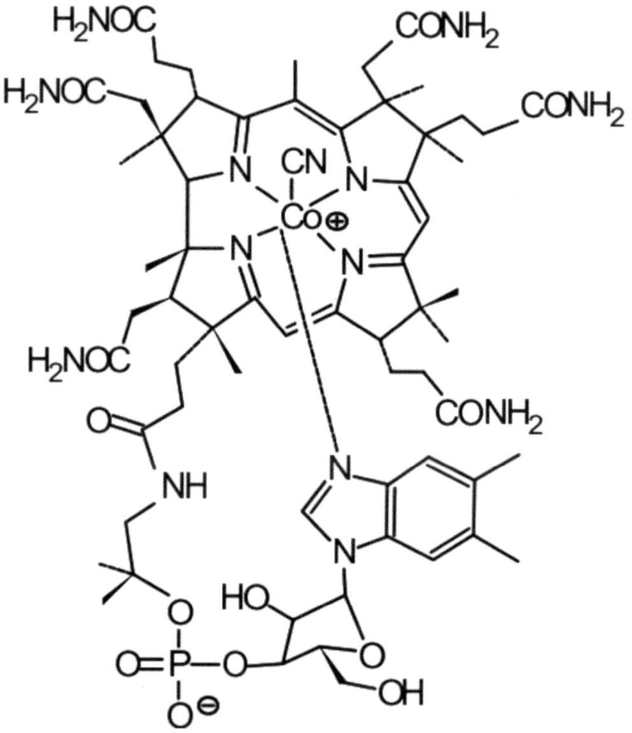

Fig. (22). Molecular structure of vitamin B_{12}.

Vitamin B_{12} can take part in two type of biochemical reactions. Firstly are the transmethylation reactions. Characteristic of these reactions is the formation of the intermediate methyl cobalamin as carrier of methyl group. In this way is biosynthesized methionine and acetate from carbon dioxide. The second group are the transformation reactions were coenzyme B_{12} participate directly. These reactions consisted of transporting hydrogen and formation of a new carbon-hydrogen bonds. This includes deamination of ethanolamine, reduction of ribonucleotides to deoxyribonucleotides, *etc* [145, 146].

A special importance has the participation of vitamin B_{12} in the biosynthesis of nucleic acids (formation of purine and pyrimidine nitrogenous bases and the reduction of ribonucleotides to deoxyribonucleotides). Vitamin B_{12} has an significant role in the forming of red blood cell at the spinal cord level, stimulates the protein and lipid anabolism, promotes cell division, maintains the integrity of the nerve cell, *etc* [146, 147].

FUTURE TRENDS AND PERSPECTIVES

Wine is complex beverage containing numerous compounds valuable for human health. The compounds from wine, a natural food product, are polyphenols, vitamins, glutathione, melatonin, serotonin, among others. Nutritional properties and health benefits of bioactive compounds from wine are of great interest around the world. Antioxidant properties, positives effects in prevention of heart diseases, and the implication in prolonging the time life are themes of major importance. Therefore, a moderate consumption of wine is recommended in a diversified and equilibrated diet as is the Mediterranean one.

CONFLICT OF INTEREST

The author confirms that he has no conflict of interest to declare for this publication.

ACKNOWLEDGMENTS

This work was supported by a grant of the Romanian National Authority for Scientific Research, CNCS – UEFISCDI, project number PN-II-ID-PCE-2-11-3-0255.

REFERENCES

[1] Spencer AF, Peynaud E. Knowing and Making Wine. John Willey & Sons 1984.

[2] Peynaud É. The Taste of Wine: The Art and Science of Wine Appreciation. John Wiley & Sons 1996.

[3] Cotea VD. Podgoriile si vinurile Romaniei. Editura Academiei Romane 2003.

[4] Jackson RS. Wine Science Principles and Applications. Burlington, MA: Academic Press 2008.

[5] Apetrei IM, Apetrei C, El Rayess Y. Characterization of red wines polyphenolics employing sensors and biosensors. In: El Rayess Y, Ed. Wine: Phenolic Composition, Classification and Health Benefits. Nova Publishers 2014; pp. 41-70.

[6] Robinson J, Ed. The Oxford Companion to Wine. 3rd ed., Oxford University Press 2006. [http://dx.doi.org/10.1093/acref/9780198609902.001.0001]

[7] Morel-Salmi C, Souquet JM, Bes M, Cheynier V. Effect of flash release treatment on phenolic

extraction and wine composition. J Agric Food Chem 2006; 54(12): 4270-6.
[http://dx.doi.org/10.1021/jf053153k] [PMID: 16756356]

[8] Cotea V D, Zănoagă C V, Cotea V V. Tratat de Oenochimie. Bucureşti: Editura Academiei Române 2009.

[9] McRae JM, Schulkin A, Kassara S, Holt HE, Smith PA. Sensory properties of wine tannin fractions: implications for in-mouth sensory properties. J Agric Food Chem 2013; 61(3): 719-27.
[http://dx.doi.org/10.1021/jf304239n] [PMID: 23289627]

[10] Davidović SM, Veljović MS, Pantelić MM, *et al.* Physicochemical, antioxidant and sensory properties of peach wine made from redhaven cultivar. J Agric Food Chem 2013; 61(6): 1357-63.
[http://dx.doi.org/10.1021/jf3043727] [PMID: 23301859]

[11] Linskens HF, Jackson JF, Eds. Wine Analysis. Berlin: Springer-Verlag 1988.
[http://dx.doi.org/10.1007/978-3-642-83340-3]

[12] Jackson RS. Wine Tasting: A Professional Handbook. Amsterdam: Elsevier 2002.

[13] Tedesco I, Russo M, Russo P, *et al.* Antioxidant effect of red wine polyphenols on red blood cells. J Nutr Biochem 2000; 11(2): 114-9.
[http://dx.doi.org/10.1016/S0955-2863(99)00080-7] [PMID: 10715597]

[14] Pereira V, Albuquerque F, Cacho J, Marques JC. Polyphenols, antioxidant potential and color of fortified wines during accelerated ageing: the Madeira Wine case study. Molecules 2013; 18(3): 2997-3017.
[http://dx.doi.org/10.3390/molecules18032997] [PMID: 23462532]

[15] Boydens C, Pauwels B, Decaluwé K, Brouckaert P, Van de Voorde J. Relaxant and antioxidant capacity of the red wine polyphenols, resveratrol and quercetin, on isolated mice corpora cavernosa. J Sex Med 2015; 12(2): 303-12.
[http://dx.doi.org/10.1111/jsm.12786] [PMID: 25443137]

[16] Landrault N, Poucheret P, Ravel P, Gasc F, Cros G, Teissedre PL. Antioxidant capacities and phenolics levels of french wines from different varieties and vintages. J Agric Food Chem 2001; 49(7): 3341-8.
[http://dx.doi.org/10.1021/jf010128f] [PMID: 11453773]

[17] Mitrou PN, Kipnis V, Thiébaut AC, *et al.* Mediterranean dietary pattern and prediction of all-cause mortality in a US population: results from the NIH-AARP Diet and Health Study. Arch Intern Med 2007; 167(22): 2461-8.
[http://dx.doi.org/10.1001/archinte.167.22.2461] [PMID: 18071168]

[18] Edlin G, Golanty E. Health and Wellness. Bartlett Publishers 2010.

[19] Dai J, Mumper RJ. Plant phenolics: extraction, analysis and their antioxidant and anticancer properties. Molecules 2010; 15(10): 7313-52.
[http://dx.doi.org/10.3390/molecules15107313] [PMID: 20966876]

[20] Lattanzio V. Phenolic Compounds: Introduction. In: Ramawat KG, Mérillon JM, Eds. Natural products. Springer-Verlag Berlin Heidelberg 2013.
[http://dx.doi.org/10.1007/978-3-642-22144-6_57]

[21] [Accessed on: Oct 2015]; Available at: https://www.weizmann.ac.il/plants/aharoni/PlantMetabolomeCourse/May092007.pdf

[22] Vickery ML, Vickery B. Secondary plant metabolism. London: MacMillan 1981. [http://dx.doi.org/10.1007/978-1-349-86109-5]

[23] Ollis WD. Recent developments in the chemistry of natural phenolic compounds: Proceedings of the plant phenolics group symposium. Amsterdam: Elsevier 2013.

[24] Moodley R, Koorbanally NA, Shahidul Islam M, Jonnalagadda SB. Structure and antioxidant activity of phenolic compounds isolated from the edible fruits and stem bark of *Harpephyllum caffrum*. J Environ Sci Health B 2014; 49(12): 938-44. [http://dx.doi.org/10.1080/03601234.2014.951578] [PMID: 25310809]

[25] Bhattacharya A, Sood P, Citovsky V. The roles of plant phenolics in defence and communication during Agrobacterium and Rhizobium infection. Mol Plant Pathol 2010; 11(5): 705-19. [PMID: 20696007]

[26] Watson RR, Ed. Polyphenols in plants Isolation, purification and extract preparation. Amsterdam: Academic Press 2014.

[27] Butler LG. Antinutritional effects of condensed and hydrolysable tannins. In: Hemingway RW, Laks PE, Eds. Plant polyphenols: Synthesis, properties and significance. New York: Plenum Press 1992. [http://dx.doi.org/10.1007/978-1-4615-3476-1_40]

[28] Matalas AL, Zampelas A, Stavrinos V. The mediterranean diet: Constituents and health promotion. CRC Press 2001. [http://dx.doi.org/10.1201/9781420042221.ch2]

[29] Gawel R, Day M, Schulkin A, Smith P. White wine phenolics – friend or foe? Aust N Z Grapegrower Winemaker 2013; 591: 67.

[30] Mercurio MD, Dambergs RG, Cozzolino D, Herderich MJ. Smith, P. A. Relationship between red wine grades and phenolics. 1. Tannin and total phenolics concentrations. J Agric Food Chem 2010; 58: 12313-9. [http://dx.doi.org/10.1021/jf103230b] [PMID: 21047137]

[31] Fernández-Pachón MS, Villaño D, Troncoso AM, García-Parrilla MC. Determination of phenolic composition of sherry and table white wines by liquid chromatography and their relation with antioxidant activity. Anal Chim Acta 2006; 563: 101-8. [http://dx.doi.org/10.1016/j.aca.2005.09.057]

[32] Cliff MA, King MC, Schlosser J. Anthocyanin, phenolic composition, colour measurement and sensory analysis of BC commercial red wines. Food Res Int 2007; 40: 92-100. [http://dx.doi.org/10.1016/j.foodres.2006.08.002]

[33] Grisa EF, Mattivi F, Ferreira EA, Vrhovsek U, Filho DW, Pedrosa RC. Bordignon- Luiz, M. T. Phenolic profile and effect of regular consumption of Brazilian red wines on *in vivo* antioxidant activity. J Food Compos Anal 2013; 31: 31-40. [http://dx.doi.org/10.1016/j.jfca.2013.03.002]

[34] Jackson RS. Wine Science: Principles and Applications. Elsevier 2014.

[35] Ribéreau-Gayon P, Glories Y, Maujean A, Dubourdieu D. Handbook of Enology, The Chemistry of Wine: Stabilization and Treatments. John Wiley & Sons 2006. [http://dx.doi.org/10.1002/0470010398]

[36] Schuster B, Herrmann K. Hydroxybenzoic and hydroxycinnamic acid derivatives in soft fruits. Phytochemistry 1985; 24: 2761-4. [http://dx.doi.org/10.1016/S0031-9422(00)80722-0]

[37] Rodriguez-Mendez ML, Apetrei C, Gay M, *et al.* Evaluation of oxygen exposure levels and polyphenolic content of red wines using an electronic panel formed by an electronic nose and an electronic tongue. Food Chem 2014; 155: 91-7. [http://dx.doi.org/10.1016/j.foodchem.2014.01.021] [PMID: 24594159]

[38] Herrmann K. Occurrence and content of hydroxycinnamic and hydroxybenzoic acid compounds in foods. Crit Rev Food Sci Nutr 1989; 28(4): 315-47. [http://dx.doi.org/10.1080/10408398909527504] [PMID: 2690858]

[39] Moreno-Arribas MV, Polo C. Wine Chemistry and Biochemistry. Springer Science & Business Media 2008.

[40] Duke JA. CRC Handbook of Medicinal Spices. CRC Press 2002. [http://dx.doi.org/10.1201/9781420040487]

[41] Buglass AJ. Handbook of Alcoholic Beverages: Technical, Analytical and Nutritional Aspects. John Wiley & Sons 2011. [http://dx.doi.org/10.1002/9780470976524]

[42] Preedy VR. Processing and Impact on Antioxidants in Beverages. Elsevier 2014.

[43] Schopp LM, Lee J, Osborne JP, Chescheir SC, Edwards CG. Metabolism of nonesterified and esterified hydroxycinnamic acids in red wines by Brettanomyces bruxellensis. J Agric Food Chem 2013; 61(47): 11610-7. [http://dx.doi.org/10.1021/jf403440k] [PMID: 24219184]

[44] Buiarelli F, Coccioli F, Merolle M, Jasionowska R, Terracciano A. Identification of hydroxycinnamic acid–tartaric acid esters in wine by HPLC–tandem mass spectrometry. Food Chem 2010; 123: 827-33. [http://dx.doi.org/10.1016/j.foodchem.2010.05.017]

[45] Singleton VL, Salgues M, Zaya J, Trousdale E. Caftaric acid disappearance and conversion to products of enzymic oxidation in grape must and wine. Am J Enol Vitic 1985; 36: 50-6.

[46] Bavaresco L, Fregoni C, Cantù E, Trevisan M. Stilbene compounds: from the grapevine to wine. Drugs Exp Clin Res 1999; 25(2-3): 57-63. [PMID: 10370866]

[47] Sun B, Ribes AM, Conceição Leandro M. Pedro Belchior. A.; Isabel Spranger, M. Stilbenes: Quantitative extraction from grape skins, contribution of grape solids to wine and variation during wine maturation. Anal Chim Acta 2006; 563: 382-90. [http://dx.doi.org/10.1016/j.aca.2005.12.002]

[48] Adrian M, Jeandet P. Effects of resveratrol on the ultrastructure of Botrytis cinerea conidia and biological significance in plant/pathogen interactions. Fitoterapia 2012; 83(8): 1345-50. [http://dx.doi.org/10.1016/j.fitote.2012.04.004] [PMID: 22516542]

[49] Terrier N, Poncet-Legrand C, Cheynier V. Flavanols, flavonols and dihydroflavonols. In: Moreno-Arribas MV, Polo MC, Eds. Wine Chemistry and Biochemistry. Springer 2009.
[http://dx.doi.org/10.1007/978-0-387-74118-5_22]

[50] Fanzone M, Zamora F, Jofré V, Assof M, Gómez-Cordovés C, Peña-Neira Á. Phenolic characterisation of red wines from different grape varieties cultivated in Mendoza province (Argentina). J Sci Food Agric 2012; 92(3): 704-18.
[http://dx.doi.org/10.1002/jsfa.4638] [PMID: 21919008]

[51] Goldberg DM, Karumanchiri A, Tsang E, Soleas GJ. Catechin and epicatechin concentrations of red wines: Regional and cultivar-related differences. Am J Enol Vitic 1998; 49: 23-34.

[52] Reynolds AG. Managing wine quality: Viticulture and wine quality. Elsevier 2010.
[http://dx.doi.org/10.1533/9781845699987]

[53] Sun B, Leandro C, Ricardo da Silva JM, Spranger I. Separation of grape and wine proanthocyanidins according to their degree of polymerization. J Agric Food Chem 1998; 46: 1390-6.
[http://dx.doi.org/10.1021/jf970753d]

[54] Fortes Gris E, Mattivi F, Antonio Ferreira E, Vrhovsek U, Curi Pedrosa R, Bordignon-Luiz MT. Proanthocyanidin profile and antioxidant capacity of Brazilian Vitis vinifera red wines. Food Chem 2011; 126: 213-20.
[http://dx.doi.org/10.1016/j.foodchem.2010.10.102]

[55] McRae JM, Kennedy JA. Wine and grape tannin interactions with salivary proteins and their impact on astringency: a review of current research. Molecules 2011; 16(3): 2348-64.
[http://dx.doi.org/10.3390/molecules16032348] [PMID: 21399572]

[56] Ma W, Guo A, Zhang Y, Wang H, Liu Y, Li H. A review on astringency and bitterness perception of tannins in wine. Trends Food Sci Technol 2014; 40: 6-19.
[http://dx.doi.org/10.1016/j.tifs.2014.08.001]

[57] Cheynier V. Polyphenols in foods are more complex than often thought. Am J Clin Nutr 2005; 81(1) (Suppl.): 223S-9S.
[PMID: 15640485]

[58] Jackisch P. Modern Winemaking. Cornell University Press 1985.

[59] Okuda T, Ito H. Tannins of constant structure in medicinal and food plants—hydrolyzable tannins and polyphenols related to tannins. Molecules 2011; 16: 2191-217.
[http://dx.doi.org/10.3390/molecules16032191]

[60] Okuda T, Mori K, Hatano T. Relationship of the structures of tannins to the binding activities with hemoglobin and methylene blue. Chem Pharm Bull (Tokyo) 1985; 33(4): 1424-33.
[http://dx.doi.org/10.1248/cpb.33.1424] [PMID: 4042219]

[61] Tsanova-Savova S, Ribarova F. Free and conjugated myricetin, quercetin, and kaempferol in bulgarian red wines. J Food Compos Anal 2002; 15: 639-45.
[http://dx.doi.org/10.1006/jfca.2002.1099]

[62] Hamidia Z, Nayeri H, Vessal M, Naderi G, Boshtam M. Myricetin, quercetin, kaempferol and galangin all increase serum paraoxonase activity. Indian Journal of Fundamental and Applied Life Sciences 2013; 3: 36-43.

[63] He F, Liang NN, Mu L, *et al.* Anthocyanins and their variation in red wines. II. Anthocyanin derived pigments and their color evolution. Molecules 2012; 17(2): 1483-519. [http://dx.doi.org/10.3390/molecules17021483] [PMID: 23442981]

[64] Wang H, Race EJ, Shrikhande AJ. Anthocyanin transformation in Cabernet Sauvignon wine during aging. J Agric Food Chem 2003; 51(27): 7989-94. [http://dx.doi.org/10.1021/jf034501q] [PMID: 14690384]

[65] Pina F, Oliveira J, de Freitas V. Anthocyanins and derivatives are more than flavylium cations. Tetrahedron 2015; 71: 3107-14. [http://dx.doi.org/10.1016/j.tet.2014.09.051]

[66] Packer L. Handbook of Antioxidants. CRC Press 2001.

[67] Amerine MA, Joslyn MA. Table wines: The technology of their production. University of California Press 1970; Vol. 2.

[68] Work HH. Wood, Whiskey and Wine: A History of Barrels. Reaktion Books 2014.

[69] Apetrei C, Apetrei IM, Nevares I, *et al.* Using an e-tongue based on voltammetric electrodes to discriminate among red wines aged in oak barrels or aged using alternative methods: Correlation between electrochemical signals and analytical parameters. Electrochim Acta 2007; 52: 2588-94. [http://dx.doi.org/10.1016/j.electacta.2006.09.014]

[70] Zhang B, Cai J, Duan CQ, Reeves MJ, He F. A review of polyphenolics in oak woods. Int J Mol Sci 2015; 16(4): 6978-7014. [http://dx.doi.org/10.3390/ijms16046978] [PMID: 25826529]

[71] Pérez-Coello MS, Sanz J, Cabezudo MD. Gas chromatographic-mass spectrometric analysis of volatile compounds in oak wood used for ageing of wines and spirits. Chromatographia 1998; 47: 427-32. [http://dx.doi.org/10.1007/BF02466474]

[72] Rappoport Z. The chemistry of phenols, 2 volume set. John Wiley & Sons 2004.

[73] Spillman PJ, Pollnitz AP, Liacopoulos D, Skouroumounis GK, Sefton MA. Accumulation of vanillin during barrel-aging of white, red, and model wines. J Agric Food Chem 1997; 45: 2584-9. [http://dx.doi.org/10.1021/jf970034z]

[74] Giménez Martínez R, López García De La Serrana H, Villalón Mir M, *et al.* Study of vanillin, syringaldehyde and gallic acid content in oak wood and wine spirit mixtures: influence of heat treatment and chip size. J Wine Res 2001; 12: 175-82. [http://dx.doi.org/10.1080/09571260120106811]

[75] Ziyatdinova G, Kozlova E, Budnikov H. Electropolymerized eugenol-MWNT-based electrode for voltammetric evaluation of wine antioxidant capacity. Electroanalysis 2015; 27: 1660-8. [http://dx.doi.org/10.1002/elan.201400712]

[76] Prida A, Chatonnet P. Impact of oak-derived compounds on the olfactory perception of barrel- aged wines. Am J Enol Vitic 2010; 61: 408-13.

[77] Pandey KB, Rizvi SI. Plant polyphenols as dietary antioxidants in human health and disease. Oxid Med Cell Longev 2009; 2(5): 270-8. [http://dx.doi.org/10.4161/oxim.2.5.9498] [PMID: 20716914]

[78] Packer L. Handbook of Antioxidants. CRC Press 2001.

[79] Moukette BM, Pieme CA, Njimou JR, Biapa CP, Marco B, Ngogang JY. *In vitro* antioxidant properties, free radicals scavenging activities of extracts and polyphenol composition of a non-timber forest product used as spice: Monodora myristica. Biol Res 2015; 48: 15-31.
[http://dx.doi.org/10.1186/s40659-015-0003-1] [PMID: 25885269]

[80] Borra SK, Gurumurthy P, Mahendra J, Jayamathi KM, Cherian CN, Chand R. Antioxidant and free radical scavenging activity of curcumin determined by using different *in vitro* and *ex vivo* models. J Med Plants Res 2013; 7: 2680-90.

[81] Liu RH. Health benefits of fruit and vegetables are from additive and synergistic combinations of phytochemicals. Am J Clin Nutr 2003; 78(3) (Suppl.): 517S-20S.
[PMID: 12936943]

[82] Watson RR, Preedy VR, Zibadi S. Polyphenols in human health and disease. Academic Press 2013.
[http://dx.doi.org/10.1007/978-1-62703-044-1]

[83] Panesar PS, Marwaha SS. Biotechnology in agriculture and food processing: Opportunities and challenges. CRC Press 2013.
[http://dx.doi.org/10.1201/b15271]

[84] Arranz S, Chiva-Blanch G, Valderas-Martínez P, Medina-Remón A, Lamuela-Raventós RM, Estruch R. Wine, beer, alcohol and polyphenols on cardiovascular disease and cancer. Nutrients 2012; 4(7): 759-81.
[http://dx.doi.org/10.3390/nu4070759] [PMID: 22852062]

[85] Apetrei IM, Rodríguez-Méndez ML, Apetrei C, Nevares I, del Alamo M, de Saja JA. Monitoring of evolution during red wine aging in oak barrels and alternative method by means of an electronic panel test. Food Res Int 2012; 45: 244-9.
[http://dx.doi.org/10.1016/j.foodres.2011.10.034]

[86] Zenebe W, Pechánová O, Bernátová I. Protective effects of red wine polyphenolic compounds on the cardiovascular system. Exp Clin Cardiol 2001; 6(3): 153-8.
[PMID: 20428452]

[87] de Rijke YB, Demacker PN, Assen NA, Sloots LM, Katan MB, Stalenhoef AF. Red wine consumption does not affect oxidizability of low-density lipoproteins in volunteers. Am J Clin Nutr 1996; 63(3): 329-34.
[PMID: 8602588]

[88] Gerós H, Chaves M, Delrot S. The biochemistry of the grape berry. Bentham Science Publishers 2012.
[http://dx.doi.org/10.2174/97816080536051120101]

[89] Vidavalur R, Otani H, Singal PK, Maulik N. Significance of wine and resveratrol in cardiovascular disease: French paradox revisited. Exp Clin Cardiol 2006; 11(3): 217-25.
[PMID: 18651034]

[90] Maffei M. Dietary Supplements of Plant Origin: A Nutrition and Health Approach; CRC Press, 2003.

[91] Henning SM, Zhang Y, Rontoyanni VG, *et al.* Variability in the antioxidant activity of dietary supplements from pomegranate, milk thistle, green tea, grape seed, goji, and acai: effects of *in vitro* digestion. J Agric Food Chem 2014; 62(19): 4313-21.

[http://dx.doi.org/10.1021/jf500106r] [PMID: 24745654]

[92] Asif M, Khodadadi E. Medicinal uses and chemistry of flavonoid contents of some common edible tropical plants. Journal of Paramedical Sciences 2013; 4: 119-38.

[93] Shahidi F, Ed. Handbook of Antioxidants for Food Preservation. Amsterdam: Elsevier 2015.

[94] Zykova TA, Zhu F, Zhai X, *et al.* Resveratrol directly targets COX-2 to inhibit carcinogenesis. Mol Carcinog 2008; 47(10): 797-805.
[http://dx.doi.org/10.1002/mc.20437] [PMID: 18381589]

[95] Masur SK, Dewal HS, Dinh TT, Erenburg I, Petridou S. Myofibroblasts differentiate from fibroblasts when plated at low density. Proc Natl Acad Sci USA 1996; 93(9): 4219-23.
[http://dx.doi.org/10.1073/pnas.93.9.4219] [PMID: 8633044]

[96] Paulo L, Oleastro M, Gallardo E, Queiroz JA, Domingues F. Antimicrobial properties of resveratrol: a review. In: Méndez-Vilas A, Ed. Science against microbial pathogens: communicating current research and technological advances. Formatex Research Center 2011.

[97] Athar M, Back JH, Tang X, *et al.* Resveratrol: a review of preclinical studies for human cancer prevention. Toxicol Appl Pharmacol 2007; 224(3): 274-83.
[http://dx.doi.org/10.1016/j.taap.2006.12.025] [PMID: 17306316]

[98] Petrovski G, Gurusamy N, Das DK. Resveratrol in cardiovascular health and disease. Ann N Y Acad Sci 2011; 1215: 22-33.
[http://dx.doi.org/10.1111/j.1749-6632.2010.05843.x] [PMID: 21261638]

[99] Gorbunov N, Petrovski G, Gurusamy N, Ray D, Kim DH, Das DK. Regeneration of infarcted myocardium with resveratrol-modified cardiac stem cells. J Cell Mol Med 2012; 16(1): 174-84.
[http://dx.doi.org/10.1111/j.1582-4934.2011.01281.x] [PMID: 21352470]

[100] Zhang X, Wang Y, Yang W, Hou X, Zou J, Cao K. Resveratrol inhibits angiotensin II-induced ERK1/2 activation by downregulating quinone reductase 2 in rat vascular smooth muscle cells. J Biomed Res 2012; 26(2): 103-9.
[http://dx.doi.org/10.1016/S1674-8301(12)60019-0] [PMID: 23554738]

[101] Troy DB, Beringer P. Remington: The science and practice of pharmacy. Lippincott Williams & Wilkins 2006.

[102] Ueda K, Kawabata R, Irie T, Nakai Y, Tohya Y, Sakaguchi T. Inactivation of pathogenic viruses by plant-derived tannins: strong effects of extracts from persimmon (*Diospyros kaki*) on a broad range of viruses. PLoS One 2013; 8(1): e55343.
[http://dx.doi.org/10.1371/journal.pone.0055343] [PMID: 23372851]

[103] Buzzini P, Arapitsas P, Goretti M, *et al.* Antimicrobial and antiviral activity of hydrolysable tannins. Mini Rev Med Chem 2008; 8(12): 1179-87.
[http://dx.doi.org/10.2174/138955708786140990] [PMID: 18855732]

[104] Juhnke J, Miller J, Hall JO, Provenza FD, Villalba JJ. Preference for condensed tannins by sheep in response to challenge infection with Haemonchus contortus. Vet Parasitol 2012; 188(1-2): 104-14.
[http://dx.doi.org/10.1016/j.vetpar.2012.02.015] [PMID: 22459112]

[105] Agustini BC, Braga de Lima D, Bordin Bonfim TM. Composition of amino acids and bioactive amines in common wines of Brazil. Acta Scientiarum 2014; 36: 225-33.

[http://dx.doi.org/10.4025/actascihealthsci.v36i2.20187]

[106] Moreno J, Peinado R. Enological Chemistry. Academic Press 2012.

[107] Ciani M, Beco L, Comitini F. Fermentation behaviour and metabolic interactions of multistarter wine yeast fermentations. Int J Food Microbiol 2006; 108(2): 239-45. [http://dx.doi.org/10.1016/j.ijfoodmicro.2005.11.012] [PMID: 16487611]

[108] Geros H, Chaves MM, Gil HM, Delrot S. Grapevine in a changing environment: A molecular and ecophysiological perspective. John Wiley & Sons 2015. [http://dx.doi.org/10.1002/9781118735985]

[109] Farkaš J. Technology and Biochemistry of Wine. Vol. 2. AMACOM Div American Mgmt Assn 1988.

[110] Peynaud E, Blouin J. The taste of wine: The art science of wine appreciation. John Wiley & Sons 1996.

[111] Robichaud JL, Noble AC. Astringency and bitterness of selected phenolics in wine. J Sci Food Agric 1990; 53: 343-53. [http://dx.doi.org/10.1002/jsfa.2740530307]

[112] Weinert DJ. Nutrition and muscle protein synthesis: a descriptive review. J Can Chiropr Assoc 2009; 53(3): 186-93. [PMID: 19714233]

[113] Silla Santos MH. Biogenic amines: their importance in foods. Int J Food Microbiol 1996; 29(2-3): 213-31. [http://dx.doi.org/10.1016/0168-1605(95)00032-1] [PMID: 8796424]

[114] Apetrei IM, Rodriguez-Mendez ML, Apetrei C, de Saja JA. Fish Freshness Monitoring Using an E-tongue Based on Polypyrrole Modified Screen-Printed Electrodes. IEEE Sens J 2013; 13: 2548-54. [http://dx.doi.org/10.1109/JSEN.2013.2253317]

[115] Konakovsky V, Focke M, Hoffmann-Sommergruber K, *et al.* Levels of histamine and other biogenic amines in high-quality red wines. Food Addit Contam Part A Chem Anal Control Expo Risk Assess 2011; 28(4): 408-16. [http://dx.doi.org/10.1080/19440049.2010.551421] [PMID: 21337238]

[116] Naila A, Flint S, Fletcher G, Bremer P, Meerdink G. Control of biogenic amines in food existing and emerging approaches. J Food Sci 2010; 75(7): R139-50. [http://dx.doi.org/10.1111/j.1750-3841.2010.01774.x] [PMID: 21535566]

[117] Lehtonen P. Determination of amines and amino acids in wine - a review. Am J Enol Vitic 1996; 47: 127-33.

[118] Martuscelli M, Arfelli G, Manetta AC, Suzzi G. Biogenic amines content as a measure of the quality of wines of Abruzzo (Italy). Food Chem 2013; 140(3): 590-7. [http://dx.doi.org/10.1016/j.foodchem.2013.01.008] [PMID: 23601412]

[119] Shalaby AR. Significance of biogenic amines in food safety and human health. Food Res Int 1996; 29: 675-90. [http://dx.doi.org/10.1016/S0963-9969(96)00066-X]

[120] Taylor SL, Stratton JE, Nordlee JA. Histamine poisoning (scombroid fish poisoning): an allergy-like

intoxication. J Toxicol Clin Toxicol 1989; 27(4-5): 225-40.
[http://dx.doi.org/10.3109/15563658908994420] [PMID: 2689658]

[121] Souza SC, Theodoro KH, Souza ÉR, da Motta S, Abreu Glória MB. Bioactive amines in brazilian wines: Types, levels and correlation with physico-chemical parameters. Braz Arch Biol Technol 2005; 48: 53-62.
[http://dx.doi.org/10.1590/S1516-89132005000100009]

[122] Kritzinger EC, Bauer FF, du Toit WJ. Role of glutathione in winemaking: a review. J Agric Food Chem 2013; 61(2): 269-77.
[http://dx.doi.org/10.1021/jf303665z] [PMID: 23240621]

[123] Ehrlich K, Viirlaid S, Mahlapuu R, *et al.* Design, synthesis and properties of novel powerful antioxidants, glutathione analogues. Free Radic Res 2007; 41(7): 779-87.
[http://dx.doi.org/10.1080/10715760701348611] [PMID: 17577738]

[124] Ugliano M, Kwiatkowski M, Vidal S, *et al.* Evolution of 3-mercaptohexanol, hydrogen sulfide, and methyl mercaptan during bottle storage of Sauvignon blanc wines. Effect of glutathione, copper, oxygen exposure, and closure-derived oxygen. J Agric Food Chem 2011; 59(6): 2564-72.
[http://dx.doi.org/10.1021/jf1043585] [PMID: 21332202]

[125] Lavigne V, Pons A, Dubourdieu D. Assay of glutathione in must and wines using capillary electrophoresis and laser-induced fluorescence detection. Changes in concentration in dry white wines during alcoholic fermentation and aging. J Chromatogr A 2007; 1139(1): 130-5.
[http://dx.doi.org/10.1016/j.chroma.2006.10.083] [PMID: 17125780]

[126] Lamprecht M. Antioxidants in Sport Nutrition. CRC Press 2014.
[http://dx.doi.org/10.1201/b17442]

[127] De Rosa SC, Zaretsky MD, Dubs JG, *et al.* N-acetylcysteine replenishes glutathione in HIV infection. Eur J Clin Invest 2000; 30(10): 915-29.
[http://dx.doi.org/10.1046/j.1365-2362.2000.00736.x] [PMID: 11029607]

[128] Nuttall SL, Martin U, Sinclair AJ, Kendall MJ. Glutathione: in sickness and in health. Lancet 1998; 351(9103): 645-6.
[http://dx.doi.org/10.1016/S0140-6736(05)78428-2] [PMID: 9500325]

[129] Huang X, Mazza G. Simultaneous analysis of serotonin, melatonin, piceid and resveratrol in fruits using liquid chromatography tandem mass spectrometry. J Chromatogr A 2011; 1218(25): 3890-9.
[http://dx.doi.org/10.1016/j.chroma.2011.04.049] [PMID: 21570690]

[130] Mercolini L, Mandrioli R, Raggi MA. Content of melatonin and other antioxidants in grape-related foodstuffs: measurement using a MEPS-HPLC-F method. J Pineal Res 2012; 53(1): 21-8.
[http://dx.doi.org/10.1111/j.1600-079X.2011.00967.x] [PMID: 22017461]

[131] Leonardo-Mendonça RC, Martinez-Nicolas A, de Teresa Galván C, *et al.* The benefits of four weeks of melatonin treatment on circadian patterns in resistance-trained athletes. Chronobiol Int 2015; 32(8): 1125-34.
[http://dx.doi.org/10.3109/07420528.2015.1069830] [PMID: 26361788]

[132] Apetrei IM, Apetrei C. Amperometric tyrosinase based biosensors for serotonin detection. Rom Biotechnol Lett 2013; 18: 8253-62.

[133] Aramouni F, Deschenes K. Methods for developing new food products: An instructional guide; DEStech publications inc, 2014.

[134] Webb ME, Marquet A, Mendel RR, Rébeillé F, Smith AG. Elucidating biosynthetic pathways for vitamins and cofactors. Nat Prod Rep 2007; 24(5): 988-1008.
[http://dx.doi.org/10.1039/b703105j] [PMID: 17898894]

[135] Kira J. Therapeutic benefits of an oral vitamin B1 derivative for human T lymphotropic virus type I-associated myelopathy/tropical spastic paraparesis (HAM/TSP). BMC Med 2013; 11: 183-5.
[http://dx.doi.org/10.1186/1741-7015-11-183] [PMID: 23945332]

[136] Sica DA. Loop diuretic therapy, thiamine balance, and heart failure. Congest Heart Fail 2007; 13(4): 244-7.
[http://dx.doi.org/10.1111/j.1527-5299.2007.06260.x] [PMID: 17673878]

[137] McCormick DB, Greene HL. Vitamins. In: Burtis VA, Ashwood ER, Eds. Tietz Textbook of Clin Chem. 2nd ed., Philadelphia: W.B. Saunders 1994.

[138] Bruno EJ Jr, Ziegenfuss TN. Water-soluble vitamins: research update. Curr Sports Med Rep 2005; 4(4): 207-13.
[http://dx.doi.org/10.1097/01.CSMR.0000306209.56939.48] [PMID: 16004830]

[139] Powers HJ. Riboflavin (vitamin B-2) and health. Am J Clin Nutr 2003; 77(6): 1352-60.
[PMID: 12791609]

[140] Rao NA, Felton SP, Huennekens FM, MacKler B. Flavin mononucleotide: the coenzyme of reduced diphosphopyridine nucleotide dehydrogenase. J Biol Chem 1963; 238: 449-55.
[PMID: 13973054]

[141] Mewies M, McIntire WS, Scrutton NS. Covalent attachment of flavin adenine dinucleotide (FAD) and flavin mononucleotide (FMN) to enzymes: the current state of affairs. Protein Sci 1998; 7(1): 7-20.
[http://dx.doi.org/10.1002/pro.5560070102] [PMID: 9514256]

[142] Sollner S, Macheroux P. New roles of flavoproteins in molecular cell biology: an unexpected role for quinone reductases as regulators of proteasomal degradation. FEBS J 2009; 276(16): 4313-24.
[http://dx.doi.org/10.1111/j.1742-4658.2009.07143.x] [PMID: 19624732]

[143] Hille R, Miller S, Palfey B. Handbook of flavoproteins: Volume 1 oxidases, dehydrogenases and related systems; walter de gruyter: berlin, 2013.

[144] Eitenmiller RR, Landen WO Jr. Vitamin Analysis for the Health and Food Sciences. CRC Press 2010.

[145] Ragsdale SW. Catalysis of methyl group transfers involving tetrahydrofolate and B(12). Vitam Horm 2008; 79: 293-324.
[http://dx.doi.org/10.1016/S0083-6729(08)00410-X] [PMID: 18804699]

[146] Banerjee R. Chemistry and Biochemistry of B12. John Wiley & Sons 1999.

[147] Greer JP. Wintrobe's Clinical Hematology. Vol. 1. Lippincott Williams & Wilkins 2009.

CHAPTER 3

Recent Advances in the Analysis of Bioactive Compounds Based on Molecular Recognition

Robert Săndulescu*, **Cecilia Cristea**, **Ede Bodoki** and **Radu Oprean**

Analytical Chemistry Department, "Iuliu Haţieganu" University of Medicine and Pharmacy, Cluj-Napoca, Romania

Abstract: Most of the biological processes, either normal or pathological, are based on molecular recognition. Molecular recognition, defined as the ability of one molecule (host) to "recognize" another molecule (guest), which presents complementarity, mainly through bonding interactions and molecular geometry, may be exploited in the analytical sciences, as long as the specific interaction produces a detectable change that can be measured by an instrument. The most impactful and acknowledged analytical use of molecular recognition has been recorded in separation sciences, namely in solid phase extraction and affinity chromatographic applications, but also in optical and electrochemical chemo- and biosensing. In living organisms the molecular recognition system is represented by natural receptors, characterized by the highest affinity and specificity, being the natural targets for toxins and mediators of physiological processes. By consequence, a large variety of bioelements, such enzymes, antibodies, nucleic acids, and cells were investigated in the last decades, being integrated in various separation and sensing devices which were successfully applied in biomedical and environmental analyses or monitoring. However, their isolation, purification, stability and cost are critical issues. Therefore, the development of artificial biomimetic receptors, such as cyclodextrins and derivatives, calixarenes, ligands (valinomycine, crown-ethers, *etc.*), aptamers, and molecularly imprinted polymers focused the attention of many research groups. The most recent achievements in the field of analytical sciences based on molecular recognition, such as separation techniques (solid phase extraction, chromatography, electrophoresis), and sensing (electrochemical, optical, *etc.*) are revised.

* **Corresponding author Robert Săndulescu**: Analytical Chemistry Department, Faculty of Pharmacy, Iuliu Haţieganu University of Medicine and Pharmacy, Cluj-Napoca, Romania; Email: rsandulescu@umfcluj.ro

Constantin Apetrei (Ed.)

Keywords: Antibodies, Aptamers, Biomimetic receptors, Calixarenes, Chiral discrimination, Cyclodextrins, DNA, Electrochemical sensors and biosensors, Guest-host interaction, Molecular recognition, Molecularly imprinted polymers, Separation methods.

MOLECULAR RECOGNITION

Most of the biological processes (normal and pathological), such as: enzyme metabolism, DNA-protein, RNA-ribosome, protein synthesis, active and passive ion transport across membranes (ion channels, ion pumps, ionophores) and receptor-substrate (antigen-antibody, sugar-lectin) binding are based on molecular recognition.

Molecular recognition is generally defined as the ability of one molecule (host) to "recognize" another molecule (guest), which presents complementarity, mainly through bonding interactions and molecular geometry. The effects of the molecular recognition, consisting in the specific interaction between host and guest molecules (such as, DNA hybridization, specific interactions between biotin and avidin or between protein A/G and antibody), through covalent (disulfide bridges between cysteine side chains) and non-covalent binding of various strength such as: hydrogen bonding, metal coordination, hydrophobic forces, van der Waals forces, π-π interactions, halogen bonding, electrostatic and electromagnetic effects, may be exploited in the analytical sciences, as long as the specific interaction produces a detectable change that can be measured by an instrument.

Some authors [1] distinguish between *static molecular recognition*, consisting in a complexation reaction between the specific recognition sites of the host molecule and a guest molecule to form a host-guest complex, and *dynamic molecular recognition*, which allow the control of the binding processes in biological systems. Thus, the binding of the first guest to the first binding site of the host affects the association constant of a second guest with a second binding site.

Perhaps the most impactful and acknowledged analytical use of molecular recognition has been recorded in separation sciences, namely in solid phase extraction based purifications of biomacromolecules and affinity chromatographic applications [2]. The advances of affinity chromatography have also stimulated immobilized ligand based methods to characterize and exploit molecular recognition, such as chromatographic separations but also optical and electrochemical chemo- and biosensing.

In spite of their outstanding sensitivity (in the range of 10^{-9}-10^{-12} M), the electrochemical methods are not selective, meaning that all the electroactive species from a given sample or matrix are non-specifically detected. Therefore, the design of specific or at least very selective interfaces between the analyte or the sample and the measuring systems was and still is the "hot aim" of a great number of research teams around the world. A great and various number of examples of such sensing entities is given by nature itself, the development and functioning of all biological systems being based on a multitude of "signal molecules" and the corresponding "natural receptors". Thus, many efforts were made in order to modify the electrode surface with an immobilized ligand in such a manner, to be able to distinguish a specific analyte in a multicomponent matrix. Due to the fact that "natural receptors" present the greatest selectivity or even specificity, a great number of papers about various sensors and biosensors were published in the last decades.

Molecular Recognition in Chiral Analysis

Differentiation between enantiomers is a fundamental phenomenon as chiral compounds interact in a stereospecific way with each other. Chiral recognition and differentiation in living organisms represents one of the most intriguing natural phenomena. This reflects the regulatory systems' (*i.e.* transporters, enzymes, receptors) high degree of selectivity, which assures in the organism a high fidelity transfer of the molecular information.

Chiral molecular recognition and enantio-differentiation can be considered as the highest form of molecular recognition, exploiting very small energetic differences between the forming transient complexes, therefore a great effort has been directed in the last decade in the understanding of the underlying fundamental mechanisms. The current state of knowledge will be briefly outlined in the following, however additional details on the mechanism of chiral separations can be found in several comprehensive reviews [3 - 5].

According to the most commonly invoked rule (three-point attachment model, TPA), first formulated by Easson and Steadman [6] to explain the stereoselective binding of chiral molecules to a protein receptor, chiral recognition implies and includes a three point interaction with a natural (*e.g.* enzyme) or artificial (*e.g.* chiral stationary phase) chiral selector. Thus for a successful discrimination between the two enantiomers a simultaneous interaction of at least three active positions of the selector with appropriate positions of one of enantiomers' molecule must occur, amongst which at least one of the interactions is stereochemically dependent [7]. In the early 80s, the idea that diastereoisomerism is the fundamental prerequisite and that the TPA is not needed in chiral recognition, was promoted.

Later on, a general criterion of the inequality of the distance matrices of the diastereomeric complexes has been introduced [8]. By this approach various models leading to chiral recognition, such one-, two- and three-point mechanisms may also be explained. Moreover, it has been recognized that interactions may rather be mediated *via* multiple points instead of single points, where for example π-π and dipole-dipole interactions are considered multipoint interactions.

Diastereomeric complex formation between selector and selectands is driven by several interactions, *e.g.*, ionic interactions, ion-dipole or dipole-dipole interactions, H-bonds, van der Waals interactions, and π-π interactions. The long-ranged, strong ionic interactions may be primarily involved in the first line of binding between selector and selectands. However, as both enantiomers of an ionized solute are able to form these interactions, they may not be stereoselective. In contrast, H-bonds and π-π interactions are short-range directional forces so that these may be primarily responsible for stereoselective interactions [5].

Although the basic idea of the simplistic and rigid TPA geometric model, based on a static interaction, continues to be accepted and promoted in similar variations for different stereospecific receptors, it has been learned that the dynamic process of selector-selectand binding involves significant conformational changes in both parties (induced-fit). Besides the attractive forces manifested in the three point attachment points, steric factors and other nonbinding or even repulsive interactions also contribute to the selectivity of the selector [3 - 5].

Additionally, usually stereospecific recognition models take into account the interactions between the chiral selector and analyte only, however chiral recognition can also be assisted by the local environment, such as an achiral mediator (*e.g.* solvent molecule) [9].

Analytical techniques capable of differentiating enantiomers are of great importance. In what concerns analytical enantioseparations, liquid and gas chromatography (TLC, HPLC, SFC, GC, SPE [10 - 14]) and electromigration techniques (EKC, MEKC, MEEKC, CEC [15, 16]) are currently the most popular techniques, although others such as sensor-related technologies (MIPs, aptamers, *etc* [17, 18].), mass spectrometry [19, 20], nuclear magnetic resonance [21] may also be successfully applied in chiral analysis.

Enantioseparations can be categorized into indirect and direct methods. In the indirect approach, the enantiomers of an analyte are derivatized with a stereoisomerically pure reagent to form diastereomers *via* covalent bonds. These diastereomers can be subsequently separated, in principle, under achiral conditions. Direct enantioseparations refer to the separation of enantiomers in a chiral environment which requires the presence of a chiral selector that is either fixed to an immobile support or added to the mobile phase (in chromatography) or the background electrolyte (in electromigration techniques). In such cases, the separation is based on the formation of transient diastereomeric complexes between the chiral selector and the analyte enantiomers in thermodynamic equilibria. Differences between the association constants of the two enantiomers with the chiral selector represent the physico-chemical basis for the stereoselective recognition and eventually for their separation.

The basic difference in direct chiral electrochemical sensing in comparison with chiral separations comes from the fact that there is only one single unitary process of separation that corresponds to one theoretical plate in chromatographic separations [22]. This clearly illustrates the challenges that must be faced in the selection of the chiral recognition layer and the employed transducer (*i.e.*

potentiometric, amperometric, voltammetric, conductimetric, *etc.*). All major groups of chiral selectors (*i.e.* cyclodextrins, crown-ethers, chiral antibiotics, MIPs, various conductive and non-conductive polymers, *etc.*) employed as chiral stationary phases in enantioseparation, as well as various specific natural or synthetic biomacromolecules (*i.e.* enzymes, antibodies, DNA, aptamers, *etc.*) are amenable to be used in the fabrication of such chiral electrochemical sensors and biosensors.

The results obtained so far, seem to be very encouraging as chiral sensors (*e.g.* ion selective sensors, polymer based voltammetric sensors, biosensors, immunosensors, aptasensors), due to their inherent cost-effective nature, residing in their simple design, manufacturing and maintenance, might find very wide applications for chiral analysis. Further details in the types of chiral selectors employed in the fabrication of enantio-selective electrochemical sensors and biosensors, as well as the potential area of application of such sensors may be found in several review articles [23 - 27].

It is considered that there is no fundamental difference between stereospecific interactions between enantiomers and a given chiral selector whether it is bound (to a stationary phase - chromatography; to the surface of a transducer - sensors and biosensors) or mobile in the background electrolyte as in electrophoretic methods, although the accessibility of the chiral selector towards certain analytes may vary in the two situations [5].

In conclusion, in chiral analysis based on enantio-selective recognition the key step is the formation of transient diastereomeric complexes between the two enantiomers of the analyte and a selector. The lifetime and stability (binding energies) of the forming complexes have to be carefully balanced in function of the chosen chiral analytical strategy (enantioseparation by chromatography/ electromigration techniques, chiral electrochemical sensing, *etc.*), as for example in chiral separations a too strong binding of the enantiomers could cause a lower enantio-selectivity. Although the popularity and achieved performances of the high performance chiral separation techniques are difficult to be matched, the achievements of the emerging alternative analytical approaches (chiral optical, piezoelectric and electrochemical sensors, surface enhanced Raman scattering,

NMR, mass spectrometry) can no longer be overlooked.

BIOLOGICAL RECOGNITION SYSTEMS

Natural Receptors

In an ideal situation the molecular recognition part of a sensor is represented by a natural receptor. Natural receptors are characterized by the highest affinity and specificity, being the natural targets for toxins and mediators of physiological processes and by consequence can be successfully applied in biomedical and environmental analyses or monitoring of these compounds.

Four classes of natural receptors (membrane molecular receptors), with different structures and functions in the cell, can be distinguished: ion channel receptors, G-protein-linked receptors, receptors with single transmembrane domain, and enzyme-linked receptors [28]. Molecular receptors are cellular, typically membrane, proteins that change their structure (conformation) after binding specific ligands (chemical species). The conformational change triggers a cellular response, like the opening of an ion channel or secreting an enzyme [29].

However, there is very little information on natural receptor-based sensors, mainly because they are very expensive, they have low stability, and the transformation of the interaction between the target molecule and the natural receptor into a measurable signal and the interpretation of the connection that exists between the signal formation and the biochemical function of the receptor are very difficult.

Bioelements

Depending on the biorecognition element, affinity sensors can be classified in immunosensors (using antibodies, antibody fragments or antigens as recognition elements), aptasensors (based on aptamers), DNA sensors (based on hybridization of DNA), peptide sensors (peptasensors, using peptides as bioelements), whole cell biosensors and molecularly imprinted polymer based sensors (including a biomimetic recognition element) [30].

Enzymes

Enzymes are molecules that potentiate biological reactions having the ability to increase reactions rate *via* specific binding with substrates, which causes conformational changes in the substrate molecules that enable the reaction to occur [31].

Enzymes were the first bioelements included in the biosensors, generally composed of an immobilized layer of enzymes deposited onto the transducer (electrode). If the properties and the activity of the enzyme are preserved after deposition, it will react and recognize the analyte. The molecular recognition process between the enzyme and the analyte induces a signal that defines the biosensor characteristics [32]. The lock and key hypothesis could explain the mechanism of the enzyme action which is highly specific. This specific catalytic reaction of the enzyme provides the ability to detect much lower limits than with normal binding techniques [33].

A crucial step, in using enzymes as sensing element in the biosensors elaboration, is the enzyme deposition at the transducer surface. In this regard, the final goal is to produce a biosensor with appropriate characteristics in terms of increased sensitivity towards the analyte, low response to interferences, extended linear range, short response time and high stability. The actual trend in this field is related to developing improved quality enzyme coatings, suitable for novel generation of advanced enzyme biosensors and biofuel cells. The enzyme electrodes will be not described in detail, in this chapter, because there are many examples reported in literature, a good review of the topic being recently published by Leca-Bouvier and Blum [207].

Antibodies

Antibodies are glycoproteins produced by specialized B lymphocyte cells of the host, as response of the immune system to the presence of foreign species called antigen. They are also known as immunoglobulins. There are five classes of immunoglobulins (IgG, IgA, IgM, IgD, IgE), IgG being the most widely spread in nature and commonly used in the fabrication immunosensors [34].

The structure of an antibody is illustrated in Fig. (**1**) IgGs are "Y"-shaped molecules constituted of two identical light chains (MW about 25000 Da) and two identical heavy chains (MW about 50000 Da), bound together by disulfide bonds and noncovalent interactions (hydrogen bonds). Both light and heavy chains have constant and variable domains: one variable domain for both light and heavy chains, V_L and V_C, and one constant domain for light chains, C_L, and three for heavy chains, C_{H1}, C_{H2}, and C_{H3}. The variable regions, V_L and V_H, are the most important for the antigen-antibody reaction, and are different among different antibodies, giving the specificity of the binding, depending on the amino acid sequences. They are the complementarity-determining regions (CDRs) and present hypervariable loops that represent the binding site to the antigen [34, 35]. The high diversity of CDRs permits to produce high specifies antibody towards many kinds of antigens [34].

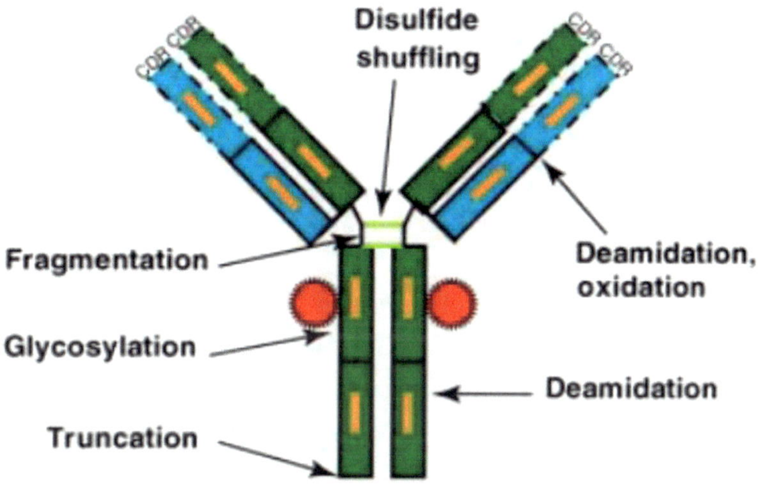

Fig. (1). The structure of an antibody [35]

IgG can be divided in an antigen binding fragment (Fab) and a non-antigen binding fragment (Fc). Antibodies are bivalent and can bind two antigens, the Fab region exhibiting two identical binding sites for the antigen. The antigen-binding region is called "paratope" and the complementary site on the antigen is called "epitope", their complementarity being based on size, shape and chemical compatibility [35]. The interaction of antibodies with their antigens is based on hydrogen bonds, ionic bonds, hydrophobic interactions and van der Waals forces, but a close fit determined by a high degree of complementarity between antibody and antigen is necessary for the non-covalent interactions to take place. In some cases the binding event is accompanied by conformational changes in antibody, antigen or both, which brings them in a closer fit but may also affect the binding affinity [35].

The strength of the binding of an antibody to its antigen is described by the equation [34]:

$$K = \frac{[Ab-Ag]}{[Ab][Ag]}$$

where K is the equilibrium constant of the interaction (affinity constant) and antibody-antigen is the resulted immuno-complex; the values of K can vary from 10^6 to $10^{12,}$ with low cross-reactivities when $K>10^8$.

Although the antibody-antigen reaction is highly sensitive, cross-reactivity can sometimes occur in the case of antigens that present identical or similar epitopes. Antibodies can bind to these similar antigens leading to false positives or interference for the immunosensor [35].

Antibodies can be produced by immunization of host animal, when a sub-lethal dose of antibody is injected to trigger the immune response. However, these procedures are complicated and time-consuming [35, 36]. The production of recombinant antibody (rAb) is shorter and allows obtaining a high diversity of antibodies. It is based on the isolation of mRNA from the B-lymphocites from the spleen of animals followed by cDNA synthesis and amplification of the variable regions by PCR. The amplified regions are then cloned in a plasmid vector. After the transformation of the host (*Escherichia coli*), different heavy and light chains can be randomly combined to obtain a DNA library with high diversity [36]. Monoclonal antibodies (mAb) are produced by identical immune cells (clones of a single parent cell). They have affinity for a single epitope, which offers high specify towards an antigen, being often used as a primary antibody in the construction of immunosensors. Polyclonal antibodies (pAb) are produced by several different immune cells as a mixture of immunoglobulins against an antigen, each of them binding a different epitope. They are in general used as secondary antibodies in immunoassays [34].

Antibody fragments can also be used in immunoassay, since antibody can be fragmented in their Fc and Fab regions using proteolytic enzyme digestion (pepsin, papain) [37].

Aptamers

Aptamers are synthetic strands of nucleic acid that can be designed to recognize amino acids, oligosaccharides, peptides, and proteins. They will be presented in this chapter due to their advantages compared to antibodies.

Aptamers are short single stranded DNA or RNA oligonucleotides that specifically bind a wide range of target molecules with high affinity and specificity [36]. They can selectively bind to small molecules, such as organic dyes, drugs, nucleotides, amino acids or to small parts of macromolecules, such as protein, polypeptides, polysaccharides [35, 38, 39]. Aptamers towards membrane bound receptors, cell surface epitopes, whole cells and even whole organisms [40] or difficult haptens such as toxins or prions [38] were also synthesized. In the binding of the aptamer to its target, hydrogen bonds, hydrophobic interaction, stacking interactions and electrostatic interactions are commonly involved [40]. Their affinity is given by their folding characteristic, aptamers being able to fold into a binding pocket upon the binding of small molecules by hydrogen bonds, or integrate in the structures of macromolecules, by complementarities driven by noncovalent interactions [38 - 40]. Aptamers have dissociation constants in the pico- to micro-molar range [38, 40] with affinity and specificity equal or superior to monoclonal antibodies [41]. Therefore they successfully substitute antibodies as recognition elements in immunoassays and immunosensors, being more stable and able to discriminate between closely related targets. They also have the advantage of avoiding the use of animals for their production since they are synthesized *in vitro* by Systematic Evolution of Ligands by Exponential Enrichment (SELEX). The method, introduced in 1990 [42, 43], starts with the synthesis of a single-stranded oligonucleotide library which contains a very large number of individual species (around 10^{15}) that increases the probability to obtain aptamers that are highly specific for a target. By PCR amplification a double stranded oligonucleotide pool is obtained, which is transcribed (for RNA) or strand separated (for DNA). The obtained oligonucleotides are incubated with the target and the aptamers that bind the target are partitioned (by filtering or by affinity chromatography), eluted and amplified by PCR to provide an enriched pool of nucleic acid. Several cycles of the SELEX process are required to obtain only a few aptamers that exhibit high specificity towards a target [38, 40].

The synthetic preparation offers them other advantages such as: stability to heat and wide range of pH and salt concentrations, affinity towards a wide range of target molecules, reduced batch-to-batch variations, they are more resistant to denaturation which makes them suitable for reusable applications and can be chemically modified to increase stability, affinity and specificity [38, 41].

Due to their inherent advantages aptamers have been widely used in a diverse range of fields such as drug delivery [44, 45], therapy [46, 47], diagnostics [48], and for various analytical applications such as immunoassay and immunosensors [49 - 51], chromatography [52], electrophoresis [53] or mass spectrometry [54].

Nucleic Acids (DNA, RNA, etc.)

DNA contains the genetic information in the cells and consists of two strands that resemble a ladder coiled into a spiral shaped the double helix. It is composed from linear array of nucleotides, each of which comprises a base plus a pentose sugar and phosphate. Four nucleotide bases are normally found in DNA: cytosine, thymine, adenine, and guanine. The information content of the DNA is embodied in the sequential arrangement of those nucleotides. This higher-order assembly containing multiple proteins bound at distinct DNA sites initiates readout of information encoded in the DNA.

The use of a nucleic acids sequence for specific diagnostic applications was developed 60 years ago and is still of interest [55]. When the nucleic acids act as the biological recognition element, highly specific affinity binding's reaction occurs between two single strands DNA (ssDNA) chains to form double stranded DNA (dsDNA). The biosensing working principle is based on recognition of the complementary strand of ssDNA to form a stable hydrogen bond between two nucleic acids in order to become dsDNA. Usually, an immobilized ssDNA is used as a probe in a bioreceptor in which the base sequence is complementary to the target of interest. Exposure of the target to the probe will result in hybridization of complementary ssDNA to form dsDNA and the biochemical reaction will be detected by the transducer and converted into a measurable signal. The literature data shows that the presence of some linker such as thiol or biotin is needed in the effort to immobilize the ssDNA onto the sensing surface [56].

Ribonucleic acid (RNA) is the other major nucleic acid besides DNA. Unlike DNA, it is single stranded containing ribose instead of deoxyribose as its sugar phosphate backbone and uracil instead of thymine in its pyrimidine bases. Like DNA, RNA can be assembled from nucleotides using DNA sequence as a template and RNA polymerase. The structure of an RNA molecule is also determined by its DNA-derived sequence. RNA is controlling how the genes are expressed to make proteins. RNA is unique in being able to store and transmit information as well as process that information. RNAs could be classified into messenger RNAs (mRNAs), which are translated into proteins, and nonprotein-coding RNAs (ncRNAs) [57]. mRNA is the short-lived intermediary in the transfer of genetic information from DNA to protein. It is transported out of the nucleus and translated into protein on the cytoplasmic ribosomes. Other categories are microRNAs (miRNAs), antisense transcripts, and transcriptional units containing a high density of stop codons and lacking any extensive open reading frame. The miRNAs are 18-24 nucleotide-long non coding RNA molecules that regulate the expression of target genes at the post-transcriptional level by either translational repression or degradation of miRNAs [58]. It has been estimated that the human genome may encode for over 1000 miRNAs, which are thought to regulate about 60% of the mammalian genes. Altered expression patterns of miRNAs have been associated with many diseases, including several types of cancer, cardiovascular disorders, diabetes, rheumatic disease and neurological disorders [59]. Due to their unique miRNA expression signatures that have been demonstrated in many pathological conditions, these molecules have been proposed as novel biomarkers in diagnostics. The miRNA profiles in clinical samples will not only signify the presence of a disease, but it can also help in their prediction and would help to choose personalized treatments. The miRNAs are currently detected in blood serum, plasma, urine, saliva, sweat, and other body fluids, their circulating levels in serum being estimated in the 200 aM to 20 pM range [60].

In the last years, miRNAs have been a hot topic in cancer research. Their function and deregulation in cancer cell apoptosis evasion is not yet fully understood. The study of miRNA regulation in tumor cell apoptosis evasion might provide new information regarding the mechanisms of cancer development and tumor cell resistance to therapy [61]. The field of circulating miRNA analysis is in its beginning and it has a great potential due to the fact that miRNAs are relatively stable molecules and can be easily accessible in a noninvasive manner [62].

Cells

The use of the living cells as the biosensing element is based on their ability to detect the intracellular and extracellular microenvironment conditions, the physiological parameters, and to produce a response through the interaction between stimulus and cell [63].

Microorganisms, such as bacteria and fungi can be used as biosensing elements for the specific detection of molecules not being influenced by the overall "state" of the surrounding environment [64]. In addition, proteins that are present in cells can also be used as bioreceptors for the detection of a specific analyte. The detection limit in the case of sensors using cells as biosensing elements is mainly determined by the natural environmental conditions in which the cell can stay alive for a long period. Therefore, the rigorous control of the physical and chemical parameters of the environment is crucial. Some parameters add major limitations to the cell based sensors, being influenced by various conditions, such as sterilization, lifetime or biocompatibility. The cell based sensor has poor selectivity compared to the microbial sensor due to the multireceptor behavior of the intact cells. However there are still some advantages when using cell as biosensing elements, the cell being less sensitive to inhibition and more tolerant of suboptimal pH and temperature values than enzymes, for example [33].

Natural receptors are able of specific *host-guest* interaction, but their fabrication is very difficult and very expensive. Therefore, besides other affinity based bioelements (DNA, antibodies or aptamers), the development of artificial biomimetic receptors focused the attention of many research groups. A large variety of cyclodextrins and derivatives, calixarenes, various ligands (valinomycine, crown-ethers, *etc.*), and molecularly imprinted polymers were studied and a great number of related papers were published in the last decade.

BIOMIMETIC RECEPTORS

Ligands (Crown Ethers, Cryptands, Valinomycin, *etc.*)

Crown ethers are cyclic oligomers of ethylene oxide that consist of a ring containing several ether groups, the most important members of this series being the tetra, the penta, and the hexamer (4, 5 and 6 ethyleneoxy units, -CH_2CH_2O-). Through the oxygen atoms which are situated so those are capable to coordinate with a cation located at the interior of the ring, crown ethers strongly bind certain cations, forming complexes. For example, 12-crown-4 has high affinity for Li^+, 15-crown-5 for Na^+, and 18-crown-6 for K^+. 21- and 18-membered diazacrown ether derivatives exhibit excellent Ca^{2+} and Mg^{2+} selectivity, being widely used in ion-selective electrodes fabrication (Fig. **2**) [65].

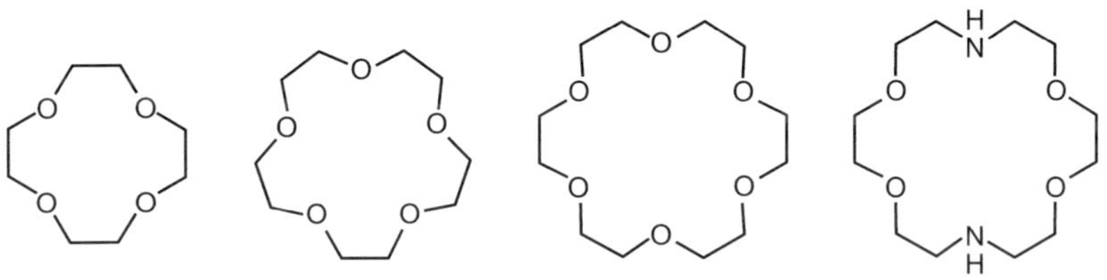

Fig. (2). Structures of 12-crown-4, 15-crown-5, 18-crown-6, dobenzo-18-crown-6, and diaza-18-crown-6.

Cryptands (Fig. **3**, left) are bi- and polycyclic multidentate synthetic ligands for a variety of cations such as: alkali and alkaline earth metals, NH_4^+, and lanthanides. In contrast to crown ethers, cryptands bind the guest ions using both nitrogen and oxygen donors and form complexes called cryptates.

Among ionophores, **valinomycin,** (Fig. **3**, right) a dodecadepsipeptide, made of twelve alternating amino acids and esters to form a macrocyclic molecule, also display a marked preference for the potassium cation over other cations. The stability constant for the potassium-valinomycin complex is 10000 times greater than for the sodium-valinomycin complex, explaining the selectivity of valinomycin for the transport of potassium ions (and not sodium ions) in biological systems and its use in potassium selective electrodes [66].

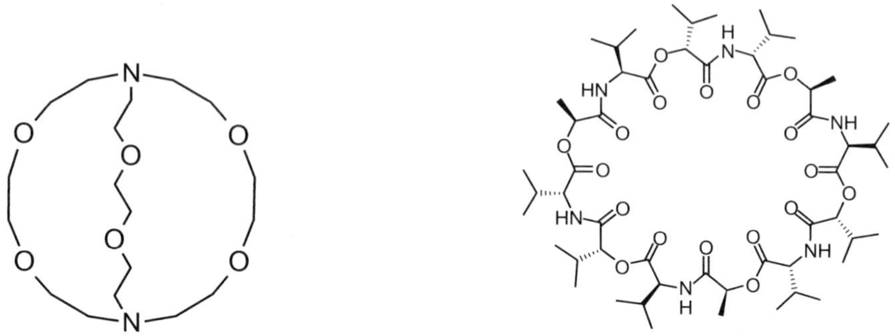

Fig. (3). Structures of 2.2.2-Cryptand (1,10-diaza-4,7,13,16,21,24-hexaoxabicyclo[8.8.8] hexacosane) and valinomycin.

Cyclodextrins

Cyclodextrins, also known as cycloamyloses, cyclomaltoses and Schardinger dextrins, were discovered by Villiers in 1891 and obtained by Schardinger from starch, by a relatively simple enzymatic conversion (cyclodextrin glucosyl transferase enzyme, CGT-ase) with *Bacilus amylobacter* and *Bacilus macerans*. Cyclodextrins are non-reducing cyclic oligosaccharides consisting of six (α-cyclodextrin), seven (β-cyclodextrin), eight (γ-cyclodextrin) or more glucopyranose units (chair conformation) linked by α-(1,4) bonds (Fig. **4**). The cyclodextrins structures were determined by X-ray crystallography showing that the "truncated cone" has the narrow side formed by the primary hydroxyl groups (C6) and the wider side by secondary hydroxyl groups (C2, C3), the cavity is lined by the hydrogen atoms and the glycoside oxygen bridges. This results in a relatively hydrophobic/nonpolar cavity compared to water and with a hydrophilic outside described as a "micro heterogeneous environment" [67 - 74].

Cyclodextrin molecules have numerous hydroxyl groups (18, 21 and 24 respectively in α-cyclodextrin, β-cyclodextrin and γ-cyclodextrin) which can be derivatized in various ways giving more than 1500 synthesized derivatives. The primary hydroxyl groups are the most reactive, readily modified to other functional groups. The secondary hydroxyl groups at carbons C2 and C3 have different reactivities such as: the reactivity of the hydroxyl group at C3 is much lower than that at C2. The hydroxyl group at C2 is the most acidic, OH at C6 is sterically favored, and OH at C3 is the least reactive (apparently because of the internal hydrogen bonds). Free electron pairs of oxygen are directed towards the interior of the cavity which produces a high electron density and lending to Lewis base characteristics. The interior cavity is non-polar in comparison with water.

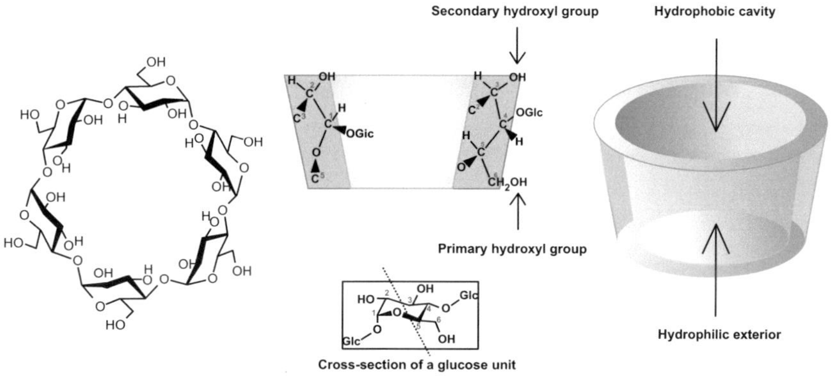

Fig. (4). Chemical structure of β-cyclodextrin [8].

The hydroxyl groups at C2 and C3 carbon atoms of the adjacent glucopyranose units form hydrogen bonds that stabilize the shape of the molecule and at the same time significantly influence the water solubility of the cyclodextrin. Through the study of the hydrogen-deuterium exchange in DMSO by NMR spectroscopy has been shown that intramolecular hydrogen bonds are formed by the secondary hydroxyl groups. As a result of investigations of γ-cyclodextrin by NMR spectroscopy in DMSO-d6 at 25°C and 80°C has been found that the α- and β-cyclodextrin protons of hydroxyl groups have lower chemical shifts indicating that the hydrogen bonds in the γ-cyclodextrin are stronger than in β-cyclodextrin and also hydrogen bonds in β-cyclodextrin are stronger than in α-cyclodextrin [75]. The mechanism of simultaneous and reversible change of directions of hydrogen bonds HO2->O3 and O3H->O2 called "flip-flop" was elucidated by X-ray studies [69]. Larger CDs (with more than 8 glucose units) have a restricted utilization because they present a "collapsed cylinder" structure, so their real cavity is even smaller than that of γ-cyclodextrin [72].

The most notable characteristic of CDs is the inclusion complexes formation with various organic and inorganic molecules (solid, liquid and gaseous compounds) in aqueous, nonaqueous and mixed media. The complex formation consist in a dimensional fit between host cavity and guest molecule when the high enthalpy water molecules from the host cavity are displaced by the more hydrophobic guest molecules leading to a nonpolar–nonpolar association and a decrease of cyclodextrin ring strain with a more stable lower energy state. The main driving forces for the inclusion complexes include Van der Waals interactions, hydrophobic interactions between the guest hydrophobic moiety and the CD cavity, hydrogen bonding between the polar groups of the guest molecules and the CDs hydroxyl groups and electrostatic interactions for ionic guests [208].

Calixarenes

Calixarenes are three-dimensional macrocycles with a truncated cone shape, consisting of 4 to 16 phenol molecules connected by methylene bridges, obtained after the hydroxyalkylation reactions of phenol and aldehydes. The phenolic groups are often substituted by *tert*-butyl in para position and the internal volume of calixarene is about 10 cubic nanometers. The name "calix" comes from the Greek, meaning vessel and the word "arena" addressing to the aromatic nuclea from their molecule. Calixarenes show one greater opening given by the larger *tert*-butyl substituents and a smaller and narrower one, due to the hydroxyl substituents [76, 77].

Due to the rotation around methylene bridges calixarenes have several chemical conformations. For example, in the case of calix [4] arene, cone conformation, partial cone conformation, and alternative structures, are all in a dynamic equilibrium. The four hydroxyl groups interact with each other through hydrogen bonds and stabilize the conformation of the cone. This conformation may be achieved either by the introduction of large substituents in place of hydroxyl groups, or on the opposite side. The p-*tert*-butyl phenol based calixarene (Fig. **5**) has such a truncated cone three-dimensional structure [78].

Calixarenes attracted much interest in recent years due to their hollow structure that allows them to form inclusion complexes with numerous neutral and charged chemical species, such as cations (alkaline ions [79], silver, lead, calcium, copper, zinc and mercury), being used to detect air pollution or water in toxicology or clinical analysis [80 - 86]. Calixarene selectivity depends on the diameter of the cavity, and the nature of the polar or non-polar functional groups on its edge.

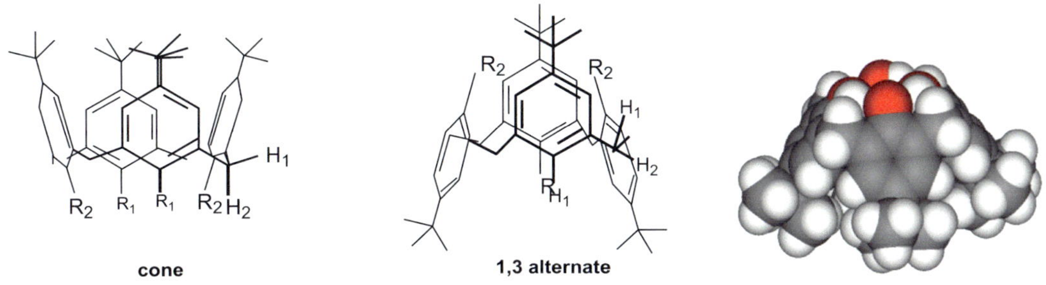

Fig. (5). p-*tert*-butyl-calix [4]arene and representation of the truncated cone three-dimensional conformation (right) [78].

Molecularly Imprinted Polymers (MIPs)

Molecularly Imprinted Polymers (MIPs) are a class of engineered materials acting as artificial receptors with a predetermined selectivity for a given target compound (analyte). The elucidation of the basic molecular principles triggered an ambitious process in the fields of polymer chemistry and material sciences in combining the nature inspired molecular recognition concept with the processability of manmade materials into a synthetic receptor that mimic natural recognition entities (biological receptors).

The complexes formed between the target molecule (template) and functional monomers (along with a cross-linking monomer) are stabilized upon polymerization into a rigid, highly cross-linked polymer. Conventionally polymerization is either photo- or thermally initiated in the presence of a radical initiator. Lately, in case of electrochemical applications involving biosensing, polymerization is performed electrochemically, which may be performed under galvanostatic, potentiometric or most frequently potentiodynamic conditions, in both nonaqueous and aqueous solutions. In the next step, upon the removal of the template, three dimensional cavities will result in the polymeric matrix, complementary in shape and chemical functionality arrangement with the template. The obtained macromolecular "mold" demonstrates a remarkable thermal and chemical stability under various experimental conditions allowing the very specific rebinding of the analyte (Fig. **6**).

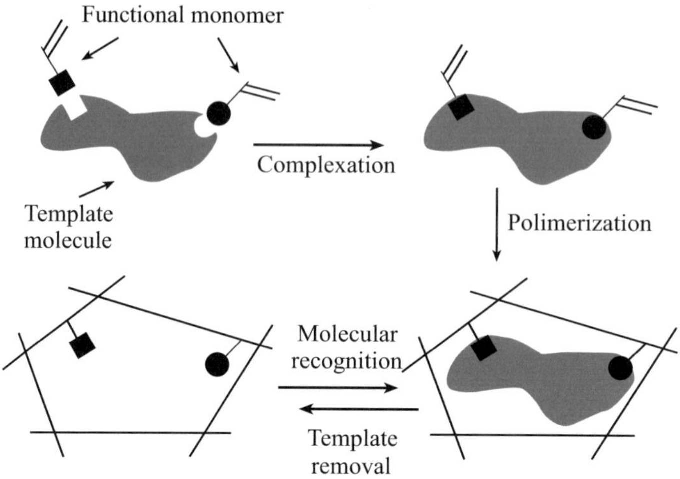

Fig. (6). Synthesis of molecularly imprinted polymers by non-covalent imprinting procedure.

The synthesis of MIPs is a relatively simple and inexpensive procedure, mixing the template molecule (either similar to the target compound - "dummy" template [87] or the target compound itself - analyte) with various functional and cross-linking monomers most often in a non-polar, aprotic, porogenic solvent.

In function of the nature of interaction prior to polymerization between the template molecule and functional monomer, three different types of molecular imprinting strategies are to be distinguished, covalent, non-covalent and metal-coordinating techniques [88]. Based on weaker interactions (hydrogen bonds, π-π interactions, van der Waals) similar to those encountered in biochemical processes, the non-covalent approach is by far the most popular one, especially because it can be applied in principle for any type of template (both small and macromolecules), but also because of the wide range of available monomers. In the case of macromolecules as templates, where bulk macromolecular imprinting leads to major inconveniences, the imprinting process needed to be adapted [89, 90]. Therefore, in contrast with the bulk imprinting, surface imprinting and epitope imprinting are able to enhance the imprinting effect for biomacromolecules, leading to higher affinity and capacity for the template, as well as to reduced non-specific binding. Additionally, they improve the binding kinetics of the macromolecule, overcoming difficulties to mass transfer and template removal in imprinted matrices.

Molecular imprinting technology became available as a technique for the creation of such receptor-like binding sites with selectivity comparable to that of affinity biomacromolecules [91]. MIPs represent robust molecular recognition elements able to mimic natural recognition entities, such as biological receptors (natural receptors or other bioelements). Additionally, they possess several advantages over their biological counterparts including low cost, ease of preparation, long term stability and shelf life, inherent reusability, high mechanical strength, durability to heat and pressure, and applicability in harsh chemical media [92, 93]. Due to their unique pre-determined selectivity towards the template molecule, these polymers have received a great deal of attention in various fields of the analytical sciences, being used either as recognition elements for highly selective sensing (*i.e.* optical, electrochemical, piezoelectric, *etc.*) [27, 94 - 96] or as highly efficient stationary phases, both for sample pretreatment (clean-up and preconcentration) and separation of complex samples by chromatographic and electrophoretic techniques [97, 98].

ANALYTICAL APPLICATIONS

Separation Techniques

The molecular recognition phenomena in the field of separation science is focused on the development of new analysis and separation concepts for applications in life science, pharmaceutical research, biotechnology and natural product isolation. Most of the molecular recognition based methods are used for chiral discrimination, where the performances of new chiral selectors are being tested by traditional separation techniques.

Solid Phase Extraction (SPE)

Traditional liquid-liquid extraction (LLE) does not fulfill current requirements in sample preparation and it has been displaced from laboratories in recent years by new extraction techniques (*e.g.* solid-phase extraction (SPE), solid-phase microextraction (SPME), stir-bar sorptive extraction (SBSE), and, more recently, matrix solid-phase dispersion (MSPD), micro SPE (MSPE) or liquid-phase microextraction (LPME)). However, all of these techniques suffer from lack of selectivity, making it necessary to optimize the typical steps involved. Unfortunately, even after careful optimization, some matrix components are co-eluted with target analytes, making it difficult to reach limits of detection (LODs) in line with the stringent regulations nowadays. To overcome this problem, molecularly imprinted SPE (MISPE) media could be a very effective alternative. The use of MIPs as selective sorbent materials allows customized sample treatment prior to the final determination. There are two types of MIPs: molecularly imprinted particle and monolith. For sample preparation, the first is generally used. They are prepared by bulk, suspension, seed, and dispersion/ precipitation polymerization methods. Other polymerization methods include grafting or coating of thin films of MIP on beaded materials or molecularly imprinting by sol-gel technique. Molecularly imprinted monoliths were divided into two categories: monoliths prepared in a column or capillary column and SPME fiber monoliths [99, 100]. A recent review focused on the application of molecularly-imprinted polymers in solid-phase microextraction (SPME) reveals 42 new applications in the field of food, environmental and drug analysis [101].

Different materials were used as functional monomers to fabricate a MIP. Li *et al.* [102] used the calixarene fiber for solid-phase microextraction of four organophosphorous pesticides in fruits and Urraca *et al.* [103] were looking for the clean-up and preconcentration of β-lactamase-resistant penicillins in milk by using molecularly imprinted solid-phase extraction (MISPE) based on 3 different monomers N-(2-aminoethyl)methacrylamide (EAMA), N,N-diethylamino ethylmethacrylate (DAEM), and 1-(4-vinylphenyl)-3-(3,5-bis(trifluromethyl) phenyl)-urea (VPU), coupled to high-performance liquid chromatography (HPLC) [103]. In order to remove *p*-nitrophenol from crude pharmaceuticals, Moghaddam and Abbasi used a MIP based on *p*-nitrophenol, methacrylic acid (functional monomer), MPTS (chemically modifier), ethylene glycol dimethacrylate (cross-linker), 2,2-azobisisobutironitril (initiator) and toluene-acetonitrile (porogenic solvent) coated on silica nanoparticle [104]. An extensive MIP database including refereed papers and reviews, conference proceedings, book chapters, theses and patents is available online [105].

Gas Chromatography (GC) and Supercritical Fluid Chromatography (SFC)

High resolution GC (HR-GC) remains as one of the most important separation technique for volatile analytes. Various novel stationary phases have been developed to achieve better resolution at higher analysis speed. By far, cyclodextrin derivatives coated onto the capillary bed are used as chiral stationary phases. Several groups reported new achievements on chiral discrimination using modified cyclodextrin stationary phases. Xiao *et al.* [106] reviewed the applications of cyclodextrin chiral stationary phases in chromatography. Several applications for chiral discrimination by GC (MS) were recently reported in the field of food analysis [107], drug analysis [108] and bioanalysis [109].

The use of SFC for chiral separations has been one of its most successful applications due to the high diffusivity and low viscosity that result from using liquid CO_2 in the mobile phase. A recent review by Plotka [110] emphasize the advantages of SFC ant its application for the analysis of pharmaceuticals and forensic drugs. Modern SFC exhibits improved precision, reproducibility, accuracy and robustness; also providing superior resolution and peak capacity compared to HPLC. Based on these findings, the use of modern chiral SFC is recommended for GMP studies of stereochemistry in pharmaceutical development and manufacturing [111]. A review of the literature covering separations of enantiomers achieved with supercritical fluids was recently published [112].

Liquid Chromatography

HPLC has achieved a great reputation in separation science because of its high speed, selectivity, efficiency, low detection and quantification limits, and reproducibility. HPLC has also gained a good reputation in chiral separations by using various types of chiral selectors. Thousands of chiral separations were reported in the last decade using different chiral separation phases (CSPs): polysaccharides, glycopeptide macrocyclic antibiotics, cyclodextrin derivatives, proteins, different types of ligands and MIPs. Several reviews are covering the topic, out of which the ones published in the last two years are summarized in Table **1**.

Table 1. The reviews covering the chiral separations by HPLC.

Analytes	Stationary Phase	Ref.
Amino acids	different CSP based on cyclodextrin, quinine derivatives, antibiotics	[113]
Primary and nonprimary amino compounds	(+)-(18-Crown-6)-2,3,11,12-tetracarboxylic	[114]
Amino acids	1,1′-Bi-2-naphthol (BINOL) derivatives	[115]
Pharmaceuticals and intermediates	helical polymers	[116]
Pharmaceutical compounds in different biological matrices	commercially available chiral stationary phases	[117]
Pharmaceuticals and intermediates	ionic liquid-functionalized stationary phases with β-cyclodextrin derivatives	[118]
Stereochemically-labile compounds, drugs	brush-type chiral stationary phases	[11]
Recombinant peptides, proteins, monoclonal antibodies, antibody-drug conjugates	wide-pore C3, C4, C8, and C18 stationary phases	[119]
Amino acids and proteins	polysaccharide-based materials, chirality-responsive polymers	[120]
Trans-stilbene oxide, benzoin and 2,2-dihydroxyl-6,6-dimethylbiphenyl	core-shell particles coated with polysaccharide chiral selectors	[121]
Mono- and di-substituted alkyl aromatics	metal-organic frameworks	[122]
Stereosynthesis	helicenes and helicene-like compounds	[123]
Pharmaceuticals in biological samples	cellulose derivatives-based chiral stationary phase	[124]
Pharmaceuticals and intermediates	immobilized polysaccharide-based chiral stationary phases	[125]
Selenoamino acids in biological samples	commercially available chiral stationary phases	[109]
Drugs	chiral monolithic stationary phases based on polysaccharides, macrocyclic glycopeptide antibiotics, cyclodextrin and proteins	[126]
Pharmaceuticals	amino acids and amino acid amides as chiral auxiliaries in cyanuric chloride based chiral derivatizing agents	[127]
Small molecules	homochiral metal-organic frameworks	[128]
Cardiovascular active compounds	optically active acids	[129]
Drugs with multiple chiral centers (labetalol, formoterol, nadolol, indenolol, nebivolol, diltiazem)	commercially available chiral stationary phases	[130]

(Table 1) contd.....

Analytes	Stationary Phase	Ref.
Amino acid and peptide	carbohydrate's macrocyclic glycopeptide antibiotics, cyclodextrins (native or derivatized), crown ethers and quinine derivatives	[108]
Chiral carboxylic acids	oligopyrrols	[131]
Carnitine in foods, dietary supplements, pharmaceutical formulations, biological tissues and body fluid	commercially available chiral stationary phases	[132]

Electrophoretic Separation

Despite of the massive use of HPLC in analytical laboratories, the electrophoretic technique became one of the most widely spread alternative. The separation of different types of isomers, especially enantiomers, is performed through molecular recognition by different kinds of selectors. If the selector is bonded to the stationary phase, the chromatographic mechanism occurs (capillary electrochromatography, CEC). By contrary, if the selector - analyte interaction occurs in the liquid phase, the electrophoretic mechanism is prevalent (capillary zone electrophoresis, CZE).

The advantages of combining separation and mass spectrometry (MS), able to provide structural information, have been demonstrated in various fields, such as environmental, pharmaceutical, pharmacological and toxicological sciences. Unfortunately, MS is not able to discriminate the molecules with the same mass and a prior separation based on molecular recognition is compulsory. Wuetrich *et al.* [133] recently review the recent developments and achievements in the field of chiral analysis by CE-MS. The use of molecularly imprinted polymers in CEC and their applications in various fields was recently reviewed by Iacob *et al.* [98]. Different reviews on enantioseparations by CE published during the last two years are summarized in Table **2**:

Table 2. Reviews covering CE enantioseparations.

Analytes	Stationary Phase	Ref.
Neuroactive molecules	cyclodextrin derivatives	[134]
Pharmaceuticals	crown ethers	[135]
Acid dissociation constant values	cyclodextrin derivatives	[136]
Drugs, metabolites in biological fluids	cyclodextrin derivatives	[137]

(Table 2) contd.....

Analytes	Stationary Phase	Ref.
β-blocker drugs	cyclodextrin derivatives, macrocyclic antibiotics, tartrate complexes, monolithic molecularly imprinted polymer, polymeric surfactants	[138]
Drugs, peptides	natural and randomly substituted cyclodextrins	[139]
Impurities and counterions of pharmaceuticals	cyclodextrin derivatives	[140]
Amino acids	cyclodextrin derivatives	[141]
Pharmaceuticals	antibiotics	[142]
Pharmaceuticals in biological fluids	cyclodextrin derivatives	[143]
Drugs and small molecules	antibiotics	[144]
Peptides	cycledextrin derivatives, MIP	[145]
Natural products	cycledextrin derivatives	[146]
Selenoamino acids in biological samples	cyclodextrin derivatives	[109]
Carnitine in foods, dietary supplements, drug formulations, biological tissues and body fluids	cycledextrin derivatives chiral selectors	[132]
Pharmaceuticals	dual and multi- cycledextrin based chiral selectors	[147]

The development of molecular recognition patterns is a promising approach in separation sciences. Combined with the recent development of the instrumental tools, the selectors' development will be one of the major challenges of the analytical chemistry.

Sensors

Sensors are devices that detect a physical, chemical or biological change and convert it into a measurable signal [148]. The general scheme of a biosensor is illustrated in Fig. (**7**).

As already defined, sensors are analytical tools capable of converting the input variable into a signal suitable for measurement [1]. Biosensors are sensors which associate a bioelement intimately connected with a transducer [149]. The operating principle is based on biological recognition. The main goal of the recognition system is to provide the sensor with a high degree of selectivity for the analyte to be measured.

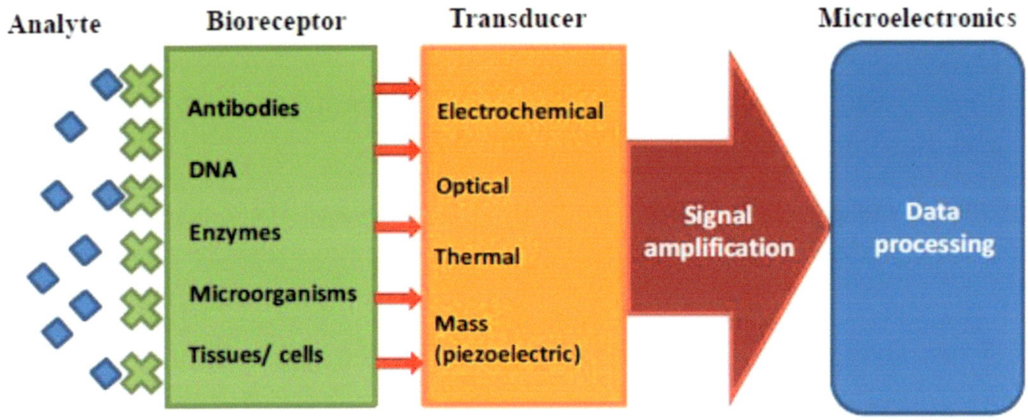

Fig. (7). Schematic representation of a biosensor.

The bioreceptor or biological recognition element is the significant part of a biosensor. A bioreceptor is responsible for binding the analyte of interest to the sensor surface for the measurement. The main component of a biosensor contains a biosensitive layer that can contain bioreceptors or may be made of bioreceptors covalently attached to the transducer. The most common forms of biosensing are: antibody-antigen interactions, nucleic acid interactions, enzyme-substrate interactions, cellular interactions (microorganisms) and in the case of synthetic bioreceptors, interactions using biomimetic materials. According to the nature of the biological recognition process, two main groups of sensors can be defined: *biocatalytic sensors*, which incorporate enzymes, microorganisms, cells, tissues and biomimetic catalysts able to recognize the analyte of interest and produce electroactive species or other detectable outcome and *affinity sensors*, mainly based on a selective interaction between the analyte and the bioelement such as antibody, nucleic acid, receptor proteins or synthetic receptors.

Many bioelements, such as: enzymes, DNA, RNA, antibodies, affibodies, microRNA, microorganisms, and living cells, were used in biosensors construction, based on different types of interactions between the bioreceptor and the analyte. The major disadvantage of all these biomolecules, represented by the poor chemical and physical stability, was the main reason for developing an alternative approach which involves the use of biomimetic receptor systems, capable of binding target molecules with affinities and specificities comparable to those of natural receptors. These chiral receptors are also able of stereoselective molecular recognition. Among the biomimetic receptors the most used are crown ethers, calixarenes, cyclodextrins and MIPs, representing a topic of current interest in supramolecular chemistry [150, 151].

In Fig. (**8**) are presented the main classes of bioreceptors and of transducers.

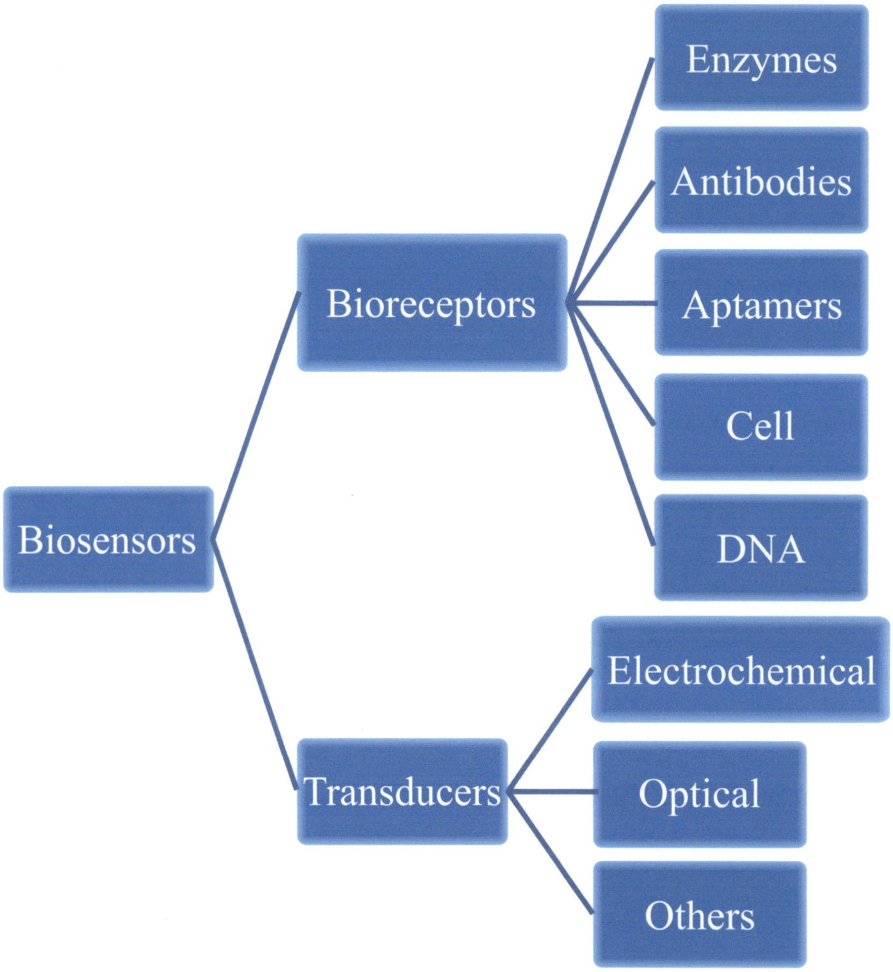

Fig. (8). Classification of biosensors after the bioreceptors and transducers type.

The enzymes and antibodies are the main classes of bioreceptors that are widely used in biosensor applications. There are several types of electrochemical transducers, as presented in Fig. (**8**). They will be briefly described in the following sections.

Electrochemical Biosensors

Electrochemical biosensors use electrochemical transducers (electrodes) capable of measuring the reduction or oxidation currents obtained after redox transformation of target analyte. Those biosensors rely on the detection of the electrical properties such as current, potential, capacitance, and impedance, which could be measured using different electrochemical methods such as amperometry, potentiometry, conductometry, voltammetry, or impedance.

Electrochemical detection is based on the chemical potential of the analyte of interest, as measured by comparison to a reference electrode. Therefore, the electrochemical response is dependent on the activity of the analyte. The resulting electrical signal is correlated with the recognition process between the target and the analyte and is proportional to the analyte concentration. Depending on the nature of the electrochemical detection method used during the biorecognition event, electrochemical biosensors are classified into five types: amperometric, potentiometric, voltammetric, conductimetric, and impedimetric.

The literature data abounds in examples of amperometric or impedimetric biosensor for a large variety of target analytes. Drugs, cancer biomarkers, pollutants (pesticides), drug of abuse and heavy metals from various matrices were analyzed with electrochemical biosensors, the detection limits being in the vicinity of ppm or ppb ranges.

Table **3** presents several examples of biosensors divided after the type of bioelement and the target analyte.

Although DNA-based biosensors represent a promising alternative for fast, simple and cheap detection of target molecules, there are some issues related to poor stability and reproducibility that hinder their competition with the currently validated methods for DNA assays (like quantitative PCR or branched-chain DNA) [171].

Table 3. Examples of electrochemical biosensors for various target analytes.

Bioelement	Type of electrode	Target analyte	LOD	Ref.
Enzymes	HRP deposited on GCE modified with PEI-SWCNT	Acetaminophen	1.36 μM	[152]
	Au/thioctic acid/EDC/NHS-biotin/ant--cortisol/cortisol/anti-cortisol with alkaline phosphatase/p-nitrophenyl phosphate	Cortisol	0.76 nM	[153]
	Mesoporous carbons-based glucose/O_2 biofuel cell with NADH	Glucose and ethanol	-	[154]
	Three-dimensional (3-D) graphene nanodots-encaged porous gold electrode with glucose oxidase and catalase	Glucose H_2O_2	30 M for glucose and 1 M for H_2O_2	[155]
	carbon paste screen printed electrodes (SPE) were coated with a ccNiR/carbon ink composite and nitrite reductase	Nitrite	1.2 μM	[156]
	Graphene based electrode materials with D-fructose dehydrogenase	D-fructose	0.6 mM	[157]
Antibodies	Protein G magnetic beads on graphite-SPE	MUC 1	0.19 nM	[158]
	3D label-free immunosensor based on crumpled graphene-Au nanocomposites	PSA	0.59 ng/mL	[159]
	Electrochemical 3D origami microfluidic paper based immunosensor	CEA CA125	0.08 pg/mL 0.06 mU/mL	[160]
	Impedimetric immunosensor based on AuNP-AgCl/polyaniline composite film	LDL cholesterol	0.34 pg/mL	[161]
	Electrochemical immunosensor based on graphite oxide	Acetaminophen	0.17 μM	[162]
Aptamers	Electrochemical aptasensors based on gold nanoparticles (AuNPs)	MUC1	EIS: 3.6 ng/mL DPV: 0.95 ng/mL	[163]
	Activated *o*-aminobenzoic acid on graphite SPE aptasensor	MUC 1	0.62 ppb	[164]
	Electrochemical aptasensor based on methylene blue and ferrocene	Thrombin	170 pM	[160]

(Table 3) contd.....

Bioelement	Type of electrode	Target analyte	LOD	Ref.
DNA	Ag^{+}-induced DNA hybridization, using Ethyl green as electroactive label/carbon paste electrode (CPE) and AuNPs-modified CPE	Ag^{+}	10^{-10} and $2.6 \cdot 10^{-11}$ M	[165]
	glassy carbon electrode modified with AuNPs, graphene and methylene blue as hybridization indicator	*Klebsiella pneumoniae*	$2 \cdot 10^{-13}$ M	[166]
	Self-assembled monolayers (SAMs) of thiolated single-stranded DNA (SH-ssDNA) on Au electrode	BRCA1	$4.6 \cdot 10^{-20}$ M	[167]
	multi-walled carbon nanotubes (MWCNT), chitosan, glutaraldehyde and DNA nanostructures (nsDNA) on graphite electrode	Toluene Naphthalene	59,5 μM 71 μM	[168]
mRNA	Electrochemical immunosensor based on graphene	Cancer biomarker mARN 21	3.12 pM	[169]
	dual-amplified duplex-specific nuclease (DSN) signal amplification strategy by coupling the enhancement of DSN and bio-bar-code conjugates on QDs (quantum dots)	Human cervical cancer cell lines (HeLa)	$4.8 \cdot 10^{-16}$ M	[170]

DNA-based biosensors use DNA strands as probes for sensing targets and two ways of recognition:

1. Hybridization of probes with DNA or RNA targets, including probes for in vitro DNA damage detection,
2. Association of probes with targets or subunits related to specific properties of the targets [172].

Two single-stranded DNA (ssDNA) with complementary sequences can hybrid together by hydrogen bonding, which is a non-covalent interaction of adenosine/thymine and cytosine/guanine bases. Based on this, DNA hybridization biosensor uses probe DNA with known sequence as the biorecognition element to detect the target DNA in the sample with various signaling mechanisms. An excellent review written by J. Chao *et al.* [171] presents several examples of DNA sensors using electrochemical transducers.

In *electrochemical impedance spectroscopy* (EIS), the sample is placed on the sensing device and a controlled alternating voltage is applied to the electrode, the current flowing through the sample being monitored. The electrical impedance resulting from the sample is calculated as the ratio of voltage over current. The resulting electrical impedance measurement has both a magnitude and a phase: a complex number. For any time-varying voltage applied, the resulting current can be in phase with the applied voltage (resistive behavior), or out of phase with it (capacitive behavior). The analytical parameters could be estimated using the radius of the semicircle obtained in Nyquist representation of the real part of the impedance plotted on the X-axis and the imaginary part plotted on the Y-axis of the chart (Fig. **9**). The main advantages of this technique consist in the capacity of real time monitoring since it is able to provide a label free or reagentless detection.

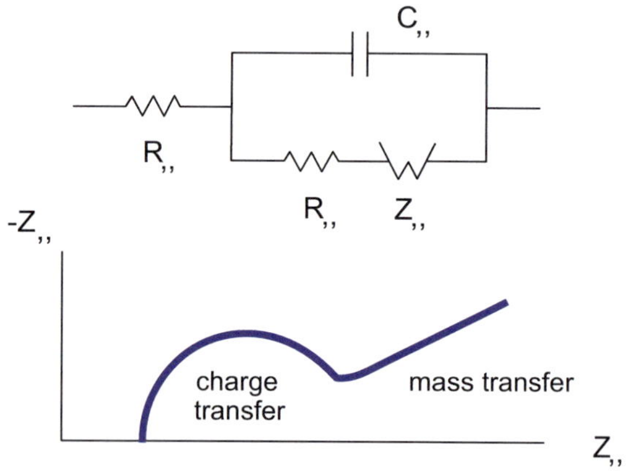

Fig. (9). Equivalent circuit and schematic Nyquist plot for an electrochemical system governed by charge and mass transfer.

EIS has been widely used by many research groups to detect cancer/tumor cells, viruses, bacteria and pathogens [163, 173, 174].

Optical Biosensors

Optical biosensors are sensitive analysis tools with various applications in biomedical, health care, pharmaceuticals, environmental monitoring, and security fields. Commonly in an optical biosensor, the transduction process induces a change in the phase, amplitude, polarization, or frequency of the input light in response to the physical or chemical change produced by the biorecognition process. Those sensors offer a large plethora of advantages such as high selectivity and specificity, remote sensing, isolation from electromagnetic interference, fast, real-time measurements, multiple channels/multi parameters detection, compact design, minimally or no-invasive for *in vivo* measurements, multiple choice for optical components and detailed chemical information on analytes [1]. The basic components of an optical biosensor are the light source, optical transmission medium (fiber, waveguide, *etc.*), the biological recognition element (enzymes, antibodies, or microbes), and optical detection system. Their classification could be done based on the detection type. Two types of detection are usually employed in optical biosensors: *Fluorescent based detection* and *Label-free detection* (such as SPR).

When low concentrations of bioanalytes are required ***fluorescence*** is one of the most sensitive spectroscopic techniques, well suited for this type of detection. In order to initiate electronic transitions in an atom or molecule, fluorescence requires an external light source (short-wavelengths), which then produces luminescence (longer wavelengths). Fluorescence-based biosensors incorporated with fluorochrome molecules are used to produce light during the biorecognition event. Since most of the biological-sensing elements and most analytes do not possess intrinsic spectral properties, the biorecognition event is transduced to optical signal by coupling fluorescence optically responsive reagents to the sensing elements.

Nucleic acid or antibodies are used to tag with fluorochrome and convert the hybridization interaction between two complementary DNA strands into an optical signal. Among the major reported drawbacks of fluorescence technology can be mentioned the complexity of instrumentation and the inability of being suitable for real time monitoring [175].

Fluorescent biosensors could be obtained by immobilizing the whole cells on the surface of a sensor layer. A common fluorescence technique used for biosensing is the sandwich assay. The analyte is selectively bound to a surface by a targeting molecule (like an antibody), which has been immobilized covalently on the surface of a well or other cell. By labeling the analyte molecule with a fluorescent tag, its surface concentration may be measured *via* highly sensitive fluorescence spectroscopy [1].

The fluorescence based biosensor could include fluorochrome molecules which are used to produce light during the biorecognition event. Since most of the biological sensing element and most analyte do not possess intrinsic spectral properties, the biorecognition event is transduced to an optical signal by coupling fluorescence in optically responsive reagents to the sensing elements [176]. In fact, fluorescence technology has been widely applied for immunochemical sensing in the medical field [177].

Surface plasmon resonance (SPR) is an optical phenomenon that provides a noninvasive, label-free means of observing binding interactions between an injected analyte and an immobilized biomolecule in real time. In the conventional SPR biosensor configuration, a thin metallic film is coated on one side of the prism, separating the sensing medium and the prism. The SPR effect is sensitive to binding of analyte because the associated increase in mass causes a proportional increase in refractive index, which is observed as a shift in the resonance angle. Optical sensors based on excitation of surface plasmons, commonly referred to as SPR sensors. SPR biosensors use surface plasmon waves (electromagnetic wave) to detect changes when the target analyte interacts with the biorecognition element on the sensor. A spectrophotometer is used to measure the absorption spectrum of the sample. Various biorecognition elements have been incorporated with SPR biosensors such as proteins, antibodies-antigens, nucleic acids, and enzymes. An important feature of SPR biosensors is that they are able to provide label-free sensing without radioactivity and fluorescence, which makes it highly attractive for real-time monitoring [178].

When free electrons in plasma experience forces applied by a variable electromagnetic field with a particular wavelength, they can oscillate in a coherent way. Large groups of electrons in the oscillation state are called plasmons. Plasmons can appear at the surface of certain metals under light irradiation, forming surface plasmons. In a very thin metal film, the plasmon region extends over the whole film thickness. In SPR model analysis, a TM (transverse magnetic) or p-polarized light causes the excitation of electron density oscillations (SPW-surface plasmon wave) at the metal-dielectric interface. When the energy as well as the momentum of both, the incident light and SPW, match, a resonance condition occurs which results in a sharp dip in the reflected light intensity.

The occurrence of plasmon resonance in metal layer is critically dependent on three factors: the angle of incidence, the wavelength of the incident radiation and the refractive index (RI) of the sample in which the evanescent wave is extended.

If the wavelength is kept constant and the angle of incidence is varied, then the sharp dip appears at a particular angle and the method is called angular interrogation. In another method, called spectral or wavelength interrogation, the angle of the incident beam is kept constant and the wavelength is varied. In this method, resonance occurs at a particular wavelength. The resonance parameter (angle or wavelength) depends on the refractive index of the dielectric medium. SPR measures changes in the refractive index near a thin metal surface in response to molecular interactions, and the RI is directly correlated to the concentration of substances near the surface (in the medium).

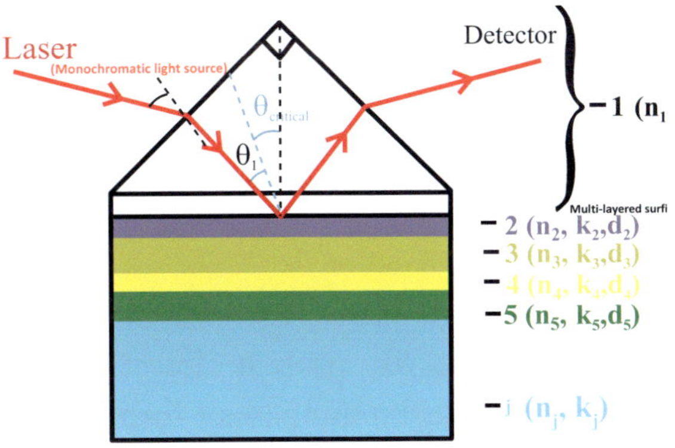

Fig. (10). Kretschmann configuration.

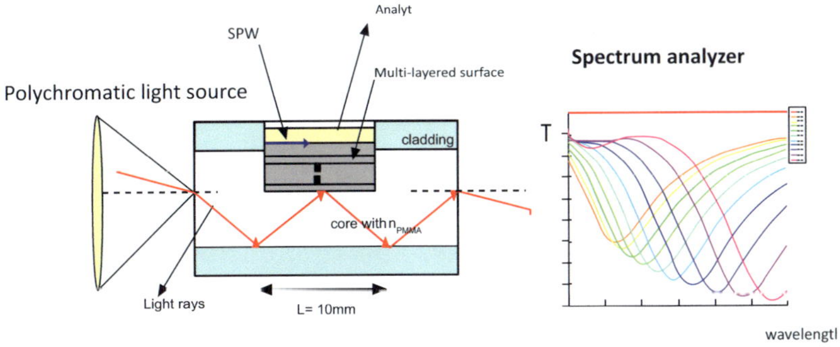

Fig. (11). SPR sensor based on POF.

SPR biosensors offer rapid screening of low concentrations of analytes in real time. To excite surface plasmons, generally, a prism is used. The most widely used prism-coupling geometry is called Kretschmann configuration (Fig. **10**). An alternative solution is an optical fiber (plastic optical fiber -POF) based SPR sensor (Fig. **11**).

This POF based sensor offers many advantages over prism based model, such as flexible optical design through the easy modeling of plastic materials, development of remote sensing setup, cost effectiveness and reduced dimensions. Furthermore, it offers the possibility of integration with optoelectronic devices (eventually leading to "lab on a chip"), to be used in continuous analysis in real-time (not just one sample at a time) and in situ monitoring. SPR technique combines rapid and real-time analysis with label-free biomarker (biomolecule or drug) recognition also allowing miniaturization for portable applications (decentralized patient monitoring).

In the last decade, over 25000 research articles (www.sciencedirect.com, October 2015) treated the subject of optical biosensors with focus on the biomedical and environmental applications.

An interesting example is a liposome-based, total internal reflection fluorescence fiber-optic biosensor (TIRF-FOB) for protein detection, which integrates a liposomal amplifier and a sandwich immunoassay format. The antibody-tagged and fluorophore-entrapped liposomes for heterogeneous detection of target molecules were designed and synthesized by Chang *et al.* [179] and the biosensor was successfully applied for the detection of the target protein IgG with a LOD of 2.0 attomoles (equivalent to 2.0 pg/mL of protein presented in 150 μL of sample solution). The advantages of this ultra-sensitive liposomal TIRF-FOB are (i) fluorescence is excited *via* evanescent waves and amplified *via* liposomes; (ii) the use of two polyclonal antibodies in the sandwich assay format increases the specificity and lowers the cost of our assay. This example has a great potential as a practical, clinical diagnostic tool in the near future.

In another study a novel arrangement of fiber-optic biosensors based on a tilted fiber Bragg grating (TFBG) coated with noble metal nanoparticles (either gold nanocages (AuNC) or gold nanospheres (AuNS) was presented. These biosensors demonstrated increased specificity and lowered detection limits for the target protein than a reference sensor without gold nanoparticles. The sensing film was fabricated by a series of thin-film and monolayer depositions to attach the gold nanoparticles to the surface of the TFBG using only covalent bonds. The minimum detected target concentrations from 90 nM for the reference sensor, to 11 pM and 8 pM respectively was obtained. This improvement of minimum detection is the result of a reduced non-specific absorption onto the gold nanoparticles (by functionalization of the external surface of the gold nanoparticles), and of an optical field enhancement due to coupling between the photonic modes of the optical fiber and the localized surface plasmon resonances (LSPR) of the gold nanoparticles [180].

Porous silicon biosensors are by far used in optical monitoring of molecular interactions. These photonic devices could greatly benefit of integration with specific microfluidic circuits that allow small volume, and thus low consumption of reagents, fast time of incubation, and highly controlled reaction conditions. The fabrication and characterization of an optical biochip, based on a porous silicon microarray of one-dimensional photonic crystals has been integrated with innovative microfluidic circuit for parallel feeding of chemical solutions. The device has been used for the label free optical monitoring of ssDNA–cDNA interaction [181].

Another sensor platform consisting of a photonic waveguide-based optical circuit used to deliver excitation light to multiple sensor windows on the platform was described by Gorin *et al.* [182]. The platform was fabricated by UV-photopatterning of photocurable hybrid organic–inorganic sol–gel materials. The sensing mechanism is based upon the detection of fluorescently labeled antibodies, in order to determine the concentration of specific analytes in a test solution. Fluorescence is excited by means of the evanescent wave in each sensor window. It was shown that the sensing properties of the platform can be dramatically enhanced by increasing the intensity of the evanescent field of the light propagating in the optical waveguide by a precise design of a high refractive index layer deposited at the waveguide surface. This work was applied for the detection of fluorescently labeled antibodies, with pg/ml detection levels. The hybrid nature of the materials developed in this study permits integration of a large range of functional groups, including organic molecules, for the immobilization of biomolecules *via* the occurrence of organic and inorganic chemical reactions. Other useful conclusion is the fact that in order to facilitate the alignment of the optical wave guides and optical fibers, three strategies have been identified, which include the integration of in-coupling Bragg gratings, prism coupling method and V-groove arrangements.

DNA detection can also be achieved by using Watson-Crick base pairing with oligonucleotides or their analogs, the so-called artificial DNAs: Peptide Nucleic Acids (PNAs), Locked Nucleic Acids, morpholino oligonucleotides, and other modified probes. Artificial DNA has shown improved properties of affinity and selectivity in target recognition compared to simple DNA probes. PNAs are nucleic acid chains that mimic DNA but with the sugar-phosphate backbone replaced by a polyamide chain composed of N-aminoethylglycine. PNAs were introduced by Nielsen *et al.* [183] and frequently coupled to SPR detection [184, 185] to enhance the biosensor sensitivity and selectivity and for the first time they were reported combined to SPR for SNPs detection in 2001 [186]. PNA-DNA hybridization is less sensitive in aqueous solution of salts, which are necessary to attenuate electrostatic repulsion in a DNA duplex. PNA is also characterized by good stability both to nucleases and peptidases, since it's synthesized, property that prevents recognition by natural enzymes, making them persistent in biological fluids. Therefore, a major drawback is that enzymatic reactions, which are often used in combination with DNA probes, are not possible using PNA substitutes. Anyway, the detection schemes involving *e.g.* DNA-ligase or DNA-polymerases cannot be performed with PNA, but they can be easily functionalized with biotin and then coupled to an enzymatic assay.

A very interesting application is about chiral boxes consisting in PNA chains containing an insert of three chiral monomers based on D-lysine. This structure was able to recognize complementary DNA oligonucleotides with sequence selectivity higher than the corresponding achiral PNA, and in particular showed greatly improved mismatch recognitionability, proving that chiral PNAs could be highly selective probes in DNA recognition of SNPs [187].

Other types of biosensors (piezoelectric, magnetoelastic, calorimetric)

The **piezoelectric biosensors** measure the changes in oscillating crystal resonance frequency which are due to bioreceptor-analyte interactions. In piezoelectric biosensors, the transducer is made of piezoelectric material (like quartz) and the biosensing material is then coated or deposited onto the piezoelectric material, which vibrates at the natural frequency. When the crystal is coated with biomolecules, the actual frequency depends on the mass of the crystal and coating [1]. The frequency can be measured with great accuracy, making possible to calculate the mass of analyte adsorbed at the crystal surface. Besides quartz, tourmaline, lithium niobate or tantalate, oriented zinc oxide or aluminium nitride can also be used as piezoelectric materials [188]. Piezoelectric transduction is suitable for portable, label-free detection and real-time biosensing. This method is highly sensitive, detection limits are close to the picogram level. Antibodies, enzymes, and antigens are used as biological elements in these devices. Such types of biosensors have been applied for detection of several gas (ammonia, methane, carbon monoxide, nitrous oxide) [189 - 192] organophosphorous compounds, pathogens [193] or for a promoter (35S) widely used in Genetically Modified Organism development [189].

The piezoelectric transducer is suitable also for DNA and protein detection with a detection limit of 1 ng cm^{-2} [194]. The piezoelectric sensor has been used in various applications such as cholera toxin detection, hepatitis B, hepatitis C, and food pathogens detection [195 - 197]. The piezoelectric transducer is a very sensitive device; LODs of 8.6 pg mL^{-1} and 25 ng mL^{-1} were obtained for hepatitis B virus DNA and for cholera toxin, respectively [198].

The **magnetoelastic biosensors** are based on magnetoelasticity instead of piezoelectricity. A magnetoelastic material changes its dimensions when it is exposed to a magnetic field. A thin strip of a magnetoelastic material forms a resonator, similar to a tuning fork. Exposing the film to a short magnetic pulse excites the film and it starts to oscillate emitting a magnetic field. The frequency and amplitude of the emitted magnetic field give information about the status of the sensor.

One of the most interesting applications of this type of sensors is the determination of blood coagulation time, essential for the anticoagulants therapy monitoring. The ribbon-like magnetoelastic sensor oscillates at a fundamental frequency, which shifts linearly as a response to the applied mass loads of changing elasticity [1].

Magnetoelastic sensors employing strips of an amorphous alloy have been in use for some time now as antitheft devices [199]. Magnetoelastic biosensors for the detection and quantification of pathogens have been studied more recently [200, 201]. In this case, an oscillating magnetic field causes a strip of amorphous alloy having a biological recognition layer on its surface to oscillate at its natural resonant frequency. The oscillation frequency is reduced as the pathogens are deposited onto the amorphous strip, allowing their detection. This is in many ways similar to the functioning of the quartz crystal microbalance (QCM). Posan *et al.* [202] used strips of amorphous Metglas 2826MB3 ribbon, cut to size and then coated with thin layers of Cr and Au. Different degrees of roughness were studied: surfaces from as-cast ribbons as well as polished surfaces. Cystamine was applied to the substrate, forming a self-assembled monolayer followed by antibodies. The effect of surface roughness was investigated for *E. coli* by measuring the frequency shift $-\Delta f$ of the sensor during bacteria capture. From the frequency shifts measured for control sensors, the authors estimated the minimum detectable concentrations to be in the range 80–400 ng/mL, the lowest values being for polished sensors and the largest for a rough sensor.

The **calorimetry based biosensing devices** rely on the general idea of generation and absorption of heat resulting in all biochemical reactions. Initially, calorimetric transduction has been employed for enzyme based sensors and has subsequently been applied in DNA/cell and immunosensors development. The principle of calorimetric biosensors is based on the measurement of the changes in temperature in the reaction between the biorecognition element and the target analyte. This change in temperature could be correlated to the amount of consumed reactants or formed products. In this type of biosensor, the heat change is measured using either a thermistor (usually metal oxide) or a thermophile (usually ceramic semiconductor).

The advantages of this type of biosensors based on thermal detection are the stability, increased sensitivity, and possibility of miniaturization. This method was used for label-free screening of biomolecule interactions and also for the rapid detection of DNA hybridization.

The literature gives several examples related to the detection of some pesticides and applications of such sensors in food analysis and biomedical field [203 - 206].

Miniaturization and inexpensive materials are needed and could be applied within this type of detection. Therefore, a paper-based microcalorimetric device was developed for the monitoring the DNA concentrations following the heat generated by the glucose oxidase. The concentration of glucose was measured using the heat generated by the glucose oxidase, which was used to convert glucose to gluconic acid and hydrogen peroxide. In biological systems, intracellular iron from ferritin reacts with hydrogen peroxide and catalyzes the formation of hydroxyl groups, which cleaves DNA. The reaction of the hydroxyl group and DNA causes DNA strand breakage. Hydrogen peroxide is a good source of hydroxyl groups, and reacts with salmon DNA in a buffer solution. The heat generated by the reaction was detected, with the paper microfluidic device (without the use of any catalyst), being proportional to the salmon DNA concentration. The calorimetric detection mechanism is a label-free method that allows the expansion of applications of enzyme-based sensors to a wide range of enzymatic reactions [204].

CONCLUSION

The process of molecular recognition has been extensively employed in the field of analytical sciences, either for separation purposes, using the biorecognition or biomimetic systems immobilized to a stationary phase or as a buffer additive for sample pretreatment, analytical separation or purification, or for sensors/ biosensors development, as a chemo- or bioreceptor intimately connected to a given transducer. Chiral molecular recognition, as the highest form of molecular recognition relying on very small energetic differences between the forming transient complexes, is also exploited for differentiation between enantiomers with numerous applications in the biomedical and environmental fields. New types of chiral selectors are introduced every year, further broadening their application range. The high selectivity of separation techniques and sensors is achieved and adapted to the target analyte by a careful selection from the large variety of natural or biomimetic recognition systems. Among the various signal transduction methodologies reported in sensor development, electrochemical techniques are widely investigated because of their potential advantages such as low-cost, miniaturization and portable instrumentation. Electrochemical detection is complementary to optical detection (*i.e.* fluorescence, the most sensitive optical technique). Label-free biosensors (including QCM and SPR) have other attractive advantages with respect to speed and simplicity of operation. Impedance is yet another rapid and inexpensive alternative technique to develop label-free biosensors for the detection of biomarkers. The continuously increasing number of papers published in the recent years, once again demonstrate the great potential and interest towards this topic.

CONFLICT OF INTEREST

The authors confirm that they have no conflict of interest to declare for this publication.

ACKNOWLEDGEMENTS

Declared none.

REFERENCES

[1] Karunakaran C, Rajkumar R, Bhargava K. Introduction to Biosensors. In: Karunakaran C, Bhargava K, Benjamin R, Eds. Biosensors and Bioelectronics. Amsterdam: Elsevier 2015; pp. 1-68. [http://dx.doi.org/10.1016/B978-0-12-803100-1.00001-3]

[2] Jones C, Patel A, Griffin S, *et al.* Current trends in molecular recognition and bioseparation. J Chromatogr A 1995; 707(1): 3-22. [http://dx.doi.org/10.1016/0021-9673(95)00466-Z] [PMID: 7633593]

[3] Sundaresan V, Abrol R. Towards a general model for protein-substrate stereoselectivity. Protein Sci 2002; 11(6): 1330-9. [http://dx.doi.org/10.1110/ps.3280102] [PMID: 12021432]

[4] Lämmerhofer M. Chiral recognition by enantioselective liquid chromatography: mechanisms and modern chiral stationary phases. J Chromatogr A 2010; 1217(6): 814-56. [http://dx.doi.org/10.1016/j.chroma.2009.10.022] [PMID: 19906381]

[5] Scriba GK. Chiral recognition mechanisms in analytical separation sciences. Chromatographia 2012; 75: 815-38.
[http://dx.doi.org/10.1007/s10337-012-2261-1]

[6] Easson LH, Stedman E. Studies on the relationship between chemical constitution and physiological action: Molecular dissymmetry and physiological activity. Biochem J 1933; 27(4): 1257-66.
[http://dx.doi.org/10.1042/bj0271257] [PMID: 16745220]

[7] Young BL, Wu L, Cooks RG. Mass spectral methods of chiral analysis. In: Chiral analysis; Busch, KW; Busch MA, Eds; Amsterdam, Elsevier BV: 2006; pp 595-660.
[http://dx.doi.org/10.1016/B978-044451669-5/50017-X]

[8] Topiol S. A general criterion for molecular recognition: implications for chiral interactions. Chirality 1989; 1(1): 69-79.
[http://dx.doi.org/10.1002/chir.530010113] [PMID: 2642036]

[9] Merten C. Chirality as probe for intermolecular interactions. Bunsenmagazin 2015; 17: 89-93.

[10] Del Bubba M, Checchini L, Lepri L. Thin-layer chromatography enantioseparations on chiral stationary phases: a review. Anal Bioanal Chem 2013; 405(2-3): 533-54.
[http://dx.doi.org/10.1007/s00216-012-6514-5] [PMID: 23161065]

[11] Cavazzini A, Marchetti N, Guzzinati R, *et al.* Enantioseparation by ultra-high-performance liquid chromatography. TrAC Trend. Anal Chem 2014; 63: 95-103.

[12] Lesellier E, West C. The many faces of packed column supercritical fluid chromatography a critical review. J Chromatogr A 2015; 1382: 2-46.
[http://dx.doi.org/10.1016/j.chroma.2014.12.083] [PMID: 25604272]

[13] Schurig V, Juza M. Analytical separation of enantiomers by gas chromatography on chiral stationary phases. Adv Chromatogr 2014; 52: 117-68.
[http://dx.doi.org/10.1201/b16668-5] [PMID: 24843907]

[14] Ulatowski F, Hamankiewicz P, Jurczak J. Enantioselective liquid-solid extraction (ELSE) an unexplored, fast, and precise analytical method. ACS Comb Sci 2015; 17(9): 488-92.
[http://dx.doi.org/10.1021/acscombsci.5b00075] [PMID: 26263300]

[15] Jáč P, Scriba GK. Recent advances in electrodriven enantioseparations. J Sep Sci 2013; 36(1): 52-74.
[http://dx.doi.org/10.1002/jssc.201200836] [PMID: 23255223]

[16] Stavrou IJ, Mavroudi MC, Kapnissi-Christodoulou CP. Chiral selectors in CE: recent developments and applications (2012-mid 2014). Electrophoresis 2015; 36(1): 101-23.
[http://dx.doi.org/10.1002/elps.201400310] [PMID: 25263224]

[17] Iacob BC, Bodoki E, Florea A, Bodoki AE, Oprean R. Simultaneous enantiospecific recognition of several β-blocker enantiomers using molecularly imprinted polymer-based electrochemical sensor. Anal Chem 2015; 87(5): 2755-63.
[http://dx.doi.org/10.1021/ac504036m] [PMID: 25630982]

[18] Feagin TA, Olsen DP, Headman ZC, Heemstra JM. High-throughput enantiopurity analysis using enantiomeric DNA-based sensors. J Am Chem Soc 2015; 137(12): 4198-206.
[http://dx.doi.org/10.1021/jacs.5b00923] [PMID: 25747268]

[19] Schug KA, Lindner W. Chiral molecular recognition for the detection and analysis of enantiomers by mass spectrometric methods. J Sep Sci 2005; 28: 1932-55.
[http://dx.doi.org/10.1002/jssc.200500232]

[20] Wu L, Vogt FG. A review of recent advances in mass spectrometric methods for gas-phase chiral analysis of pharmaceutical and biological compounds. J Pharm Biomed Anal 2012; 69: 133-47.
[http://dx.doi.org/10.1016/j.jpba.2012.04.022] [PMID: 22579598]

[21] Wenzel TJ. Chiral derivatizing agents, macrocycles, metal complexes, and liquid crystals for enantiomer differentiation in NMR spectroscopy. Top Curr Chem 2013; 341: 1-68.
[http://dx.doi.org/10.1007/128_2013_433] [PMID: 23595365]

[22] Trojanowicz M, Kaniewska M. Electrochemical chiral sensors and biosensors. Electroanalysis 2009; 21: 3-5, 229-238.
[http://dx.doi.org/10.1002/elan.200804382]

[23] Stefan RI, van Staden JF, Aboul-Enein HY. Molecular recognition in chiral discrimination. Cryst Eng 2001; 4: 113-8.
[http://dx.doi.org/10.1016/S1463-0184(00)00047-2]

[24] Trojanowicz M. Electrochemical and piezoelectric enantioselective sensors and biosensors. Anal Lett 2005; 38: 523-47.
[http://dx.doi.org/10.1081/AL-200050157]

[25] Izake EL. Chiral discrimination and enantioselective analysis of drugs: an overview. J Pharm Sci 2007; 96(7): 1659-76.
[http://dx.doi.org/10.1002/jps.20820] [PMID: 17221858]

[26] Sharma R, Ragavan KV, Thakur MS, Raghavarao KS. Recent advances in nanoparticle based aptasensors for food contaminants. Biosens Bioelectron 2015; 74: 612-27.
[http://dx.doi.org/10.1016/j.bios.2015.07.017] [PMID: 26190473]

[27] Tiwari MP, Prasad A. Molecularly imprinted polymer based enantioselective sensing devices: a review. Anal Chim Acta 2015; 853: 1-18.
[http://dx.doi.org/10.1016/j.aca.2014.06.011] [PMID: 25467446]

[28] Haga T. Receptor biochemistry. In: Meyers RA, Ed. Molecular biology and biotechnology, a comprehensive desk reference. New York: VCH 1995; pp. 789-93.

[29] Subrahmanyam S, Piletsky SA, Turner AP. Application of natural receptors in sensors and assays. Anal Chem 2002; 74(16): 3942-51.
[http://dx.doi.org/10.1021/ac025673+] [PMID: 12199559]

[30] Matharu Z, Bandodkar AJ, Gupta V, Malhotra BD. Fundamentals and application of ordered molecular assemblies to affinity biosensing. Chem Soc Rev 2012; 41(3): 1363-402.
[http://dx.doi.org/10.1039/C1CS15145B] [PMID: 22105315]

[31] Ackaah-Gyasi NA, Patel P, Zhang Y, Simpson BK. Current and future uses of enzymes in food processing. In: Yada RY, Ed. Improving and tailoring enzymes for food quality and functionality. Amsterdam: Elsevier B.V. 2015; pp. 103-22.
[http://dx.doi.org/10.1016/B978-1-78242-285-3.00005-3]

[32] Ammam M. Electrochemical and electrophoretic deposition of enzymes: principles, differences and application in miniaturized biosensor and biofuel cell electrodes. Biosens Bioelectron 2014; 58: 121-31.
[http://dx.doi.org/10.1016/j.bios.2014.02.030] [PMID: 24632138]

[33] Perumal V, Hashim U. Advances in biosensors: Principle, architecture and applications. J Appl Biomed 2014; 12: 1-15.
[http://dx.doi.org/10.1016/j.jab.2013.02.001]

[34] Fowler JM, Wong DK, Halsall HB, Heineman WR. Recent developments in electrochemical immunoassays and immunosensors. 2008.
[http://dx.doi.org/10.1016/B978-012373738-0.50007-6]

[35] Cristea C, Florea A, Tertis M, Sandulescu R. Immunosensors. In: Rinken T, Ed. Biosensors - Micro and Nanoscale Applications. 165-202.

[36] Hock B. Antibodies for immunosensors. Anal Chim Acta 1997; 347: 177-86.
[http://dx.doi.org/10.1016/S0003-2670(97)00167-0]

[37] Gopinath SC, Tang TH, Citartan M, Chen Y, Lakshmipriya T. Current aspects in immunosensors. Biosens Bioelectron 2014; 57: 292-302.
[http://dx.doi.org/10.1016/j.bios.2014.02.029] [PMID: 24607580]

[38] Tombelli S, Minunni M, Mascini M. Analytical applications of aptamers. Biosens Bioelectron 2005; 20(12): 2424-34.
[http://dx.doi.org/10.1016/j.bios.2004.11.006] [PMID: 15854817]

[39] Nezlin R. Aptamers in immunological research. Immunol Lett 2014; 162(2 Pt B): 252-5.
[http://dx.doi.org/10.1016/j.imlet.2014.10.001] [PMID: 25447403]

[40] Šmuc T, Ahn IY, Ulrich H. Nucleic acid aptamers as high affinity ligands in biotechnology and biosensorics. J Pharm Biomed Anal 2013; 81-82: 210-7.
[http://dx.doi.org/10.1016/j.jpba.2013.03.014] [PMID: 23666257]

[41] Ling Z, Wang M, Wang J, Ye Z. Application of biosensor surface immobilization methods for aptamer. Chin J Anal Chem 2011; 39(3): 432-8.
[http://dx.doi.org/10.1016/S1872-2040(10)60429-X]

[42] Ellington AD, Szostak JW. In vitro selection of RNA molecules that bind specific ligands. Nature 1990; 346(6287): 818-22.
[http://dx.doi.org/10.1038/346818a0] [PMID: 1697402]

[43] Tuerk C, Gold L. Systematic evolution of ligands by exponential enrichment: RNA ligands to bacteriophage T4 DNA polymerase. Science 1990; 249(4968): 505-10.
[http://dx.doi.org/10.1126/science.2200121] [PMID: 2200121]

[44] Ninomiya K, Yamashita T, Kawabata S, Shimizu N. Targeted and ultrasound-triggered drug delivery using liposomes co-modified with cancer cell-targeting aptamers and a thermosensitive polymer. Ultrason Sonochem 2014; 21(4): 1482-8.
[http://dx.doi.org/10.1016/j.ultsonch.2013.12.023] [PMID: 24418100]

[45] Sundaram P, Wower J, Byrne ME. A nanoscale drug delivery carrier using nucleic acid aptamers for extended release of therapeutic. Nanomedicine (Lond) 2012; 8(7): 1143-51.

[PMID: 22321384]

[46] Liang HR, Hu GQ, Li L, Gao YW, Yang ST, Xia XZ. Aptamers targeting rabies virus-infected cells inhibit street rabies virus *in vivo*. Int Immunopharmacol 2014; 21(2): 432-8. [http://dx.doi.org/10.1016/j.intimp.2014.03.020] [PMID: 24735817]

[47] Shigdar S, Qiao L, Zhou SF, *et al.* RNA aptamers targeting cancer stem cell marker CD133. Cancer Lett 2013; 330(1): 84-95. [http://dx.doi.org/10.1016/j.canlet.2012.11.032] [PMID: 23196060]

[48] Ding F, Guo S, Xie M, *et al.* Diagnostic applications of gastric carcinoma cell aptamers *in vitro* and *in vivo*. Talanta 2015; 134: 30-6. [http://dx.doi.org/10.1016/j.talanta.2014.09.036] [PMID: 25618637]

[49] Zelada-Guillén GA, Blondeau P, Rius FX, Riu J. Carbon nanotube-based aptasensors for the rapid and ultrasensitive detection of bacteria. Methods 2013; 63(3): 233-8. [http://dx.doi.org/10.1016/j.ymeth.2013.07.008] [PMID: 23872060]

[50] Meini N, Farre C, Chaix C, Kherrat R, Dzyadevych S, Jaffrezic-Renault N. A sensitive and selective thrombin impedimetric aptasensor based on tailored aptamers obtained by solid-phase synthesis. Sensor. Actuat. Biol Chem 2012; 166-167: 715-20.

[51] Xu H, Gorgy K, Gondran C, *et al.* Label-free impedimetric thrombin sensor based on poly(pyrrole-nitrilotriacetic acid)-aptamer film. Biosens Bioelectron 2013; 41: 90-5. [http://dx.doi.org/10.1016/j.bios.2012.07.044] [PMID: 22959014]

[52] Han B, Zhao C, Yin J, Wang H. High performance aptamer affinity chromatography for single-step selective extraction and screening of basic protein lysozyme. J Chromatogr B Analyt Technol Biomed Life Sci 2012; 903: 112-7. [http://dx.doi.org/10.1016/j.jchromb.2012.07.003] [PMID: 22841745]

[53] Zhou ZM, Feng Z, Zhou J, *et al.* Capillary electrophoresis-chemiluminescence detection for carcino-embryonic antigen based on aptamer/graphene oxide structure. Biosens Bioelectron 2015; 64: 493-8. [http://dx.doi.org/10.1016/j.bios.2014.09.050] [PMID: 25299985]

[54] Dick LW Jr, McGown LB. Aptamer-enhanced laser desorption/ionization for affinity mass spectrometry. Anal Chem 2004; 76(11): 3037-41. [http://dx.doi.org/10.1021/ac049860e] [PMID: 15167780]

[55] Liu A, Wang K, Weng S, *et al.* Development of electrochemical DNA biosensors. TrAC. Trends Analyt Chem 2012; 37: 101-11. [http://dx.doi.org/10.1016/j.trac.2012.03.008]

[56] Cagnin S, Caraballo M, Guiducci C, *et al.* Overview of electrochemical DNA biosensors: new approaches to detect the expression of life. Sensors (Basel) 2009; 9(4): 3122-48. [http://dx.doi.org/10.3390/s90403122] [PMID: 22574066]

[57] Kwok CK, Balasubramanian S. Targeted detection of g-quadruplexes in cellular RNAs. Angew Chem Int Ed Engl 2015; 54(23): 6751-4. [http://dx.doi.org/10.1002/anie.201500891] [PMID: 25907625]

[58] Ambros V. The functions of animal microRNAs. Nature 2004; 431(7006): 350-5. [http://dx.doi.org/10.1038/nature02871] [PMID: 15372042]

[59] Labib M, Berezovski MV. Electrochemical sensing of microRNAs: avenues and paradigms. Biosens Bioelectron 2015; 68: 83-94.
[http://dx.doi.org/10.1016/j.bios.2014.12.026] [PMID: 25562735]

[60] Keller A, Leidinger P, Lange J, *et al.* Multiple sclerosis: microRNA expression profiles accurately differentiate patients with relapsing-remitting disease from healthy controls. PLoS One 2009; 4(10): e7440.
[http://dx.doi.org/10.1371/journal.pone.0007440] [PMID: 19823682]

[61] Pileczki V, Cojocneanu-Petric R, Maralani M, Berindan Neagoe I, Săndulescu R. MicroRNAs as regulators of apoptosis mechanisms in cancer. Clujul Med 2016; 89(1): 50-5.
[http://dx.doi.org/10.15386/cjmed-512]

[62] Cristea C, Florea A. Immunosensors. In: Rinken T, Intech Rijeka, Eds. Biosensors - Micro and nanoscale applications. 2015; pp. 165-202.

[63] Wang TH, Hui GH, Deng SP. A novel sweet taste cell-based sensor. Biosens Bioelectron 2010; 26(2): 929-34.
[http://dx.doi.org/10.1016/j.bios.2010.06.049] [PMID: 20691579]

[64] Acha V, Andrews T, Huang Q, Sardar DK, Hornsby PJ. Tissue-based biosensors. In: Zourob M, Ed. Recognition Receptors in Biosensors. New York: Springer 2010; pp. 365-82.
[http://dx.doi.org/10.1007/978-1-4419-0919-0_9]

[65] Suzuki K, Watanabe K, Matsumoto Y, *et al.* Design and synthesis of calcium and magnesium ionophores based on double-armed diazacrown ether compounds and their application to an ion sensing component for an ion-selective electrode. Anal Chem 1995; 67(2): 324-34.
[http://dx.doi.org/10.1021/ac00098a016]

[66] Safiulina D, Veksler V, Zharkovsky A, Kaasik A. Loss of mitochondrial membrane potential is associated with increase in mitochondrial volume: physiological role in neurones. J Cell Physiol 2006; 206(2): 347-53.
[http://dx.doi.org/10.1002/jcp.20476] [PMID: 16110491]

[67] Del Valle ME. Cyclodextrins and their uses. Process Biochem 2004; 39(9): 1033-46.
[http://dx.doi.org/10.1016/S0032-9592(03)00258-9]

[68] Abd-Elgawad R, Shimaa E. Electrochemistry of cyclodextrin inclusion complexes of pharmaceutical compounds. Open Chem Biomed Methods J 2010; 3: 74-85.
[http://dx.doi.org/10.2174/1875038901003010074]

[69] Dodziuk H. Cyclodextrins and their complexes; Wiley-VCH Verlag GmbH & Co. KGaA: Weinheim, 2006.

[70] Szejtli J, Osa T. Comprehensive supramolecular chemistry, vol. 3; Cyclodextrins; Elsevier Science Ltd.: Oxford, 1996.

[71] Szejtli J. Past, present and future of cyclodextrin research. Pure Appl Chem 2004; 10(76): 1825-45.

[72] Szejtli J. Introduction and general overview of cyclodextrin chemistry. Chem Rev 1998; 98(5): 1743-54.
[http://dx.doi.org/10.1021/cr970022c] [PMID: 11848947]

[73] Connors KA. The stability of cyclodextrin complexes in solution. Chem Rev 1997; 97(5): 1325-58. [http://dx.doi.org/10.1021/cr960371r] [PMID: 11851454]

[74] Available from: http://www.chemiedidaktik.uni-wuppertal.de/disido_cy/cyen/info/nav/nav_info_en.html.

[75] Popovici I, Lupuleasa D. Tehnologie farmaceutică 3rd ed. 2011; 1. Polirom: Iaşi pp. 290-292.

[76] McMahon G, O'Malley S, Nolan K, Diamond D. Important calixarene derivatives - their synthesis and applications. ARKIVOC 2003; 7: 23-31.

[77] Diamond D, Nolan K. Calixarenes: designer ligands for chemical sensors. Anal Chem 2001; 73(1): 22A-9A. [http://dx.doi.org/10.1021/ac012376g] [PMID: 11195503]

[78] Available from: http://chemphys.u-strasbg.fr/mpb/teach/calixarenes/calixarenes.html.

[79] Yagi K, Khoo S, Sugawara M, *et al.* Channel mimetic sensing membranes for alkali metal cations based on oriented monolayers of calixarene esters. J Electroanal Chem 1996; 401: 65-79. [http://dx.doi.org/10.1016/0022-0728(95)04293-8]

[80] Thompson C, Field S, Jones A, Kan M, Hall C, Nicholson G. Development of selective calixarene for uranium. WM'02 Conference. Tucson, AZ, USA, February 24-28, 2002.

[81] Mulchandani A, Sadik OA. Chemical and biological sensors for environmental analysis ACS Symposium Series no. 762. [http://dx.doi.org/10.1021/bk-2000-0762.fw001]

[82] Katz JJ, Seaborg GT, Morss LR. The chemistry of the actinide elements. 2nd ed., London: Chapman and Hall 1986. [http://dx.doi.org/10.1007/978-94-009-4077-2]

[83] Vicens J, Böhmer V. Calixarenes, a versatile class of macrocyclic compounds. London: Kluwer Academic Publishers 1991.

[84] Arnaud-Neu F, Schwing-Weill MJ. Calixarenes, new selective molecular receptors. Synth Met 1997; 90: 157-64. [http://dx.doi.org/10.1016/S0379-6779(98)80001-5]

[85] McKittrick T, Diamond D, Marrs DJ, O'Hagan P, Anthony McKervey M. Calcium-selective electrode based on a calix[4]arene tetraphosphine oxide. Talanta 1996; 43(7): 1145-8. [http://dx.doi.org/10.1016/0039-9140(96)01888-7] [PMID: 18966592]

[86] Gupta VK, Jain AK, Khurana U, Singh LP. PVC-based neutral carrier and organic exchanger membranes as sensors for the determination of Ba^{2+} and Sr^{2+}. Sens. Actuat. Biol Chem 1999; 55: 201-11.

[87] Feás X, Seijas JA, Vázquez-Tato MP, Regal P, Cepeda A, Fente C. Syntheses of molecularly imprinted polymers: Molecular recognition of cyproheptadine using original print molecules and azatadine as dummy templates. Anal Chim Acta 2009; 631(2): 237-44. [http://dx.doi.org/10.1016/j.aca.2008.10.054] [PMID: 19084632]

[88] Ge Y, Turner AP. Too large to fit? Recent developments in macromolecular imprinting. Trends Biotechnol 2008; 26(4): 218-24.

[http://dx.doi.org/10.1016/j.tibtech.2008.01.001] [PMID: 18295919]

[89] Lv Y, Tan T, Svec F. Molecular imprinting of proteins in polymers attached to the surface of nanomaterials for selective recognition of biomacromolecules. Biotechnol Adv 2013; 31(8): 1172-86. [http://dx.doi.org/10.1016/j.biotechadv.2013.02.005] [PMID: 23466364]

[90] Ding X, Heiden PA. Recent developments in molecularly imprinted nanoparticles by surface imprinting techniques. J Mol Recognit 2014; 27(6): 297-401. [PMID: 24700625]

[91] Sharma PS, Pietrzyk-Le A, D'Souza F, Kutner W. Electrochemically synthesized polymers in molecular imprinting for chemical sensing. Anal Bioanal Chem 2012; 402(10): 3177-204. [http://dx.doi.org/10.1007/s00216-011-5696-6] [PMID: 22302165]

[92] Yan H, Row KH. Characteristic and synthetic approach of molecularly imprinted polymer. Int J Mol Sci 2006; 7: 155-78. [http://dx.doi.org/10.3390/i7050155]

[93] Moreira FT, Dutra RA, Noronha JP, Sales MG. Electrochemical biosensor based on biomimetic material for myoglobin detection. Electrochim Acta 2013; 107: 481-7. [http://dx.doi.org/10.1016/j.electacta.2013.06.061]

[94] Chillawar RR, Tadi KK, Motghare RV. Voltammetric techniques at chemically modified electrodes. J Anal Chem 2015; 70(4): 399-418. [http://dx.doi.org/10.1134/S1061934815040152]

[95] Huynh TP, Kutner W. Molecularly imprinted polymers as recognition materials for electronic tongues. Biosens Bioelectron 2015; 74: 856-64. [http://dx.doi.org/10.1016/j.bios.2015.07.054] [PMID: 26233642]

[96] Wackerlig J, Lieberzeit PA. Molecularly imprinted polymer nanoparticles in chemical sensing - Synthesis, characterisation and application. Sens. Actuat. Biol Chem 2015; 207: 144-57.

[97] Cheong WJ, Yang SH, Ali F. Molecular imprinted polymers for separation science: a review of reviews. J Sep Sci 2013; 36(3): 609-28. [http://dx.doi.org/10.1002/jssc.201200784] [PMID: 23281278]

[98] Iacob BC, Bodoki E, Oprean R. Recent advances in capillary electrochromatography using molecularly imprinted polymers. Electrophoresis 2014; 35(19): 2722-32. [http://dx.doi.org/10.1002/elps.201400253] [PMID: 25042043]

[99] Martin-Esteban A. Molecularly-imprinted polymers as a versatile, highly selective tool in sample preparation. TrAC Trend. Anal Chem 2013; 45: 169-81.

[100] Haginaka J. Molecularly imprinted polymers as affinity-based separation media for sample preparation. J Sep Sci 2009; 32(10): 1548-65. [http://dx.doi.org/10.1002/jssc.200900085] [PMID: 19472278]

[101] Sarafraz-Yazdi A, Razavi N. Application of molecularly-imprinted polymers in solid-phase microextraction techniques. TrAC Trend Anal Chem 2015; 73: 81-90. [http://dx.doi.org/10.1016/j.trac.2015.05.004]

[102] Li JW, Wang YL, Yan S, Li XJ, Pan SY. Molecularly imprinted calixarene fiber for solid-phase microextraction of four organophosphorous pesticides in fruits. Food Chem 2016; 192: 260-7.

[http://dx.doi.org/10.1016/j.foodchem.2015.07.018] [PMID: 26304345]

[103] Urraca JL, Chamorro-Mendiluce R, Orellana G, Moreno-Bondi MC. Molecularly imprinted polymer beads for clean-up and preconcentration of β-lactamase-resistant penicillins in milk. Anal Bioanal Chem 2016; 408(7): 1843-54.
[http://dx.doi.org/10.1007/s00216-015-8941-6] [PMID: 26342308]

[104] Hashemi-Moghaddam H, Abbasi F. Synthesis of molecularly imprinted polymers coated on silica nanoparticles for removal of p-nitrophenol from crude pharmaceuticals. Pharm Chem J 2015; 49: 280-6.
[http://dx.doi.org/10.1007/s11094-015-1270-4]

[105] Available from: http://mipdatabase.com

[106] Xiao Y, Ng SC, Tan TT, Wang Y. Recent development of cyclodextrin chiral stationary phases and their applications in chromatography. J Chromatogr A 2012; 1269: 52-68.
[http://dx.doi.org/10.1016/j.chroma.2012.08.049] [PMID: 22959844]

[107] Zhan J, Li J, Liu D, *et al.* A simple method for the determination of organochlorine pollutants and the enantiomers in oil seeds based on matrix solid-phase dispersion. Food Chem 2016; 194: 319-24.
[http://dx.doi.org/10.1016/j.foodchem.2015.07.067] [PMID: 26471561]

[108] Dołowy M, Pyka A. Application of TLC, HPLC and GC methods to the study of amino acid and peptide enantiomers: a review. Biomed Chromatogr 2014; 28(1): 84-101.
[http://dx.doi.org/10.1002/bmc.3016] [PMID: 23946167]

[109] Chen B, He M, Zhong C, Hu B. Chiral speciation of selenoamino acids in biological samples. J Chromatogr A 2014; 1363: 62-70.
[http://dx.doi.org/10.1016/j.chroma.2014.07.098] [PMID: 25130085]

[110] Plotka JM, Biziuk M, Morrison C, Namiesnik J. Pharmaceutical and forensic drug applications of chiral supercritical fluid chromatography. TrAC. Trends Analyt Chem 2014; 56: 74-89.
[http://dx.doi.org/10.1016/j.trac.2013.12.012]

[111] Hicks MB, Regalado EL, Tan F, Gong X, Welch CJ. Supercritical fluid chromatography for GMP analysis in support of pharmaceutical development and manufacturing activities. J Pharm Biomed Anal 2016; 117: 316-24.
[http://dx.doi.org/10.1016/j.jpba.2015.09.014] [PMID: 26412720]

[112] West C. Enantioselective separations with supercritical fluids - review. Curr Anal Chem 2014; 10: 99-120.
[http://dx.doi.org/10.2174/1573411011410010009]

[113] Tanwar S, Bhushan R. Enantioresolution of amino acids: a decade's perspective, prospects and challenges. Chromatographia 2015; 78: 1113-34.
[http://dx.doi.org/10.1007/s10337-015-2933-8]

[114] Hyun MH. Development of HPLC chiral stationary phases based on (+)-(18-crown-6)-2,3,1-,12-tetracarboxylic acid and their applications. Chirality 2015; 27(9): 576-88.
[http://dx.doi.org/10.1002/chir.22484] [PMID: 26237013]

[115] Yu S, Pu L. Recent progress on using BINOLs in enantioselective molecular recognition. Tetrahedron 2015; 71: 745-72.

[http://dx.doi.org/10.1016/j.tet.2014.11.007]

[116] Okamoto Y. Precision synthesis, structure and function of helical polymers. Proc Jpn Acad, Ser B, Phys Biol Sci 2015; 91(6): 246-61.
[http://dx.doi.org/10.2183/pjab.91.246] [PMID: 26062738]

[117] Lipka E, Vaccher C. Quantitative analysis of drugs in biological matrices by HPLC hyphenated to fluorescence detection. Bioanalysis 2015; 7(6): 743-62.
[http://dx.doi.org/10.4155/bio.15.20] [PMID: 25871590]

[118] Zhang M, Mallik AK, Takafuji M, Ihara H, Qiu H. Versatile ligands for high-performance liquid chromatography: An overview of ionic liquid-functionalized stationary phases. Anal Chim Acta 2015; 887: 1-16.
[http://dx.doi.org/10.1016/j.aca.2015.04.022] [PMID: 26320781]

[119] Fekete S, Schappler J, Veuthey JL, Guillarme D. Current and future trends in UHPLC. TrAC. Trends Analyt Chem 2014; 63: 2-13.
[http://dx.doi.org/10.1016/j.trac.2014.08.007]

[120] Ding P, Chang B, Qing G, Sun T. New approach for chiral separation: From polysaccharide-based materials to chirality-responsive polymers. Sci China Chem 2014; 57: 1492-506.
[http://dx.doi.org/10.1007/s11426-014-5206-8]

[121] Hayes R, Ahmed A, Edge T, Zhang H. Core-shell particles: preparation, fundamentals and applications in high performance liquid chromatography. J Chromatogr A 2014; 1357: 36-52.
[http://dx.doi.org/10.1016/j.chroma.2014.05.010] [PMID: 24856904]

[122] Yusuf K, Aqel A, ALOthman Z. Metal-organic frameworks in chromatography. J Chromatogr A 2014; 1348: 1-16.
[http://dx.doi.org/10.1016/j.chroma.2014.04.095] [PMID: 24845827]

[123] Hoffmann N. Photochemical reactions applied to the synthesis of helicenes and helicene-like compounds. J Photochem Photobiol Photochem Rev 2014; 19: 1-19.
[http://dx.doi.org/10.1016/j.jphotochemrev.2013.11.001]

[124] Weng X, Bao Z, Luo F, Su B, Yang Y, Ren Q. Progress in preparation and applications of cellulose derivatives-based chiral stationary phase. Progr Chem 2014; 26: 415-23.

[125] Shen J, Ikai T, Okamoto Y. Synthesis and application of immobilized polysaccharide-based chiral stationary phases for enantioseparation by high-performance liquid chromatography. J Chromatogr A 2014; 1363: 51-61.
[http://dx.doi.org/10.1016/j.chroma.2014.06.042] [PMID: 24997110]

[126] Al-Othman ZA, Al-Warthan A, Ali I. Advances in enantiomeric resolution on monolithic chiral stationary phases in liquid chromatography and electrochromatography. J Sep Sci 2014; 37(9-10): 1033-57.
[http://dx.doi.org/10.1002/jssc.201301326] [PMID: 24634395]

[127] Batra S, Bhushan R. Amino acids as chiral auxiliaries in cyanuric chloride-based chiral derivatizing agents for enantioseparation by liquid chromatography. Biomed Chromatogr 2014; 28(11): 1532-46.
[http://dx.doi.org/10.1002/bmc.3305] [PMID: 25137358]

[128] Peluso P, Mamane V, Cossu S. Homochiral metal-organic frameworks and their application in

chromatography enantioseparations. J Chromatogr A 2014; 1363: 11-26. [http://dx.doi.org/10.1016/j.chroma.2014.06.064] [PMID: 25001329]

[129] Némethy A. Cardiovascular active compounds - Preparation and the study of chiral aspects of derivatives of the aryloxyaminopropanol type, (Zlúčeniny s účinkom na KVS - Príprava a štúdium chiralnych vlastností derivátov typu aryloxyaminopropanolov). Farm Obz 2014; 83: 65-8.

[130] Al-Othman ZA, Al-Warthan A, Alam SD, Ali I. Enantio-separation of drugs with multiple chiral centers by chromatography and capillary electrophoresis. Biomed Chromatogr 2014; 28(11): 1514-24. [http://dx.doi.org/10.1002/bmc.3259] [PMID: 25067822]

[131] Setsune JI. Helical chirality control of oligopyrrols. J Synt Org Chem 2014; 72: 280-91. [http://dx.doi.org/10.5059/yukigoseikyokaishi.72.280]

[132] Dąbrowska M, Starek M. Analytical approaches to determination of carnitine in biological materials, foods and dietary supplements. Food Chem 2014; 142: 220-32. [http://dx.doi.org/10.1016/j.foodchem.2013.06.137] [PMID: 24001835]

[133] Wuethrich A, Haddad PR, Quirino JP. Chiral capillary electromigration techniques-mass spectrometry-hope and promise. Electrophoresis 2014; 35(1): 2-11. [http://dx.doi.org/10.1002/elps.201300377] [PMID: 24265218]

[134] Sánchez-López E, Montealegre C, Crego AL, Marina ML. Recent contributions of capillary electrophoresis to neuroscience. TrAC. Trends Analyt Chem 2015; 67: 82-99. [http://dx.doi.org/10.1016/j.trac.2014.12.008]

[135] Mohammadzadeh Kakhki R, Assadi H. Capillary electrophoresis analysis based on crown ethers. J Incl Phenom Macrocycl Chem 2015; 81: 1-12. [http://dx.doi.org/10.1007/s10847-014-0419-1]

[136] Nowak P, Woźniakiewicz M, Kościelniak P. Application of capillary electrophoresis in determination of acid dissociation constant values. J Chromatogr A 2015; 1377: 1-12. [http://dx.doi.org/10.1016/j.chroma.2014.12.032] [PMID: 25543298]

[137] Escuder-Gilabert L, Martín-Biosca Y, Medina-Hernández MJ, Sagrado S. Cyclodextrins in capillary electrophoresis: recent developments and new trends. J Chromatogr A 2014; 1357: 2-23. [http://dx.doi.org/10.1016/j.chroma.2014.05.074] [PMID: 24947884]

[138] Ma Y, Zhang H, Chen H, Chen X. Recent developments in chiral analysis of β-blocker drugs by capillary electromigration techniques. Electrophoresis 2014; 35(23): 3345-54. [http://dx.doi.org/10.1002/elps.201400363] [PMID: 25224837]

[139] Řezanka P, Navrátilová K, Řezanka M, Král V, Sýkora D. Application of cyclodextrins in chiral capillary electrophoresis. Electrophoresis 2014; 35(19): 2701-21. [http://dx.doi.org/10.1002/elps.201400145] [PMID: 24853583]

[140] Stěpánová S, Kašička V. Determination of impurities and counterions of pharmaceuticals by capillary electromigration methods. J Sep Sci 2014; 37(15): 2039-55. [http://dx.doi.org/10.1002/jssc.201400266] [PMID: 24838601]

[141] Giuffrida A, Maccarrone G, Cucinotta V, Orlandini S, Contino A. Recent advances in chiral separation of amino acids using capillary electromigration techniques. J Chromatogr A 2014; 1363: 41-50. [http://dx.doi.org/10.1016/j.chroma.2014.08.041] [PMID: 25171947]

[142] Dixit S, Park JH. Application of antibiotics as chiral selectors for capillary electrophoretic enantioseparation of pharmaceuticals: a review. Biomed Chromatogr 2014; 28(1): 10-26. [http://dx.doi.org/10.1002/bmc.2950] [PMID: 23780664]

[143] Sánchez-Hernández L, Guijarro-Diez M, Marina ML, Crego AL. New approaches in sensitive chiral CE. Electrophoresis 2014; 35(1): 12-27. [http://dx.doi.org/10.1002/elps.201300355] [PMID: 24132619]

[144] Domínguez-Vega E, Pérez-Fernández V, Crego AL, García MA, Marina ML. Recent advances in CE analysis of antibiotics and its use as chiral selectors. Electrophoresis 2014; 35(1): 28-49. [http://dx.doi.org/10.1002/elps.201300347] [PMID: 24395661]

[145] Kašička V. Recent developments in capillary and microchip electroseparations of peptides (2011-2013). Electrophoresis 2014; 35(1): 69-95. [http://dx.doi.org/10.1002/elps.201300331] [PMID: 24255019]

[146] Tubaon RM, Rabanes H, Haddad PR, Quirino JP. Capillary electrophoresis of natural products: 2011-2012. Electrophoresis 2014; 35(1): 190-204. [http://dx.doi.org/10.1002/elps.201300473] [PMID: 24395664]

[147] Müllerová L, Dubský P, Gaš B. Twenty years of development of dual and multi-selector models in capillary electrophoresis: a review. Electrophoresis 2014; 35(19): 2688-700. [http://dx.doi.org/10.1002/elps.201400149] [PMID: 24946108]

[148] Eggins BR. Chemical sensors and biosensors. Chichester, West Sussex, England: John Wiley & Sons 2002.

[149] Hulanicki A, Glab S, Ingman F. Chemical sensors definitions and classification. Pure Appl Chem 1991; 63: 1247-50. [http://dx.doi.org/10.1351/pac199163091247]

[150] Vo-Dinh T, Cullum B. Biosensors and biochips: advances in biological and medical diagnostics. Fresenius J Anal Chem 2000; 366(6-7): 540-51. [http://dx.doi.org/10.1007/s002160051549] [PMID: 11225766]

[151] Shahgaldian P, Pieles U. Cyclodextrin derivatives as chiral supramolecular receptors for enantioselective sensing. Sensors (Basel Switzerland) 2006; 6: 593-615. [http://dx.doi.org/10.3390/s6060593]

[152] Tertiş M, Florea A, Săndulescu R, Cristea C. Carbon based electrodes modified with horseradish peroxidase immobilized in conducting polymers for acetaminophen analysis. Sensors (Basel) 2013; 13(4): 4841-54. [http://dx.doi.org/10.3390/s130404841] [PMID: 23580052]

[153] Sun K, Ramgir N, Bhansali S. An immunoelectrochemical sensor for salivary cortisol measurement. Sens. Actuat. Biol Chem 2008; 133: 533-7.

[154] Zhou M, Deng L, Wen D, Shang L, Jin L, Dong S. Highly ordered mesoporous carbons-based glucose/O_2 biofuel cell. Biosens Bioelectron 2009; 24(9): 2904-8. [http://dx.doi.org/10.1016/j.bios.2009.02.028] [PMID: 19321330]

[155] Wang J, Zhu H, Xu Y, *et al.* Graphene nanodots encaged 3-D gold substrate as enzyme loading platform for the fabrication of high performance biosensors. Sens. Actuat. Biol Chem 2015; 220:

1186-95.

[156] Monteiro T, Rodrigues PR, Gonçalves AL, *et al.* Construction of effective disposable biosensors for point of care testing of nitrite. Talanta 2015; 142: 246-51. [http://dx.doi.org/10.1016/j.talanta.2015.04.057] [PMID: 26003719]

[157] Šakinytė I, Barkauskas J, Gaidukevič J, Razumienė J. Thermally reduced graphene oxide: The study and use for reagentless amperometric D-fructose biosensors. Talanta 2015; 144: 1096-103. [http://dx.doi.org/10.1016/j.talanta.2015.07.072] [PMID: 26452933]

[158] Florea A, Ravalli A, Cristea C, Săndulescu R, Marrazza G. An optimized bioassay for mucin1 detection in serum samples. Electroanalysis 2015; 27(7): 1594-601. [http://dx.doi.org/10.1002/elan.201400689]

[159] Zhou F, Lu M, Wang W, Bian ZP, Zhang JR, Zhu JJ. Electrochemical immunosensor for simultaneous detection of dual cardiac markers based on a poly(dimethylsiloxane)-gold nanoparticles composite microfluidic chip: a proof of principle. Clin Chem 2010; 56(11): 1701-7. [http://dx.doi.org/10.1373/clinchem.2010.147256] [PMID: 20852134]

[160] Sinha B, Ramulu TS, Kim KW, Venu R, Lee JJ, Kim CG. Planar Hall magnetoresistive aptasensor for thrombin detection. Biosens Bioelectron 2014; 59: 140-4. [http://dx.doi.org/10.1016/j.bios.2014.03.021] [PMID: 24727201]

[161] Chung D, Kim K, Choi S. Electrochemical DNA biosensor based on avidin-biotin conjugation for influenza virus (type A) detection. Appl Surf Sci 2011; 257(22): 9390-6. [http://dx.doi.org/10.1016/j.apsusc.2011.06.015]

[162] Tertiş M, Hosu O, Fritea L, *et al.* A novel label-free immunosensor based on activated graphene oxide for acetaminophen detection. Electroanalysis 2015; 27: 638-47. [http://dx.doi.org/10.1002/elan.201400583]

[163] Florea A, Taleat Z, Cristea C, Mazloum-Ardakani M, Săndulescu R. Label free MUC1 aptasensors based on electrodeposition of gold nanoparticles on screen printed electrodes. Electrochem Commun 2013; 33: 127-30. [http://dx.doi.org/10.1016/j.elecom.2013.05.008]

[164] Taleat Z, Cristea C, Marrazza G, Mazloum-Ardakani M, Săndulescu R. Electrochemical immunoassay based on aptamer–protein interaction and functionalized polymer for cancer biomarker detection. J Electroanal Chem 2014; 717-718: 119-24. [http://dx.doi.org/10.1016/j.jelechem.2014.01.015]

[165] Ebrahimi M, Raoof JB, Ojani R. Novel electrochemical DNA hybridization biosensors for selective determination of silver ions. Talanta 2015; 144: 619-26. [http://dx.doi.org/10.1016/j.talanta.2015.07.020] [PMID: 26452869]

[166] Pan HZ, Yu HW, Wang N, *et al.* Electrochemical DNA biosensor based on a glassy carbon electrode modified with gold nanoparticles and graphene for sensitive determination of *Klebsiella pneumoniae* carbapenemase. J Biotechnol 2015; 214: 133-8. [http://dx.doi.org/10.1016/j.jbiotec.2015.09.015] [PMID: 26387446]

[167] Benvidi A, Dehghani Firouzabadi AD, Dehghan Tezerjani M, Moshtaghiun SM, Mazloum-Ardakani M, Ansarin A. A highly sensitive and selective electrochemical DNA biosensor to diagnose breast

cancer. J Electroanal Chem 2015; 750: 57-64.
[http://dx.doi.org/10.1016/j.jelechem.2015.05.002]

[168] Gayathri SB, Kamaraj P, Arthanareeswari M, Devikala S. DNA nanostructures based biosensor for the determination of aromatic compounds. Biosens Bioelectron 2015; 72: 191-6.
[http://dx.doi.org/10.1016/j.bios.2015.05.002] [PMID: 25982727]

[169] Kilic T, Erdem A, Erac Y, Seydibeyoglu MO, Okur S, Ozsoz M. Electrochemical detection of a cancer biomarker mir-21 in cell lysates using graphene modified sensors. Electroanalysis 2014; 26: 1-16.

[170] Li XM, Wang LL, Luo J, Wei QL. A dual-amplified electrochemical detection of mRNA based on duplex-specific nuclease and bio-bar-code conjugates. Biosens Bioelectron 2015; 65: 245-50.
[http://dx.doi.org/10.1016/j.bios.2014.10.051] [PMID: 25461165]

[171] Chao J, Zhu D, Zhang Y, Wang L, Fan C. DNA nanotechnology-enabled biosensors. Biosens Bioelectron 2016; 76: 68-79.
[http://dx.doi.org/10.1016/j.bios.2015.07.007] [PMID: 26212206]

[172] Teles FR, Fonseca LP. Trends in DNA biosensors. Talanta 2008; 77(2): 606-23.
[http://dx.doi.org/10.1016/j.talanta.2008.07.024]

[173] Diouani MF, Helali S, Hafaid I, Hassen WM, Snoussi MA, Ghram A. Miniaturized biosensor for avian influenza virus detection. Mater Sci Eng C 2008; 28: 580-3.
[http://dx.doi.org/10.1016/j.msec.2007.10.043]

[174] Ohno R, Ohnuki H, Wang H, *et al.* Electrochemical impedance spectroscopy biosensor with interdigitated electrode for detection of human immunoglobulin A. Biosens Bioelectron 2013; 40(1): 422-6.
[http://dx.doi.org/10.1016/j.bios.2012.07.052] [PMID: 22917917]

[175] Ramanathan A, Pape L, Schwartz DC. High-density polymerase-mediated incorporation of fluorochrome-labeled nucleotides. Anal Biochem 2005; 337(1): 1-11.
[http://dx.doi.org/10.1016/j.ab.2004.09.043] [PMID: 15649370]

[176] Daly CJ, McGrath JC. Fluorescent ligands, antibodies, and proteins for the study of receptors. Pharmacol Ther 2003; 100(2): 101-18.
[http://dx.doi.org/10.1016/j.pharmthera.2003.08.001] [PMID: 14609715]

[177] Abdulhalim I, Zourob M, Lakhtakia A. Overview of optical biosensing techniques. In: Marks RS, Lowe CR, Cullen DC, Weetall HH, Karube I, Eds. Handbook of Biosensors and Biochips. Weinheim: Wiley 2007.

[178] Mol NJ, Fischer MJ. Surface plasmon resonance: a general introduction. In: Mol NJ, Fischer MJ, Eds. Surface Plasmon Resonance Methods and Protocols. New York: Springer 2010; pp. 1-14.
[http://dx.doi.org/10.1007/978-1-60761-670-2_1]

[179] Chang YF, Fu C, Chen YT, *et al.* Use of liposomal amplifiers in total internal reflection fluorescence fiber-optic biosensors for protein detection. Biosens Bioelectron 2016; 77: 1201-7.
[http://dx.doi.org/10.1016/j.bios.2015.10.012] [PMID: 26595485]

[180] Lepinay S, Staff A, Ianoul A, Albert J. Improved detection limits of protein optical fiber biosensors coated with gold nanoparticles. Biosens Bioelectron 2014; 52: 337-44.
[http://dx.doi.org/10.1016/j.bios.2013.08.058] [PMID: 24080213]

[181] De Stefano L, Orabona E, Lamberti A, Rea I, Rendina I. Microfluidics assisted biosensors for label-free optical monitoring of molecular interactions, Sens. Actuat. Biol Chem 2013; 179: 157-62.

[182] Oubaha M, Gorin A, McDonagh C, Duffy B, Copperwhite R. Development of a multianalyte optical sol–gel biosensor for medical diagnostic, Sens. Actuat. Biol Chem 2015; 221: 96-103.

[183] Nielsen PE, Egholm M, Berg RH, Buchardt O. Sequence-selective recognition of DNA by strand displacement with a thymine-substituted polyamide. Science 1991; 254(5037): 1497-500. [http://dx.doi.org/10.1126/science.1962210] [PMID: 1962210]

[184] D'Agata R, Spoto G. Artificial DNA and surface plasmon resonance. Artif DNA PNA XNA 2012; 3(2): 45-52. [http://dx.doi.org/10.4161/adna.21383] [PMID: 22821257]

[185] D'Agata R, Spoto G. Surface plasmon resonance imaging for nucleic acid detection. Anal Bioanal Chem 2013; 405(2-3): 573-84. [http://dx.doi.org/10.1007/s00216-012-6563-9] [PMID: 23187826]

[186] Ermini ML, Mariani S, Scarano S, Minunni M. Bioanalytical approaches for the detection of single nucleotide polymorphisms by surface plasmon resonance biosensors. Biosens Bioelectron 2014; 61: 28-37. [http://dx.doi.org/10.1016/j.bios.2014.04.052] [PMID: 24841091]

[187] Corradini R, Feriotto G, Sforza S, Marchelli R, Gambari R. Enhanced recognition of cystic fibrosis W1282X DNA point mutation by chiral peptide nucleic acid probes by a surface plasmon resonance biosensor. J Mol Recognit 2004; 17(1): 76-84. [http://dx.doi.org/10.1002/jmr.646] [PMID: 14872540]

[188] Tombelli S, Minunni M, Mascini M. Piezoelectric biosensors: strategies for coupling nucleic acids to piezoelectric devices. Methods 2005; 37(1): 48-56. [http://dx.doi.org/10.1016/j.ymeth.2005.05.005] [PMID: 16216523]

[189] Ogimoto Y, Selyanchyna R, Takahara N, Wakamatsu S, Lee SW. Detection of ammonia in human breath using quartz crystal microbalance sensors with functionalized mesoporous SiO_2 nanoparticle films. Sens. Actuat. Biol Chem 2015; 215: 428-36.

[190] Xie G, Sun P, Yan X, Du X, Jiang Y. Fabrication of methane gas sensor by layer-by-layer self-assembly of polyaniline/PdO ultra thin films on quartz crystal microbalance. Sens. Actuat. Biol Chem 2010; 145: 373-7.

[191] Özbek C, Okur S, Mermer Ö, Kurt M, Sayine S, Yılmaz M. Effect of Fe doping on the CO gas sensing of functional calixarene molecules measured with quartz crystal microbalance technique. Sens. Actuat. Biol Chem 2015; 215: 464-70.

[192] Wetchakun K, Samerjai T, Tamaekong N, *et al.* Semiconducting metal oxides as sensors for environmentally hazardous gases. Sens. Actuat. Biol Chem 2011; 160: 580-91.

[193] Wong YY, Ng SP, Ng MH, Si SH, Yao SZ, Fung YS. Immunosensor for the differentiation and detection of Salmonella species based on a quartz crystal microbalance. Biosens Bioelectron 2002; 17(8): 676-84. [http://dx.doi.org/10.1016/S0956-5663(02)00030-1] [PMID: 12052353]

[194] Tukkiniemia K, Rantala A, Nirschl M, Pitzer D, Huber T, Schreiter M. Fully integrated FBAR sensor matrix for mass detection. Procedia Chem 2009; 1(1): 1051-4. [http://dx.doi.org/10.1016/j.proche.2009.07.262]

[195] Somiya M, Yoshimoto N, Iijima M, *et al.* Targeting of polyplex to human hepatic cells by bio-nanocapsules, hepatitis B virus surface antigen L protein particles. Bioorg Med Chem 2012; 20(12): 3873-9. [http://dx.doi.org/10.1016/j.bmc.2012.04.031] [PMID: 22591855]

[196] Skládal P, dos Santos Riccardi C, Yamanaka H, da Costa PI. Piezoelectric biosensors for real-time monitoring of hybridization and detection of hepatitis C virus. J Virol Methods 2004; 117(2): 145-51. [http://dx.doi.org/10.1016/j.jviromet.2004.01.005] [PMID: 15041211]

[197] Guo X, Lin CS, Chen SH, Ye R, Wu VC. A piezoelectric immunosensor for specific capture and enrichment of viable pathogens by quartz crystal microbalance sensor, followed by detection with antibody-functionalized gold nanoparticles. Biosens Bioelectron 2012; 38(1): 177-83. [http://dx.doi.org/10.1016/j.bios.2012.05.024] [PMID: 22683250]

[198] Bragazzi NL, Amicizia D, Panatto D, Tramalloni D, Valle I, Gasparin R. Quartz-crystal microbalance (QCM) for public health: An overview of its applications. Adv. protein chem., in press (Available online 26 September 2015). [http://dx.doi.org/10.1016/bs.apcsb.2015.08.002]

[199] Herzer G. Magnetic materials for electronic article surveillance. J Magn Magn Mater 2003; 254-255: 598-602. [http://dx.doi.org/10.1016/S0304-8853(02)00930-7]

[200] Grimes CA, Roy SC, Rani S, Cai Q. Theory, instrumentation and applications of magnetoelastic resonance sensors: a review. Sensors (Basel) 2011; 11(3): 2809-44. [http://dx.doi.org/10.3390/s110302809] [PMID: 22163768]

[201] Shen W, Mathison LC, Petrenko VA, Chin BA. Design and characterization of a magnetoelastic sensor for the detection of biological agents. J Phys D Appl Phys 2010; 43: 01500413,14.

[202] Possan AL, Menti C, Beltrami M, Santos AD, Roesch-Ely M, Missell FP. Effect of surface roughness on performance of magnetoelastic biosensors for the detection of Escherichia coli. Mater Sci Eng C Mater Biol Appl 2016; 58: 541-7. [http://dx.doi.org/10.1016/j.msec.2015.08.029] [PMID: 26478342]

[203] Nath N, Chilkoti A. A colorimetric gold nanoparticle sensor to interrogate biomolecular interactions in real time on a surface. Anal Chem 2002; 74(3): 504-9. [http://dx.doi.org/10.1021/ac015657x] [PMID: 11838667]

[204] Maskow T, Wolf K, Kunze W, Enders S, Harms H. Rapid analysis of bacterial contamination of tap water using isothermal calorimetry. Thermochim Acta 2012; 543: 273-80. [http://dx.doi.org/10.1016/j.tca.2012.06.002]

[205] Kirchner P, Oberlander J, Friedrich P, *et al.* Realisation of a calorimetric gas sensor on polyimide foil for applications in aseptic food industry. Sens. Actuat. Biol Chem 2012; 170: 60-6.

[206] Davaji B, Lee CH. A paper-based calorimetric microfluidics platform for bio-chemical sensing. Biosens Bioelectron 2014; 59: 120-6.

[http://dx.doi.org/10.1016/j.bios.2014.03.022] [PMID: 24713542]

[207] Leca-Bouvier B, Blum L. Enzyme for Biosensing application. In: Zourob M, Ed. Recognition Receptors in Biosensors. New York: Springer 2010; pp. 177-220. [http://dx.doi.org/10.1007/978-1-4419-0919-0_4]

[208] Fritea L, Tertiş M, Săndulescu R. Applications of cyclodextrins in biomedical electrochemistry. In: Ramirez FG, Ed. Cyclodextrins: Synthesis, Chemical applications and Role in drug delivery. Hauppage, NY: Nova Science Publishers, Inc. 2015; pp. 45-83.

CHAPTER 4

Electrochemical Sensors for the Detection of Antioxidants

María Luz Rodríguez-Méndez[1,*], **Celia García-Hernández**[1], **Cristina Medina-Plaza**[1], **Cristina García-Cabezón**[2], **David Paniagua-Albillos**[1], **Silvia Rodríguez-Sánchez**[1] and **José Antonio de Saja-Sáez**[3]

[1] *Department of Inorganic Chemistry, Engineers School, Universidad de Valladolid, 47011 Valladolid, Spain.*

[2] *Department of Materials Science, Engineers School, Universidad de Valladolid, 47011 Valladolid, Spain.*

[3] *Department of Condensed Matter Physics, Faculty of Sciences, Universidad de Valladolid, 47011 Valladolid, Spain*

Abstract: The Mediterranean diet has largely demonstrated its health benefits which are related to the intake of antioxidants. Besides the health benefits, antioxidants influence in the organoleptic characteristics, the quality and the preservation of foods and beverages.

Although a variety of methods have been described to detect antioxidants and to evaluate the antioxidant capacity of foods, field experiments still represent a challenge. In this context, electrochemical techniques can represent an advantage thanks to higher sensitivity, short experimental times and inherent portability. In this chapter, the electrochemical sensors dedicated to the analysis of antioxidants will be presented. Particular attention will be paid to voltammetric sensors. The methods used to improve their selectivity and sensitivity by modifying the surface using chemical or biological materials will be described. The new developments obtained thanks to new nanomaterials and nanostructured films will be evidenced. Finally, the advantages of multisensor systems based on electrochemical sensors will be described.

[*] **Corresponding author María Luz Rodríguez-Méndez:** Department of Inorganic Chemistry, Industrial Engineers School, Paseo del Cauce, 59, 47011, Valladolid, Spain; Tel: +34 983423540; Fax: +34 983423310; E-mail: mluz@eii.uva.es

Constantin Apetrei (Ed.)

Keywords: Antioxidant, Biosensor, Food, Langmuir-Blodgett, Nanoparticle, Phthalocyanine, Sensor.

INTRODUCTION

Over the last years the action of free radicals in living organisms has been extensively studied. Their destructive action can be reduced by antioxidants, a broad group of active biological substances which are widespread in the nature [1, 2]. An antioxidant is defined as a substance that, in contact with an oxidizable substrate, protects that substrate from oxidation, protecting then the organism from harmful effects of oxidative stress. It has been demonstrated that the consumption of antioxidants decreases the risk of cardiovascular diseases [2 - 4]. The main antioxidants in food include substances such as ascorbic acid (vitamin C), vitamin E and urate. In addition, other antioxidants are present in lower concentration such as beta-carotene, glutathione and a variety of polyphenols [5]. The chemical structure of polyphenols is specially adapted to their antioxidant action (hydrogen or electron donor, or capture of free radicals). Many polyphenols are better antioxidants than vitamin E or C.

The habitual and moderate consumption of fruits and vegetables such as oranges or grapes and beverages such as fruit juices or wines, can produce beneficial health effects due to their content of phenolic compounds, particularly quercetin and resveratrol. Epidemic studies have related the intake of polyphenols to a decrease of the coronary cardiopathy or the cancer risk in humans [6]. Polyphenols also affect the quality and organoleptic characteristic of foods and their contribution to the flavor is extremely important. They also contribute to the antioxidant capacity of foods. The contribution of each component to the global antioxidant capacity depends on its antioxidant power, its concentration and the interaction with other components [7].

Different methods have been developed to assess the antioxidant capacity of foods [8]. However, up to now, a single method has not been recognized as the most adequate, and the results obtained depend on the method used. The antioxidant capability of a complex mixture, is not only given by the addition of the antioxidant capacities of each one of its components; it also depends on the synergic and inhibitory effects of each compound with the rest. The methods to assess the antioxidant activity are usually based on the evaluation of the capability of an antioxidant or of a mixture of antioxidants to inhibit the oxidative damage caused by an oxidizing agent to a substrate. Any test for the evaluation of the antioxidant capacity needs an appropriate substrate to monitor the inhibition of the oxidation, an initiator of the oxidation (free radical) and an appropriate measure of the endpoint of the oxidation. This measure can be carried out by chemical, instrumental or sensorial methods.

When approaching the study of the antioxidant activity of foods and beverages, it has been recommended to use more than one method. The reason is that each method gives different information: certain antioxidants do not react with certain oxidizing species, but they do react with others. As a consequence different values can be obtained with different methods. The antioxidant activity can be evaluated by means of the measure of the absorbance capacity of the radical oxygen (ORAC) [9] and by means of the trolox equivalent antioxidant capacity (TEAC) (Trolox is a soluble analogue of vitamin E) [10]. On the other hand, the measure of phenols using the Folin-Ciocalteu method, based on a redox reaction, is also an accepted method to measure the antioxidant activity of foods in function of their Polyphenolic content [11, 12]. ORAC and TEAC are used to measure the antioxidant capacity, while Folin is an indicator of the antioxidant capacity in function of the global content of polyphenols in foods. Taking into account that the antioxidant activity of many foods and beverages is mainly related to the Polyphenolic content, this method is preferred by some authors. The evaluation of the total polyphenol index (TPI) measuring the absorbance at 280 nm is also recommended, because this method evaluates the reducing power of wines [13]. The knowledge of the phenolic profile of foods and beverages and their relation with the antioxidant capacity are of great interest to evaluate the modifications that foods can undergo. All these techniques have been developed for the analysis of foods at the laboratory level. They are not suitable for routine, on line industrial or field analysis because they require long periods of sample preparation, analysis and interpretation.

Natural antioxidants are usually electroactive compounds. This electroactivity can be exploited for electrochemical detection. Similarly, the antioxidant capacity can be easily assessed by electrochemical techniques [14 - 17].

Moreover, electrochemical techniques have often been considered advantageous due to their higher sensitivity, and because they do not require sophisticated and expensive equipment. The combination of electrochemical detectors with flow injection systems speed up the time of analysis and open the possibility of miniaturization and automation [14]. For this reason, electrochemical analysis has been widely used to assess the activity of a variety of antioxidants.

Next section will describe the most relevant electroanalytical methodologies for detection of individual antioxidants or the total antioxidant capacity in several kinds of matrices.

ELECTROCHEMICAL SENSORS FOR THE DETECTION OF ANTIOXIDANTS

The number of electrochemical sensors dedicated to the detection of antioxidants is quite large. The simplest sensors are bare electrode surfaces (carbon, platinum, ITO glass, gold, *etc.*) but the selectivity and sensitivity can be improved by modifying the electrode surfaces. Such modification can be carried out using a large variety of chemical or biological materials (*i.e.* enzymes). Moreover, a variety of techniques can be used to modify the electrodes, leading to a large number of different electrodes [16]. It is also important to keep in mind that in the last years new developments in the field of nanotechnology have helped to improve the sensitivity of the electrochemical sensors [18].

Different electrochemical approaches have been used for the direct determination of the total or individual antioxidants present in beverages/food. Cyclic voltammetry (CV) is the most widely used electrochemical technique although differential pulse voltammetry (DPV) and square wave voltammetry (SWV) have also been extensively employed [19 - 21]. Using voltammetry, an antioxidant can be detected as a redox peak whose position depends on its chemical nature, and the intensity is proportional to its concentration. In complex mixtures such as foods, that contain more than one antioxidant, the antioxidant capacity can be evaluated by measuring the area under the broad peak produced by the overlap of several oxidation peaks [16]. Additionally, the peak potential (Ep) is an indication of the reducing power. Many examples can be found in the literature dedicated to the analysis of individual antioxidants or the antioxidant capacity in a variety of foods. Here we have selected some examples that illustrate the state of the art in this field. For instance, using simple electrodes (graphite, platinum or ITO glass) the antioxidants present in teas [22, 23], natural juices [20], wines [24, 25], or onions [26] have been analyzed.

Chemically modified electrodes (CME) have also been widely used for the detection of antioxidants. In such sensors, the electrode surface is covered with a sensing material that by means of interfacial interactions or faradaic (electron-transfer) reactions can facilitate the charge-transfer reactions between the electrode and the solution. CME can thus reduce the oxidation potential of the antioxidant while increasing the intensity of the response [27]. In other cases, the chemical modifiers are redox mediators that shuttle electrons from the antioxidant to the electrode surface reducing the potential at which the oxidation takes place. The electron mediator immobilized at the electrode surface is first reduced while it oxidizes chemically the antioxidant. The reduced mediator is then electrochemically oxidized at the electrode surface, regenerating the mediator. The catalytic cycle thus established, increases the anodic current observed increasing the sensitivity of the determination. The change in the oxidation potential is also an advantage because it can increase the selectivity of the electrodes. For instance, ascorbic acid can be directly detected using bare electrodes. However, the presence of compounds such as dopamine that are oxidized at similar potentials than ascorbic acid, can interfere in the analysis. Electrodes modified with cationic surfactants make more difficult the oxidation of dopamine, causing a shift of the oxidation potential to more positive values improving thus the peak separation [28].

Many sensing materials have been used as chemical modifiers (carbon nanotubes, graphene, nanoparticles, porphyrins, phthalocyanines or conducting polymers, among others). In addition, the modifiers can be deposited on the surfaces using a variety of methods that include physical adsorption, covalent attachment, electropolymerization, incorporation into an electrode matrix (*i.e.* carbon paste electrodes), or entrapment in a polymeric or organic matrix. Miniaturized electrodes can be prepared by screen-printing technologies [29]. Finally, sophisticated nanostructured sensors can be prepared using the Langmuir-Blodgett (LB), self-assembling (SAM) or the electrostatic Layer by Layer (LbL) technique [21]. In the next paragraphs, some examples of electrodes used to detect antioxidants will be presented.

The simplest modification method consists in the adsorption of the sensing material on the surface. Carbon paste electrodes (CPE) can also be easily prepared by incorporating the modifier inside a carbon paste prepared by mixing graphite powder and a binder (*e.g.* parafilm oil, nujol, silicon) [25]. Ionic liquids can totally or partially replace the classical binders in carbon pastes because they exhibit high ionic conductivity and a wide potential window. In an original work, vegetables oils have been used as electroactive binder material of carbon paste electrodes. The electrochemical response of such electrodes immersed in a variety of electrolytic aqueous solutions, has been exploited to discriminate oils [30].

Carbon nanotubes (CNT) or graphene can be easily adsorbed on graphite surfaces. They are excellent modifiers, not only because they are promoters of the electron transfer and have an excellent electrocatalytic activity, but also because they increase the surface area, so they have been used to test many antioxidants in sub-nM concentrations [31, 32]. CNTs can also be mixed with graphite and nujol to obtain CPEs with high electron transfer rates for phenolic redox couples [33].

Carbon pastes modified with MWCNTs-thionine-nafion carbon composite [34] or cobalt 5-nitrosalophen, a well-known mediator of the oxidation of ascorbic acid [35] show improved limits of detection with respect of CPE. The catalytic properties of a CPE modified with copper immobilized in a polyester resin has been successfully used to detect catechin in teas [36].

Electrodeposition is a versatile technique and can be used to obtain polymers from a variety of monomers and under different electropolymerization conditions. Electrodes modified with electrodeposited o-aminophenol catalyze the oxidation of ascorbic acid found in fruit juices [37]. Electropolymerization of aspartic acid [38] catalyzes the oxidation of catechin in teas. Electrodes covered with electrodeposited conducting polymers such as polyaniline or polypyrrole can be used as electrocatalytic materials for a variety of phenols [39, 40].

Electrodes modified with transition metal complexes, have the ability to catalyze the oxidation or reduction of solved compounds by lowering the potential required for the catalyzed redox systems, when compared with the unmodified electrode. Metallophthalocyanines have been widely applied as electrocatalytic materials for the detection of a variety of compounds [41]. Using phthalocyanine-based CPEs, phenols have been successfully detected. These electrodes have been particularly useful to detect antioxidants present in wines and to assess their antioxidant capacity [42 - 46]. Good results have also been obtained using Screen printed electrodes modified with phthalocyanines [47].

The selectivity of voltammetric sensors can be improved by means of molecular recognition. Porous compounds such as cyclodextrins can form complexes with phenols and their selectivity is related to the size of the cavities. Using this strategy, catechin has been successfully detected in several beverages [48]. Selectivity can also been improved by using molecularly imprinted polymers (MIPs). One of the difficulties with using MIPs as the recognition element in electrochemical sensors is their lack of conductivity. This drawback can be solved by using hybrid materials combining a catalytic MIP, capable of oxidizing the template, with an electrically conducting polymer. Using this strategy detection of catechol has demonstrated to be specific [49].

Certain properties of nanoparticles, mainly gold nanoparticles (AuNPs) (*e.g.*, conductivity, catalytic and photocatalytic activity) make them applicable in fabrication of electrochemical sensors [50]. Electrodes modified with gold nanoparticles can detect catechol and hydroquinone in tea samples with satisfactory results comparable to those obtained using chromatography [51].

A completely different approach is based on the correlation between the reducing activity (antioxidant properties) of phenolic compounds and their ability to obtain gold nanoparticles by reduction of gold salts or to promote the enlargement of AuNPs [52]. Solutions of ferricyanide/ferrocyanide were used as probes to test the growth of AuNPs. The growth of the Au seeds induced by the reducing activity of flavonoids or phenolic compounds decreased the redox activity of the probe, making possible the detection of phenols in herbal samples.

Combinations of electrocatalytic materials can improve drastically the performance of the sensors. Many examples can be found in the literature. In fact, most of the papers published recently in the field of electrochemical sensors for the detection of phenols are based on combinations of materials. In a recent example, AuNPs were self-assembled onto the surfaces of functionalized multi-walled carbon nanotubes (p-MWCNs) sheets. The nanocomposites were applied to simultaneous determination of flavonoids quercetin and rutin [53]. AuNPs combined with graphene has also been used to detect phenols [54]. In another example combinations of conducting polymers and nanoparticles have been found to have superior performance relative to conventional polymers due to the synergy between both components [55, 56].

As it has been presented in previous paragraphs, nanotechnology provides new nanomaterials (nanoparticles, nanocarbons, *etc.*) with improved electrocatalytic properties. But, nanotechnology can offer another advantage. Traditional sensing materials can be deposited as nanostructured films using Self Assembling (SAM), LbL or the LB techniques. Nanostructured films have the advantage of their enhanced number of active sites that produce an increase in the intensity, and lower the detection limits. Nanostructured films also show faster kinetics.

The spontaneous self-assembling of alkyl thiols on Au surfaces can be used to immobilize molecules. For instance, a nickel(II) complex was immobilized in a mercaptopropionic acid monolayer. This complex was able to oxidize catechin on the interface, which was subsequently reduced on the electrode surface [57].

LB and LbL films based on phthalocyanines have been frequently used to analyze phenols. Detection limits calculated for mono- di- and tri-phenols are one order of magnitude lower than those found in CPE sensors and can reach 10^{-6} M [58, 59]. Similarly, electrodes formed by LuPc2 nanowires prepared by means of electrophoretic techniques show an enhanced electrocatalytic activity when exposed to solutions of phenols [60]. Nanoparticles in LB films produced an increase in the sensitivity towards phenols, and detection limits of 10^{-7} M are attained [61]. Such films can detect simultaneously organic acids without interferences.

Another advantage of nanostructured films is that the control of the molecular architectures afforded by these techniques has led to a variety of devices where synergy is achieved by combining electroactive materials, including organic-inorganic hybrids. For instance, LB films of phthalocyanines combined with nanoparticles [62] or carbon nanotubes [63] show highly sensitive responses towards phenols or citric acid. In the latest case, the synergy is promoted by the pi-pi stacking between both components. Similar detection limits were obtained by combining phthalocyanines with silver nanoparticles in LbL films [64, 65].

ELECTROCHEMICAL ENZYMATIC BIOSENSORS FOR ANTIOXIDANT EVALUATION

During the last years, many efforts have been carried out to develop biosensors specific for certain antioxidants. The incorporation of enzymes in the sensing layer of electrochemical sensors can help to improve the selectivity of the sensing devices thanks to the specificity enzyme-substrate. During the reaction enzymes consume oxygen (*e.g.* oxidases), or produce hydrogen peroxide or produce the reduced form of the NAD(P)H (*e.g.* dehydrogenases). These transformations can be measured by electrochemical methods.

The enzymes used for the detection of antioxidants are oxidase enzymes such as polyphenol oxidases [66] tyrosinase [67 - 70], laccase [71 - 74], horseradish peroxidase [75] or ascorbate oxidase [76, 77].

Amperometric or voltammetric biosensors based on the enzyme tyrosinase and laccase have been widely used to phenol determination due to their high sensitivity. Tyrosinase is a di-copper enzyme that catalyzes the oxidation of monophenols (hydroxylase activity) by the molecular oxygen to o-diphenols, which are subsequently oxidized to o-quinones species (oxidase activity). The detection and quantification of the reaction within the biosensor can be based on various different principles, such as detection of oxygen consumption, direct reduction of liberated o-quinone or reduction of the o-quinone. In order to improve the electron transfer between the enzyme and the electrode, redox mediators can be used. Examples of electron mediators are hexacyanoferrate [78], conducting polymers [79, 80] or phthalocyanines [81, 82].

The adequate immobilization of enzymes on the electrode surface is a critical step to preserve the enzymatic activity. Enzymes have been immobilized using a range of techniques that include immobilization onto carbonaceous electrodes *via* the cross-linking step by glutaraldehyde [83, 84] or immobilization of enzymes in

polymeric matrixes [80, 85] among others. The LbL and the LB techniques are of special interest for enzyme immobilization because using these methods enzymes can be immobilized in lipidic layers with structures similar to those of the biological membranes. This biomimetic environment can help to preserve the functionality of the enzyme by conserving the conformation and the orientation of the enzyme [81, 86 - 88].

The role of the molecular interactions in the electrocatalytic properties in biomimetic systems has been studied by combining enzymes with different electron mediators and the existence of synergistic effects has been evidenced. Detection limits as low as 10^{-8} M towards phenols can be attained using these biosensors.

ELECTRONIC TONGUES DEDICATED TO THE ANALYSIS OF ANTIOXIDANTS

The selectivity of sensors can be further improved by using combinations of sensors with cross-selectivity, instead of single sensors. This is the principle of the so-called electronic tongues (ET) which try to emulate the way in which human senses can perceive tastes [89]. According to the IUPAC, an electronic tongue is a multisensor system, which consists of a number of low-selective sensors and cross-sensitivity to different species in solution, and an appropriate method of pattern recognition and/or multivariate calibration for data processing [90]. Sensors used in electronic tongues can use several measurement principles including mass, optical or electrochemical transduction. The most used ETs dedicated to the detection of antioxidants are those based on arrays of electrochemical sensors including potentiometric [91, 92], voltammetric [44, 93] and impedimetric [94] sensors.

ETs are holistic systems that provide global information about a sample instead of information of particular components. ETs can be used in qualitative applications.

In this case, ETs are used to address the classification or discrimination of samples based on its phenolic content by means of chemometric tools such as Principal Component Analysis (PCA). On the other hand, the signals provided by the sensor array, are correlated with the chemical composition of the sample. For this reason, ETs can also be used for the quantitative determination of phenolic compounds trough statistical tools such as Partial Least Squares (PLS1) [95 - 97].

The quality of the correlations is improved by introducing in the array biosensors (bio-Etongue) [90]. In the case of bio-Etongues, most of the works have been dedicated to the wine industry and correlations with individual phenolic acids such as gallic and caffeic acids or groups of antioxidants *e.g.* catechins, anthocyanins or tannins content have been established [81, 86, 98]. Good correlations have been found with the chemical parameters classically used for the analysis of the antioxidant capacity such as the Folin-Ciocalteu index (FC) and the UV Polyphenol Index (I_{280}).

Due to the good correlations found with antioxidant content of foods E-tongues might be considered an attractive alternative to traditional chemical analysis, with the advantage of its low-cost, portability or fast-response. Special attention needs to be paid to the development of an array of sensors with cross-selectivity, specifically adapted to the analysis of antioxidants.

CONFLICT OF INTEREST

The authors confirm that they have no conflict of interest to declare for this publication.

ACKNOWLEDGEMENTS

Financial support by the MINECO and FEDER (grant CICYT-AGL2012-33535) and Junta de Castilla y León (VA-032U13) is gratefully acknowledged. CMP would also like to thank the University of Valladolid for the grant (PIF-UVa).

REFERENCES

[1] Ames BN, Shigenaga MK, Hagen TM. Oxidants, antioxidants, and the degenerative diseases of aging. Proc Natl Acad Sci USA 1993; 90(17): 7915-22. [http://dx.doi.org/10.1073/pnas.90.17.7915] [PMID: 8367443]

[2] Dai J, Mumper RJ. Plant phenolics: extraction, analysis and their antioxidant and anticancer properties. Molecules 2010; 15(10): 7313-52.
[http://dx.doi.org/10.3390/molecules15107313] [PMID: 20966876]

[3] Roleira FM, Tavares-da-Silva EJ, Varela CL, *et al.* Plant derived and dietary phenolic antioxidants: anticancer properties. Food Chem 2015; 183: 235-58.
[http://dx.doi.org/10.1016/j.foodchem.2015.03.039] [PMID: 25863633]

[4] Zulueta A, Caretti A, Signorelli P, Ghidoni R. Resveratrol: A potential challenger against gastric cancer. World J Gastroenterol 2015; 21(37): 10636-43.
[http://dx.doi.org/10.3748/wjg.v21.i37.10636] [PMID: 26457023]

[5] Rice-Evans CA, Miller NJ, Paganga G. Antioxidant properties of phenolic compounds. Trends Plant Sci 1997; 2(4): 152-9.
[http://dx.doi.org/10.1016/S1360-1385(97)01018-2]

[6] Yang CS, Landau JM, Huang MT, Newmark HL. Inhibition of carcinogenesis by dietary polyphenolic compounds. Annu Rev Nutr 2001; 21: 381-406.
[http://dx.doi.org/10.1146/annurev.nutr.21.1.381] [PMID: 11375442]

[7] Sies H. Oxidative stress: oxidants and antioxidants. Exp Physiol 1997; 82(2): 291-5.
[http://dx.doi.org/10.1113/expphysiol.1997.sp004024] [PMID: 9129943]

[8] Laguerre M, Decker EA, Lecomte J, Villeneuve P. Methods for evaluating the potency and efficacy of antioxidants. Curr Opin Clin Nutr Metab Care 2010; 13(5): 518-25.
[http://dx.doi.org/10.1097/MCO.0b013e32833aff12] [PMID: 20601864]

[9] Cao G, Alessio HM, Cutler RG. Oxygen-radical absorbance capacity assay for antioxidants. Free Radic Biol Med 1993; 14(3): 303-11.
[http://dx.doi.org/10.1016/0891-5849(93)90027-R] [PMID: 8458588]

[10] Re R, Pellegrini N, Proteggente A, Pannala A, Yang M, Rice-Evans C. Antioxidant activity applying an improved ABTS radical cation decolorization assay. Free Radic Biol Med 1999; 26(9-10): 1231-7.
[http://dx.doi.org/10.1016/S0891-5849(98)00315-3] [PMID: 10381194]

[11] Everette JD, Bryant QM, Green AM, Abbey YA, Wangila GW, Walker RB. Thorough study of reactivity of various compound classes toward the Folin-Ciocalteu reagent. J Agric Food Chem 2010; 58(14): 8139-44.
[http://dx.doi.org/10.1021/jf1005935] [PMID: 20583841]

[12] Antolovich M, Prenzler PD, Patsalides E, McDonald S, Robards K. Methods for testing antioxidant activity. Analyst (Lond) 2002; 127(1): 183-98.
[http://dx.doi.org/10.1039/b009171p] [PMID: 11827390]

[13] Rivero-Pérez MD, Muñiz P, Gonzalez-Sanjosé ML. Antioxidant profile of red wines evaluated by total antioxidant capacity, scavenger activity, and biomarkers of oxidative stress methodologies. J Agric Food Chem 2007; 55(14): 5476-83.
[http://dx.doi.org/10.1021/jf070306q] [PMID: 17579427]

[14] Wang J. Portable electrochemical systems. Trends Analyti Chem. 2002; 21(4): 226.
[http://dx.doi.org/10.1016/S0165-9936(02)00402-8]

[15] Blasco AJ, Crevillén AG, González MC, Escarpa A. Direct electrochemical sensing and detection of natural antioxidants and antioxidant capacity *in vitro* systems. Electroanalysis 2007; 19(22): 2275-86. [http://dx.doi.org/10.1002/elan.200704004]

[16] Barroso MF, de-los-Santos-Álvarez N, Delerue-Matos C, Oliveira MB. Towards a reliable technology for antioxidant capacity and oxidative damage evaluation: electrochemical (bio)sensors. Biosens Bioelectron 2011; 30(1): 1-12. [http://dx.doi.org/10.1016/j.bios.2011.08.036] [PMID: 21963095]

[17] Ivanova AV, Gerasimova EL, Brainina KhZ. Potentiometric study of antioxidant activity: development and prospects. Crit Rev Anal Chem 2015; 45(4): 311-22. [http://dx.doi.org/10.1080/10408347.2014.910443] [PMID: 25830410]

[18] Medina-Plaza C, Rodríguez-Méndez ML, Sutter P, Tong X, Sutter E. Nanoscale Au-In alloy-oxide core-shell particles as electrocatalysts for efficient hydroquinone detection. J Phys Chem 2015; 119(44): 25100-7. [http://dx.doi.org/10.1021/acs.jpcc.5b07960]

[19] Intarakamhang S, Leson C, Schuhmann W, Schulte A. A novel automated electrochemical ascorbic acid assay in the 24-well microtiter plate format. Anal Chim Acta 2011; 687(1): 1-6. [http://dx.doi.org/10.1016/j.aca.2010.11.023] [PMID: 21241840]

[20] Pisoschi AM, Cheregi MC, Danet AF. Total antioxidant capacity of some commercial fruit juices: electrochemical and spectrophotometrical approaches. Molecules 2009; 14(1): 480-93. [http://dx.doi.org/10.3390/molecules14010480] [PMID: 19158657]

[21] Casilli S, de Luca M, Apetrei C, *et al.* Langmuir-Blodgett and Langmuir-Schaefer films of homoleptic and heteroleptic phthalocyanine complexes as voltammetric sensors: Applications to the study of antioxidants. Appl Surf Sci 2005; 246(4): 304-12. [http://dx.doi.org/10.1016/j.apsusc.2004.11.002]

[22] Piljac-Žegarac J, Valek L, Stipčević T, Martinez S. Electrochemical determination of antioxidant capacity of fruit tea infusions. Food Chem 2010; 121(3): 820-5. [http://dx.doi.org/10.1016/j.foodchem.2009.12.090]

[23] Buratti S, Scampicchio M, Giovanelli G, Mannino S. A low-cost and low-tech electrochemical flow system for the evaluation of total phenolic content and antioxidant power of tea infusions. Talanta 2008; 75(1): 312-6. [http://dx.doi.org/10.1016/j.talanta.2007.11.014] [PMID: 18371884]

[24] Makhotkina O, Kilmartin PA. The use of cyclic voltammetry for wine analysis: determination of polyphenols and free sulfur dioxide. Anal Chim Acta 2010; 668(2): 155-65. [http://dx.doi.org/10.1016/j.aca.2010.03.064] [PMID: 20493292]

[25] Apetrei C, Apetrei IM, Nevares I, *et al.* Using an e-tongue based on voltammetric electrodes to discriminate among red wines aged in oak barrels or aged using alternative methods: correlation between electrochemical signals and analytical parameters. Electrochim Acta 2007; 52(7): 2588-94. [http://dx.doi.org/10.1016/j.electacta.2006.09.014]

[26] Zielinska D, Wiczkowski W, Piskula MK. Determination of the relative contribution of quercetin and its glucosides to the antioxidant capacity of onion by cyclic voltammetry and spectrophotometric

methods. J Agric Food Chem 2008; 56(10): 3524-31.
[http://dx.doi.org/10.1021/jf073521f] [PMID: 18454541]

[27] Durst RA, Bäumner AJ, Murray RW, Buck RP, Andrieux P. Chemically modified electrodes: recommended terminology and definitions. Pure Appl Chem 1997; 69(6): 1317-23.
[http://dx.doi.org/10.1351/pac199769061317]

[28] Luo LQ, Li QX, Ding YP, Zhang Y, Shen X. Docosyltrimethylammonium chloride modified glassy carbon electrode for simultaneous determination of dopamine and ascorbic acid. J Solid State Electrochem 2010; 14(7): 1311-6.
[http://dx.doi.org/10.1007/s10008-009-0936-9]

[29] Sakthinathan S, Palanisamy S, Chen SM, Wu PS, Yao L, Lou BS. Electrochemical detection of phenol in industrial pollutant absorbed molecular sieves by electrochemically activated screen printed carbon electrode. Int J Electrochem Sci 2015; 10: 3319-28.

[30] Apetrei C, Rodríguez-Méndez ML, de Saja JA. Modified carbon paste electrodes for the discrimination of vegetable oils. Sens Actuators B Chem 2005; 111: 403-9.
[http://dx.doi.org/10.1016/j.snb.2005.03.041]

[31] Yang LJ, Tang C, Xiong HY, Zhang XH, Wang SF. Electrochemical properties of catechin at a single-walled carbon nanotubes-cetylramethylammonium bromide modified electrode. Bioelectrochemistry 2009; 75(2): 158-62.
[http://dx.doi.org/10.1016/j.bioelechem.2009.03.009] [PMID: 19383571]

[32] Du HJ, Ye JS, Zhang JQ, Huang XD, Yu CZ. Graphene nanosheets modified glassy carbon electrode as a highly sensitive and selective voltammetric sensor for rutin. Electroanalysis 2010; 22(20): 2399-406.
[http://dx.doi.org/10.1002/elan.201000110]

[33] Apetrei C, Apetrei IM, De Saja JA, Rodríguez-Mendez ML. Carbon paste electrodes made from different carbonaceous materials: application in the study of antioxidants. Sensors (Basel) 2011; 11(2): 1328-44.
[http://dx.doi.org/10.3390/s110201328] [PMID: 22319354]

[34] Shahrokhian S, Zare-Mehrjardi HR. Application of thionine-nafion supported on multi-walled carbon nanotube for preparation of a modified electrode in simultaneous voltammetric detection of dopamine and ascorbic acid. Electrochim Acta 2007; 52(22): 6310-7.
[http://dx.doi.org/10.1016/j.electacta.2007.04.023]

[35] Shahrokhian S, Zare-Mehrjardi HR. Cobalt salophen-modified carbon-paste electrode incorporating a cationic surfactant for simultaneous voltammetric detection of ascorbic acid and dopamine. Sens Actuators B Chem 2007; 121: 530-7.
[http://dx.doi.org/10.1016/j.snb.2006.04.088]

[36] Freitas KH, Fatibello O. Carbon composite electrode modified with copper(II) phosphate immobilized in a polyester resin for voltammetric determination of catechin in teas. Anal Lett 2010; 43(13): 2091-104.
[http://dx.doi.org/10.1080/00032711003687112]

[37] Civit L, Nassef HM, Fragoso A, O'Sullivan CK. Amperometric determination of ascorbic acid in real samples using a disposable screen-printed electrode modified with electrografted o-aminophenol film.

J Agric Food Chem 2008; 56(22): 10452-5.
[http://dx.doi.org/10.1021/jf802536k] [PMID: 18950182]

[38] Wang XG, Fan YJ. Fast detection of catechin in tea beverage using a poly-aspartic acid film based sensor. Mikrochim Acta 2010; 169(1–2): 173-9.
[http://dx.doi.org/10.1007/s00604-010-0335-z]

[39] Li CM, Sun CQ, Chen W, Pan L. Electrochemical thin film deposition of polypyrrole on different substrates. Surf Coat Tech 2005; 198(1-3): 474-7.
[http://dx.doi.org/10.1016/j.surfcoat.2004.10.065]

[40] Zhang L, Zhang C, Lian J. Electrochemical synthesis of polyaniline nano-networks on p-aminobenzene sulfonic acid functionalized glassy carbon electrode Its use for the simultaneous determination of ascorbic acid and uric acid. Biosens Bioelectron 2008; 24(4): 690-5.
[http://dx.doi.org/10.1016/j.bios.2008.06.025] [PMID: 18706798]

[41] Sharma PS, Pietrzyk-Le A, D'Souza F, Kutner W. Electrochemically synthesized polymers in molecular imprinting for chemical sensing. Anal Bioanal Chem 2012; 402(10): 3177-204.
[http://dx.doi.org/10.1007/s00216-011-5696-6] [PMID: 22302165]

[42] Rodríguez-Méndez ML, Gay M, de Saja JA. New insights into sensors based on radical bisphthalocyanines. J Porphyr Phthalocyanines 2009; 13(11): 1159-67.
[http://dx.doi.org/10.1142/S1088424609001509]

[43] Rodríguez-Méndez ML, Parra V, Apetrei C, Gay M, Prieto N, de Saja JA. E-tongue based on voltammetric electrodes modified with electroactive materials. Applications. Mikrochim Acta 2008; 163(1): 23-31.
[http://dx.doi.org/10.1007/s00604-007-0907-8]

[44] Cetó X, Apetrei C, del Valle M, Rodríguez-Mendez ML. Evaluation of red wines antioxidant capacity by means of a voltammetric e-tongue with an optimized sensor array. Electrochim Acta 2014; 120(20): 180-6.
[http://dx.doi.org/10.1016/j.electacta.2013.12.079]

[45] Parra V, Hernando T, Rodríguez-Méndez ML, de Saja JA. Electrochemical sensor array made from bisphthalocyanine modified carbon paste electrodes for discrimination of red wines. Electrochim Acta 2004; 49(28): 5177-85.
[http://dx.doi.org/10.1016/j.electacta.2004.06.031]

[46] Rodríguez-Méndez ML, Apetrei C, de Saja JA. Evaluation of the polyphenolic content of extra virgin olive oils using an array of voltammetric sensors. Electrochim Acta 2008; 53(20): 5867-72.
[http://dx.doi.org/10.1016/j.electacta.2008.04.006]

[47] Matemadombo F, Apetrei C, Nyokong T, Rodríguez-Méndez ML, de Saja JA. Comparison of carbon screen printed and disk electrodes in the detection of antioxidants using CoPc derivatives. Sens Actuators B Chem 2012; 166-167: 457-66.
[http://dx.doi.org/10.1016/j.snb.2012.02.088]

[48] El-Hady D, El-Maali N. Selective square wave voltammetric determination of (+)-catechin in commercial tea samples using beta-cyclodextrin modified carbon paste electrode. Microchimica Acta 161(1): 225-31.

[49] Lakshmi D, Bossi A, Whitcombe MJ, *et al.* Electrochemical sensor for catechol and dopamine based on a catalytic molecularly imprinted polymer-conducting polymer hybrid recognition element. Anal Chem 2009; 81(9): 3576-84.
[http://dx.doi.org/10.1021/ac802536p] [PMID: 19354259]

[50] Saha K, Agasti SS, Kim C, Li X, Rotello VM. Gold nanoparticles in chemical and biological sensing. Chem Rev 2012; 112(5): 2739-79.
[http://dx.doi.org/10.1021/cr2001178] [PMID: 22295941]

[51] Wang G, He X, Zhou F, *et al.* Application of gold nanoparticles/TiO_2 modified electrode for the electrooxidative determination of catechol in tea samples. Food Chem 2012; 135(2): 446-51.
[http://dx.doi.org/10.1016/j.foodchem.2012.04.139] [PMID: 22868112]

[52] Ma X, Qian W. Phenolic acid induced growth of gold nanoshells precursor composites and their application in antioxidant capacity assay. Biosens Bioelectron 2010; 26(3): 1049-55.
[http://dx.doi.org/10.1016/j.bios.2010.08.055] [PMID: 20863682]

[53] Yola ML, Atar N. A novel voltammetric sensor based on gold nanoparticles involved in p-aminothiophenol functionalized multi-walled carbon nanotubes: Application to the simultaneous determination of quercetin and rutin. Electrochim Acta 2014; 119: 24-31.
[http://dx.doi.org/10.1016/j.electacta.2013.12.028]

[54] Penu R, Obreja AC, Patroi D, Diaconu M, Radu GL. Graphene and gold nanoparticles based reagentless biodevice for phenolic endocrine disruptors monitoring. Microchem J 2015; 121: 130-5.
[http://dx.doi.org/10.1016/j.microc.2015.03.002]

[55] Rajesh , Ahuja T, Kumar D. Recent progress in the development of nano-structured conducting polymers/nanocomposites for sensor applications. Sens Actuators B 2009; 136: 275-86.

[56] García-Hernández C, García-Cabezón C, Medina-Plaza C, *et al.* Electrochemical behavior of polypyrrol/AuNP composites deposited by different electrochemical methods: sensing properties towards catechol. Beilstein J Nanotechnol 2015; 6: 2052-61.
[http://dx.doi.org/10.3762/bjnano.6.209] [PMID: 26665076]

[57] Moccelini SK, Fernandes SC, de Camargo TP, Neves A, Vieira IC. Self-assembled monolayer of nickel(II) complex and thiol on gold electrode for the determination of catechin. Talanta 2009; 78(3): 1063-8.
[http://dx.doi.org/10.1016/j.talanta.2009.01.038] [PMID: 19269473]

[58] Apetrei C, Casilli S, de Luca M, *et al.* Spectroelectrochemical characterisation of Langmuir-Shaeffer films of heteroleptic phthalocyanine complexes. Potential applications. Colloids Surf A Physicochem Eng Asp 2006; 284-285: 574-82.
[http://dx.doi.org/10.1016/j.colsurfa.2005.10.069]

[59] Alessio P, Pavinatto FJ, Oliveira ON Jr, De Saja Saez JA, Constantino CJ, Rodríguez-Méndez ML. Detection of catechol using mixed Langmuir-Blodgett films of a phospholipid and phthalocyanines as voltammetric sensors. Analyst (Lond) 2010; 135(10): 2591-9.
[http://dx.doi.org/10.1039/c0an00159g] [PMID: 20714515]

[60] Martín MG, Rodríguez-Méndez ML, de Saja JA. Films of lutetium bisphthalocyanine nanowires as electrochemical sensors. Langmuir 2010; 26(24): 19217-24.

[http://dx.doi.org/10.1021/la1025894] [PMID: 21082798]

[61] Medina-Plaza C, García-Cabezón C, García-Hernández C, *et al.* Analysis of organic acids and phenols of interest in the wine industry using Langmuir-Blodgett films based on functionalized nanoparticles. Anal Chim Acta 2015; 853: 572-8. [http://dx.doi.org/10.1016/j.aca.2014.10.046] [PMID: 25467505]

[62] Medina-Plaza C, Furini LN, Constantino CJ, de Saja JA, Rodrí Guez-Mendez ML. Synergistic electrocatalytic effect of nanostructured mixed films formed by functionalised gold nanoparticles and bisphthalocyanines. Anal Chim Acta 2014; 851: 95-102. [http://dx.doi.org/10.1016/j.aca.2014.08.049] [PMID: 25440670]

[63] Apetrei C, Nieto M, Rodríguez-Méndez ML, de Saja JA. Development of lutetium bisphthalocyanine/carbon nanotube Langmuir-Blodgett films. Sensing properties. J Porphyr Phthalocyanines 2011; 15(9): 1-10.

[64] Fernandes EG, Brazaca LC, Rodríguez-Mendez ML, Saja JA, Zucolotto V. Immobilization of lutetium bisphthalocyanine in nanostructured biomimetic sensors using the LbL technique for phenol detection. Biosens Bioelectron 2011; 26(12): 4715-9. [http://dx.doi.org/10.1016/j.bios.2011.05.032] [PMID: 21704506]

[65] Alessio P, Rodríguez-Méndez ML, De Saja Saez JA, Constantino CJ. Iron phthalocyanine in non-aqueous medium forming layer-by-layer films: growth mechanism, molecular architecture and applications. Phys Chem Chem Phys 2010; 12(16): 3972-83. [http://dx.doi.org/10.1039/b922242c] [PMID: 20379489]

[66] Tan Y, Kan J, Shengqi L. Amperometric biosensor for catechol using electrochemical template process. Sens Actuators B Chem 2011; 152(2): 285-91. [http://dx.doi.org/10.1016/j.snb.2010.12.021]

[67] Cortina-Puig M, Muñoz-Berbel X, Calas-Blanchard C, Marty JL. Diazonium-functionalized tyrosinase-based biosensor for the detection of tea polyphenols. Mikrochim Acta 2010; 171(1): 187-93. [http://dx.doi.org/10.1007/s00604-010-0425-y]

[68] El Kaoutit M, Naranjo-Rodriguez I, Temsamani KR, Hidalgo-Hidalgo de Cisneros JL. The sonogel-carbon materials as basis for development of enzyme biosensors for phenols and polyphenols monitoring: A detailed comparative study of three immobilization matrixes. Biosens Bioelectron 2007; 22(12): 2958-66. [http://dx.doi.org/10.1016/j.bios.2006.12.008] [PMID: 17215118]

[69] Singh RP. A catechol biosensor based on a gold nanoparticles encapsulated-dendrimer. Analyst (Lond) 2011; 136(6): 1216-21. [http://dx.doi.org/10.1039/c0an00601g] [PMID: 21240422]

[70] Wang Y, Hasebe Y. Acridine orange-induced signal enhancement effect of tyrosinase-immobilized carbon-felt-based flow biosensor for highly sensitive detection of monophenolic compounds. Anal Bioanal Chem 2011; 399(3): 1151-62. [http://dx.doi.org/10.1007/s00216-010-4369-1] [PMID: 21042905]

[71] ElKaoutit M, Naranjo-Rodriguez I, Temsamani KR, Hernández-Artiga MP, Bellido-Milla D, Cisneros JL. A comparison of three amperometric phenoloxidase-Sonogel-Carbon based biosensors for

determination of polyphenols in beers. Food Chem 2008; 110(4): 1019-24.
[http://dx.doi.org/10.1016/j.foodchem.2008.03.006] [PMID: 26047296]

[72] Di Fusco M, Tortolini C, Deriu D, Mazzei F. Laccase-based biosensor for the determination of polyphenol index in wine. Talanta 2010; 81(1-2): 235-40.
[http://dx.doi.org/10.1016/j.talanta.2009.11.063] [PMID: 20188914]

[73] Ibarra-Escutia P, Gomez JJ, Calas-Blanchard C, Marty JL, Ramirez-Silva MT. Amperometric biosensor based on a high resolution photopolymer deposited onto a screen-printed electrode for phenolic compounds monitoring in tea infusions 2010.
[http://dx.doi.org/10.1016/j.talanta.2010.03.017]

[74] Tan Y, Deng W, Ge B, Xie Q, Huang J, Yao S. Biofuel cell and phenolic biosensor based on acid-resistant laccase-glutaraldehyde functionalized chitosan-multiwalled carbon nanotubes nanocomposite film. Biosens Bioelectron 2009; 24(7): 2225-31.
[http://dx.doi.org/10.1016/j.bios.2008.11.026] [PMID: 19153037]

[75] Santos AS, Pereira AC, Sotomayor MD, Tarley CR, Duran N, Kubota LT. Determination of phenolic compounds based on co-immobilization of methylene blue and HRP on multi-wall carbon nanotubes. Electroanalysis 2007; 19(5): 549-54.
[http://dx.doi.org/10.1002/elan.200603759]

[76] Chauhan N, Dahiya T. Priyanka, Pundi, C.S. Fabrication of an amperometric ascorbate biosensor using egg shell membrane bound *Lagenaria siceraria* fruit ascorbate oxidase. J Mol Catal, B Enzym 2010; 67(1-2): 66-71.
[http://dx.doi.org/10.1016/j.molcatb.2010.07.007]

[77] Wang X, Watanabe H, Uchiyama S. Amperometric l-ascorbic acid biosensors equipped with enzyme micelle membrane. Talanta 2008; 74(5): 1681-5.
[http://dx.doi.org/10.1016/j.talanta.2007.09.008] [PMID: 18371836]

[78] Bonakdar M, Vilchez JL, Mottola HA. Bioamperometric sensors for phenol based on carbon paste electrodes. J Electroanal Chem 1989; 266(1): 47-55.
[http://dx.doi.org/10.1016/0022-0728(89)80214-1]

[79] Wang P, Liu M, Kan JQ. Amperometric phenol biosensor based on polyaniline. Sens Actuators B Chem 2009; 140(2): 577-84.
[http://dx.doi.org/10.1016/j.snb.2009.05.005]

[80] Apetrei C, Rodríguez-Méndez ML, de Saja JA. Amperometric tyrosinase based biosensor using an electropolymerized phosphate-doped polypyrrole film as an immobilization support. Application for detection of phenolic compounds. Electrochim Acta 2011; 56(24): 8919-25.
[http://dx.doi.org/10.1016/j.electacta.2011.07.127]

[81] Medina-Plaza C, Revilla G, Muñoz R, *et al.* Electronic tongue formed by sensors and biosensors containing phthalocyanines as electron mediators. Application to the analysis of red grapes. J Porphyr Phthalocyanines 2014; 18: 76-86.
[http://dx.doi.org/10.1142/S1088424613501137]

[82] Apetrei C, Alessio P, Constantino CJ, *et al.* Biomimetic biosensor based on lipidic layers containing tyrosinase and lutetium bisphthalocyanine for the detection of antioxidants. Biosens Bioelectron 2011; 26(5): 2513-9.

[http://dx.doi.org/10.1016/j.bios.2010.10.047] [PMID: 21123042]

[83] Sanz VC, Mena ML, González-Cortés A, Yáñez-Sedeño P, Pingarrón JM. Development of a tyrosinase biosensor based on gold nanoparticles-modified glassy carbon electrodes - Application to the measurement of a bioelectrochemical polyphenols index wines. Anal Chim Acta 2005; 528(1): 1-8. [http://dx.doi.org/10.1016/j.aca.2004.10.007]

[84] Zhao J, Zhi J, Zhou Y, Yan W. A tyrosinase biosensor based on ZnO nanorod clusters/nanocrystalline diamond electrodes for biosensing of phenolic compounds. Anal Sci 2009; 25(9): 1083-8. [http://dx.doi.org/10.2116/analsci.25.1083] [PMID: 19745534]

[85] Yildiz HB, Toppare L, Gursel YH, Yagci Y. Immobilization of polyphenol oxidase in conducting graft copolymers and determination of phenolic amount in red wines with enzyme electrodes. Enzyme Microb Technol 2006; 39(4): 945-8. [http://dx.doi.org/10.1016/j.enzmictec.2006.01.026]

[86] Medina-Plaza C, de Saja JA, Rodríguez-Mendez ML. Bioelectronic tongue based on lipidic nanostructured layers containing phenol oxidases and lutetium bisphthalocyanine for the analysis of grapes. Biosens Bioelectron 2014; 57: 276-83. [http://dx.doi.org/10.1016/j.bios.2014.02.023] [PMID: 24594595]

[87] Cabaj J, Sołoducho J, Nowakowska-Oleksy A. Langmuir-Blodgett Film Based Biosensor for Estimation of Phenol Derivatives. Sens Actuators B Chem 2010; 143: 508-15. [http://dx.doi.org/10.1016/j.snb.2009.09.047]

[88] Schmidt TF, Caseli L, Viitala T, Oliveira ON Jr. Enhanced activity of horseradish peroxidase in Langmuir-Blodgett films of phospholipids. Biochim Biophys Acta 2008; 1778(10): 2291-7. [http://dx.doi.org/10.1016/j.bbamem.2008.05.012] [PMID: 18585999]

[89] Śliwińska M, Wiśniewska P, Dymerski T, Namieśnik J, Wardencki W. Food analysis using artificial senses. J Agric Food Chem 2014; 62(7): 1423-48. [http://dx.doi.org/10.1021/jf403215y] [PMID: 24506450]

[90] Vlasov Y, Legin A, Rudnitskaya A, Di Natale C, D'Amico A. Nonspecific sensor arrays "electronic tongue" for chemical analysis of liquids. Pure Appl Chem 2005; 77(11): 1965-83. [http://dx.doi.org/10.1351/pac200577111965]

[91] Rudnitskaya A, Rocha SM, Legin A, Pereira V, Marques JC. Evaluation of the feasibility of the electronic tongue as a rapid analytical tool for wine age prediction and quantification of the organic acids and phenolic compounds. The case-study of Madeira wine. Anal Chim Acta 2010; 662(1): 82-9. [http://dx.doi.org/10.1016/j.aca.2009.12.042] [PMID: 20152269]

[92] Artigas J, Jimenez C, Dominguez C, Minguez S, Gonzalo A, Alonso J. Development of a multiparametric analyser based on ISFET sensors applied to process control in the wine industry. Sens Actuators B Chem 2003; 89: 199-204. [http://dx.doi.org/10.1016/S0925-4005(02)00464-1]

[93] Gay-Martin M, De Saja JA, Munoz R, Rodríguez-Méndez ML. Multisensor system based on bisphthalocyanine nanowires for the detection of antioxidants 2012; 68: 88-94. [http://dx.doi.org/10.1016/j.electacta.2012.02.042]

[94] Olivati CA, Riul A Jr, Balogh DT, Oliveira ON Jr, Ferreira M. Detection of phenolic compounds using

impedance spectroscopy measurements. Bioprocess Biosyst Eng 2009; 32(1): 41-6.
[http://dx.doi.org/10.1007/s00449-008-0218-4] [PMID: 18408956]

[95] Kirsanov D, Mednova O, Vietoris V, Kilmartin PA, Legin A. Towards reliable estimation of an "electronic tongue" predictive ability from PLS regression models in wine analysis. Talanta 2012; 90: 109-16.
[http://dx.doi.org/10.1016/j.talanta.2012.01.010] [PMID: 22340124]

[96] Rodriguez-Mendez ML, Apetrei C, Gay M, *et al.* Evaluation of oxygen exposure levels and polyphenolic content of red wines using an electronic panel formed by an electronic nose and an electronic tongue. Food Chem 2014; 155: 91-7.
[http://dx.doi.org/10.1016/j.foodchem.2014.01.021] [PMID: 24594159]

[97] Verrelli G, Lvova L, Paolesse R, Di Natale CD, Amico A. Metalloporphyrin - based electronic tongue: an application for the analysis of Italian white wines. Sensors (Basel) 2007; 7(11): 2750-62.
[http://dx.doi.org/10.3390/s7112750]

[98] Cetó X, Céspedes F, Pividori MI, Gutiérrez JM, del Valle M. Resolution of phenolic antioxidant mixtures employing a voltammetric bio-electronic tongue. Analyst (Lond) 2012; 137(2): 349-56.
[http://dx.doi.org/10.1039/C1AN15456G] [PMID: 22102984]

CHAPTER 5

Biosensors for Characterisation of Bioactive Compounds from Wines

Bogdan Bucur[a], **Ioana Silvia Hosu**[b,c] and **Alina Vasilescu**[d,*]

[a] *Centre of Bioanalysis, National Institute of Research and Development for Biological Sciences, 296, Splaiul Independentei, Bucharest 060031, Romania*

[b] *National Institute for R&D in Chemistry and Petrochemistry, Splaiul. Independentei 202, 060021, Bucharest, Romania.*

[c] *Institute of Electronics, Microelectronics and Nanotechnology (IEMN), UMR-CNRS 8520, Université Lille 1, Avenue Poincaré—BP 60069, 59655 Villeneuve d'Ascq, France*

[d] *International Centre of Biodynamics, 1B Intrarea Portocalelor, sector 6, 060101, Bucharest, Romania*

Abstract: Biosensors are promising analytical tools for the determination and characterisation of bioactive compounds in wines, providing fast, sensitive and cost-effective detection, assisting in the assessment of quality, sensory and safety attributes of wines. In this chapter, some of the most representative, recent examples of biosensors for the detection of polyphenols, biogenic amines, mycotoxins and allergens in wine are discussed, emphasizing the wealth of biosensing strategies and the particularities of wine as a complex analytical matrix.

Keywords: Allergens, Biogenic amines, Biosensor, Mycotoxins, Polyphenols, Wine.

INTRODUCTION

Bioactive compounds are generally perceived as those compounds that can exert a beneficial effect on living cells or organisms, as per a widely accepted definition [1].

* **Corresponding author Alina Vasilescu:** International Centre of Biodynamics, 1B Intrarea Portocalelor, sector 6, 060101, Bucharest, Romania; E-mail: avasilescu@biodyn.ro

There is however, a second line of thinking among researchers who adopted a broader definition of "bioactive" to include any type of effect on living cells, being that beneficial or toxic. In line with this later definition naming bioactive "a compound which has the capability and the ability to interact with one or more component(s) of the living tissue by presenting a wide range of probable effects" [2], this chapter will summarize the current information regarding the use of biosensors for the detection and characterisation of four groups of bioactive compounds in wine: (poly)phenolic compounds, biogenic amines, mycotoxins and allergens.

Biosensors are analytical tools which combine the specificity of biorecognition using an enzyme, antibody, DNA sequence, whole cell, peptide *etc.* with a sensitive physical transducer (electrochemical, optical, mass-sensitive, *etc.*) in order to detect a target molecule. The biorecognition element is immobilized on the transducer by adsorption, entrapment, covalent binding, cross linking, *etc.* and the biorecognition event is translated into an electrical signal, as illustrated in Fig. (**1**).

Biosensors have been intensively researched in the last decades as potential alternatives to conventional methods in the food, biomedical and environmental fields based on their promise for fast, sensitive, specific, portable, low-cost analysis and compatibility with miniaturization.

Wine analysis was among the practical applications explored with biosensors, which were used either for monitoring the progress of fermentation processes, to verify relevant quality attributes, to assess the sensory characteristics or for classifying wine samples. Due to the complexity of wine matrix, the development of biosensors for wine poses specific analytical challenges, particularly related to the presence of possible interfering compounds. A few examples of biosensors for the detection of bioactive compounds in wine are given in Table **1** and some illustrative biosensors are discussed in more detail below, for each group of bioactive compounds.

Detection of Phenolic Compounds

Wines are a rich source of polyphenolic antioxidants that could act as powerful scavengers of free radicals species formed during metabolic processes in human cells. High levels of free radicals are triggers of oxidative stress, associated with various diseases [16]. Natural antioxidants have a huge potential for alleviating a number of medical conditions linked to oxidative stress induced by free radicals, as indicated by several ongoing clinical trials.

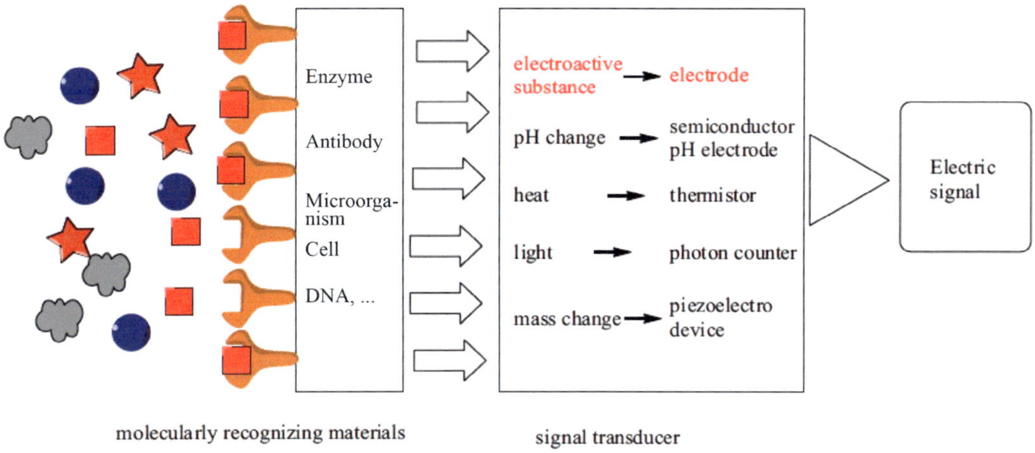

Fig. (1). Schematic diagram for the principal of the biosensors. Mechanism of the biosensor and the detection method. Figure reproduced from [109] by permission of INTECHOPEN (www.intechopen.com) and under the terms and conditions of the Creative Commons Attribution license (http://creativecommons.org/licenses/by-nc-sa/3.0/).

Table 1. Examples of recent biosensors for (poly)phenolic compounds in wine.

Bioactive compound/ matrix	Biological and chemical elements/transduction system	Specificity/Mode of action	Figures of merit	Reference
Phenol	Microbial cells/ Electrochemical Transducer	*Pseudomonas putida* immobilised on gold electrodes.	Linear range: $0.5x10^{-6}$ to $6x10^{-6}$ M Sensitivity: $5.03x10^{-3}$ A M^{-1}	[3]
Phenol	QD (CdS or Cu_2S) -MWCNTs and bacteria/ Electrochemical transducer	*Acaligense* spp. catalyses the conversion of phenol to catechol and o-quinone, and the QD-MWCNTs act as electron mediator of these redox reactions.	LR: $0.5x10^{-3}$ – $5x10^{-3}$ M (CdS) $0.7x10^{-3}$ - $10x10^{-3}$ M (Cu_2S)	[5]

(Table 1) contd.....

Bioactive compound/ matrix	Biological and chemical elements/transduction system	Specificity/Mode of action	Figures of merit	Reference
Catechol, rosmarinic acid, caffeic acid, chlorogenic acid, and gallic acid	Laccase and Nafion/ Electrochemical transducer	Laccase catalysed oxidation of phenols/electrochemical reduction of quinones	Gallic acid: LR: $2x10^{-6}$ - $7x10^{-6}$ M, LOD: $1.5x10^{-6}$ M, S: 1×10^{-6} M^{-1}. Rosmarinic acid: LR: $7x10^{-7}$ - $1.5x10^{-6}$ M, LOD: $1.19x10^{-7}$ M, S: $190x10^{-6}$ M^{-1}.	[4]
Catechin and caffeic acid	Laccase immobilised on derivatised polyethersulfone membrane/ Electrochemical Transducer	Laccase immobilised on derivatised polyethersulfone membrane, enzyme-catalysed oxidation of phenols to quinones/ electrochemical reduction of quinones.	Catechin: LR: $1x10^{-5}$ - $6x10^{-5}$ M, S: 0.02443mA.M^{-1} Caffeic acid: LR: $1x10^{-5}$ - $8x10^{-5}$ M, S: 0.02401mA.M^{-1}	[6]
Catechol	Sol-gel gold nanoparticles (Sol-AuNPs) and Tyrosinase (Tyr)/ Electrochemical Transducer	Tyrosinase-catalysed oxidation of catechol to benzoquinone	LR: $1x10^{-6}$ - $6x10^{-6}$ M, LOD: $3x10^{-7}$ M S: $1x10^{-2}$ A M^{-1}.	[7]
Quercetin	Ag Nps in ionic liquid (IL) 1-butyl-3-methylimidazolium bis(trifluoromethylsulfonyl) imide (BMI.Tf_2N), Laccase and β-cycLODextrin modified with epichlorohydrin/ Electrochemical Transducer	Laccase catalysed oxidation of quercetin, and the reduction back to quercetin is measured by SWV.	LR: $4.99x10^{-7}$ - $7.40x10^{-6}$ M, LOD: $(3.7\pm0.4)x10^{-8}$ M	[8]
Resveratrol, coumaric acid, caffeic and ascorbic acid	Guanine, glucose oxidase and Fe@Fe_2O_3 core-shell, (GOX) / Electrochemical Transducer	(GOX) catalysed the generation of H2O2, further decomposed to·OH. Free radicals damage guanine and antioxidants protect against damage/. Electrochemical oxidation of guanine.	Antioxidant activity decreased in the order ascorbic acid> resveratrol> coumaric acid> caffeic acid	[35]

(Table 1) contd.....

Bioactive compound/ matrix	Biological and chemical elements/transduction system	Specificity/Mode of action	Figures of merit	Reference
Caffeic acid, chlorogenic acid, epicatechin	Laccase, MoS_2-graphene quantum dots-SPCE/ Electrochemical transducer	Laccase-catalysed oxidation of polyphenols to quinones/ electrochemical reduction of quinones	Caffeic acid: LR:0.38-100 μM LOD: 0.32 μM Chlorogenic acid: LR: 0.38–8.26 μM LOD: 0.19 μM Epicatechin: Linear range: 2.86–100.00 μM LOD: 2.04 μM	[9]
Resveratrol	Peroxidase basic isoenzymes/ Electrochemical Transducer	Enzyme catalysed oxidation of resveratrol in the presence of H_2O_2/electrochemical detection of H_2O_2 on ferrocene-modified carbon paste electrode	LR: $1x10^{-6}$ -$2.35x10^{-5}$ M, LOQ: 0,83 μM, S: $(2.31\pm0.05)\times10^{6}$ nA M^{-1}.	[10]
Total phenolic content expressed as L-DOPA	Tyrosinase and 3-Methyl-2-benzothiazolinone hydrazone hydrochloride hydrate (MBTH)/ Optical transducer	Paper bioactive sensor acts as catalyst of phenolic conversion to quinones and the formation of coloured adduct.	LR: 0 - $0.5x10^{-3}$ M, LOD: 5 $x10^{-6}$ M S: $9.5x10^{-3}$ M^{-1}.	[11]
Phenol in waste water	Laccase/ Optical transducer based on adsorption at 400 nm.	Enzyme immobilized on polyacrylamide films	LR: $1.09x10^{-4}$ M - $2.5x10^{-3}$ M, LOD: 3.27 $x10^{-5}$ M S: $5.4x10^{1}$ M^{-1}	[12]
Wine, Pentagalloyl glucose (PGG)	α-amylase(AMY)/ SPR Optical Transducer	Specific interaction of PGG with AMY (immobilized of glass substrate with Au disks)	LR: $0.5x10^{-6}$ -$154x10^{-6}$ M, S: 0.360 ± 0.02 nm×M^{-6}	[13].
Catechol	Tyrosinase /Optical SPR transducer	Change in resonance wavelength due to the tyrosinase-catalysed oxidation of catechol to quinone	LOD: $1x10^{-4}$ M	[14]

(Table 1) contd.....

Bioactive compound/ matrix	Biological and chemical elements/transduction system	Specificity/Mode of action	Figures of merit	Reference
Fisetin	ds DNA-abasic aptamer/Optical Fluorescent Transducer.	Double stranded (ds) DNA, creates a binding pocket very specific for fisetin/Enhanced fluorescence of fisetin upon binding	LOD: $5x10^{-8}$ M .	[15]

LR: linear range; LOD: Limit of detection;S: sensitivity; LOQ: limit of quantitation; QD: quantum dots; MWCNT: multiwall carbon nanotubes; SWV: Square Wave voltammetry; SPCE: screen-printed carbon electrodes; SPR: Surface Plasmon Resonance.

The physiological relevance of antioxidant consumption to complement the action of endogenous enzymes in the organisms (*i.e.* superoxide dismutase, catalase and peroxidase) for controlling free radicals remains yet to be demonstrated, according to European Food Safety Agency [17] and United States Department of Agriculture [18], considering the limited bioavailability of polyphenols and their transformation in human body.

A glass of red wine contains about 100 mg polyphenols, including flavonoid and nonflavonoid compounds (stilbenes, phenolic acids, and tannins [19, 20], see classification in Fig. (**1**), all contributing to the high antioxidant capacity of wine. Moderate consumption of red wine is believed to lower the risk of death from coronary heart disease [21] by preventing the oxidation of low density lipoprotein, LDL [22]. For example, tyrosol prevents injuries associated with ischemia/reperfusion [23] while resveratrol was reported to possess anti-cancer [24], anti-inflammatory [25], and anti-aging activity [26]. Its role in cardiovascular, cerebral and metabolic diseases was recently reviewed [20].

Polyphenols from wine intervene in various types of transformations induced by reactive oxygen and nitrogen species (*e.g.* hydrogen peroxide, hydroxyl radicals, peroxynitrite) such as protein nitration and oxidation, DNA damage or lipid peroxidation [27 - 29].

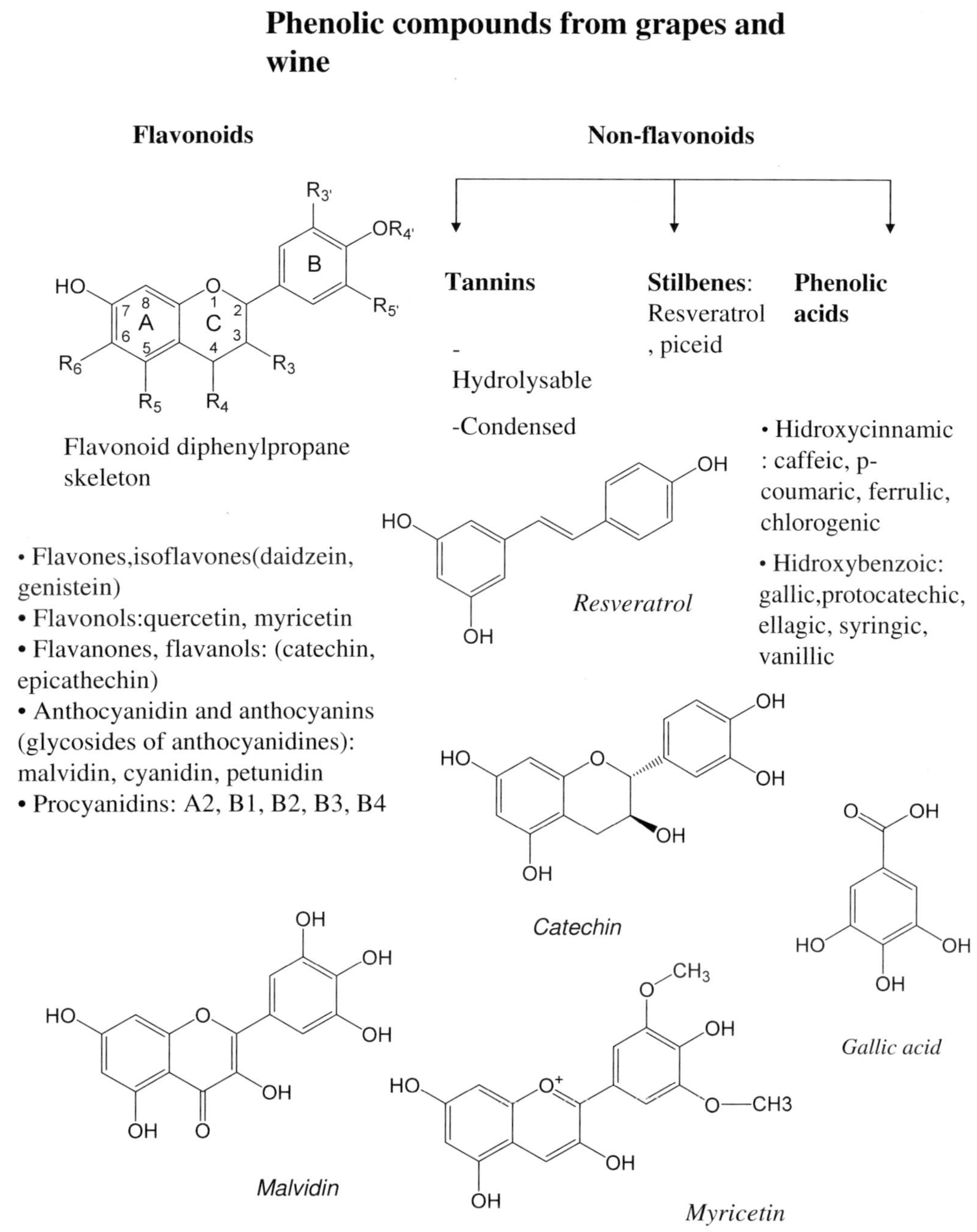

Fig. (2). Phenolic compounds from grapes and wine. [Reprinted with permission from (110], copyright American Chemical Society, 2014).

Flavonoids are secondary metabolites with a common skeleton of phenyl-benzo-γ-pyran (C6-C3-C6), (Fig. **2**). The large group of flavonoids in wine includes flavan-3-ols (catechin, epicatechin), anthocyanidins (*e.g.* malvidin) and their glycosylated forms, anthocyanins, and flavonols (such as quercetin and rutin). Besides flavonoids, phenolic acids from wine are also powerful antioxidants [30]. They include hydroxycinnamic acids such as caffeic acid and hydroxybenzoic acids, from which gallic is the most important and is also frequently used as a standard in antioxidant assays. Stilbenes represent a class of compounds with a common 1,2-diphenylethylene backbone, the most famous representative of this group being resveratrol (3,5,4-trihydroxystilbene). Stilbenes have shown extraordinary potential in the biomedical field, several studies attributing to resveratrol aging-delaying effects [31, 32], and significant potential in the fight against cardiovascular diseases and some types of cancer [33, 34]. Resveratrol (Fig. **2**) is synthesized in grapes by stilbene-synthase [33] as response to fungal infections- *e.g.* by *Botrytis cinerea*. Hydroxylated derivatives of stilbenes are called "stilbenoids", while oligomers formed by the oxidative dimerization of resveratrol are named "viniferins". Stilbenes are more prevalent in red wines as compared to the white and rose ones and were detected in wines from all over the world, in concentrations ranging from an average 0.6 mg/L in Greek wines to 12.9 mg/L in Cabernet Sauvignon wines in the Idaho Valley region in the US [35].

Besides their antioxidant role, polyphenols are strongly associated with sensory characteristics of wines, for example anthocyanins are responsible for the red colour in wine [36], epicatechin and tannins are associated with astringency [37] *etc*.

Detection of phenolic compounds from wine with biosensors rely on enzyme, whole cells or DNA-modified sensors. Various sensor architectures have been proposed for analysing either individual, groups or total phenolic compounds (i), for determining the total antioxidant capacity (iii) or for developing bioelectronic tongues, *i.e.* sets of several biosensors coupled with chemometric analysis, used to discriminate wines (iii).

The most used enzymes in the detection of phenolic compounds are: laccase, tyrosinase and peroxidase. Tyrosinase catalyses the oxidation of phenolic compounds with at least one free *ortho*-position. Two reactions occur consecutively: the o-hydroxylation of monophenols to o-diphenols and the subsequent oxidation of o-diphenols to o-quinones. The laccase biosensors, on the other hand, exploit the one-electron oxidation of polyphenols catalysed by the enzyme, leading to quinones *via* an intermediary aryl-oxy radical [38]. Laccase is able to catalyse the oxidation of various aromatic compounds [38]. Tyrosinase and laccase biosensors allow to determine the total polyphenol content of wines, with high sensitivity and better selectivity than the classic Folin-Ciocalteu test [38]. Moreover, increased sensitivity was achieved with flow-based detection systems as compared to tests performed in batch.

DNA-based sensors were typically used to assess the total antioxidant capacity of phenolic compounds in wines, based on the protective effect of phenolic antioxidants against DNA damage by free radicals [39].

Several representative biosensors for phenolic compounds in wines with electrochemical and optical detection are summarized in Table **1**, while selected biosensors (including based on more rarely used detection modes) are discussed in more detail below. Also included in Table **1** are some biosensing concepts that have been developed for other applications but could be potentially adapted for wine analysis.

Biosensors for Polyphenols with Electrochemical Detection

Most biosensors used for the detection of bioactive components in wine rely on electrochemical detection, which provides the advantage of sensitivity doubled by simplicity, low cost, potential for portability and miniaturization. Electrochemical methods allow studying the differences between various polyphenolic compounds, to capture global, characteristic electrochemical behaviour of wines (*e.g.* by recording intensity *versus* potential curves that presents wine-specific features), to measure the total polyphenolic content and to determine the total antioxidant capacity [40].

Several reviews have been published recently on biosensors for antioxidants in wine and food in general [38, 41 - 43].

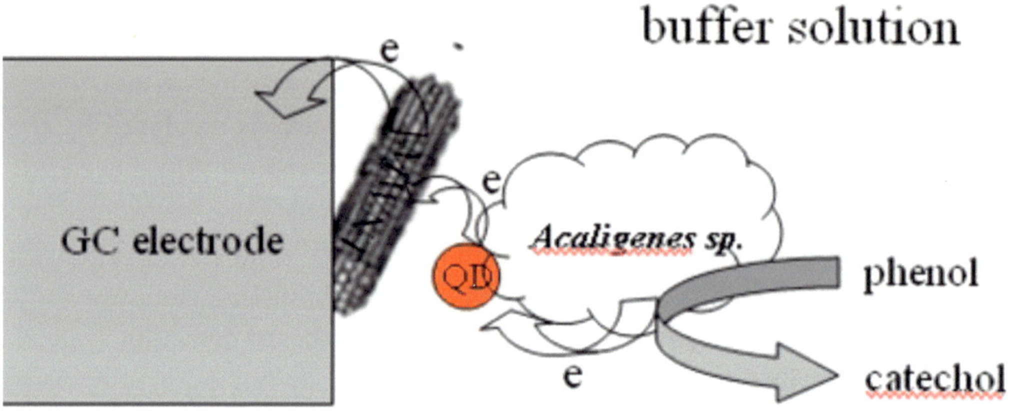

Fig. (3). Mechanism of the electron transfer for the *Acaligense spp* /QD-MWCN/GCE. Figures reproduced from [4] by permission of MDPI, Basel, Switzerland (www.mdpi.com/journal/sensors) and under the terms and conditions of the Creative Commons Attribution license (http://creativecommons.org/licenses/by/3.0/).

An interesting biosensor combined large bacterial cells of *Acaligense spp* as specific recognition element with nanomaterial-modified electrode [4]. The bacteria play the role of phenol oxidases, converting phenol into quinone. Glassy carbon electrodes have been modified with CdS or Cu_2S quantum nanodots – multi wall carbon nanotubes (Fig. **3**). The biosensor was applied for the determination of the total phenolic content in red wines [4]. The authors proposed a mechanism for the electron transfer on the surface of this biosensor in presence of phenol and bacteria as illustrated in Fig. (**3**).

Another electrochemical method using a glassy carbon electrode modified with a guanine composite membrane, covered with core-shell $Fe@Fe_2O_3$ and glucose oxidase, was used to assess the antioxidant capacity of four phenolic antioxidants: ascorbic acid, resveratrol, coumaric acid and caffeic acid. The results were in good agreement with the literature [35]. Fig. (**4**) illustrates the principle of this biosensor. Three different response modes can be observed, depending on the presence of antioxidant. In the absence of glucose, guanine is electrochemically oxidized around 0.8 V (Fig. **4a**). The signal recorded by Differential Pulse Voltammetry with undamaged guanine represents the initial, reference analytical response (i_0). In the presence of glucose, the enzymatic reaction catalysed by glucose oxidase ("Gox") produces hydrogen peroxide and eventually hydroxyl radicals, which attack and damage guanine. Consequently, the analytical signal recorded by DPV decreases, as only a part of the guanine - the undamaged one - will be oxidized on the electrode (Fig. **4b**). Finally, upon addition of an antioxidant compound to the glucose solution, the antioxidant scavenges the free radicals and protects guanine from damage. Consequently, the biosensor response increases again (i_t), (Fig. **4c**). The antioxidant capacity de-fined as AOT % = $i_t/i_0 \times 100$ of ascorbic acid, resveratrol, coumaric acid and caffeic acid was 73.5%, 71.6%, 64.4% and 60.7%, respectively. This biosensor still awaits application in real samples.

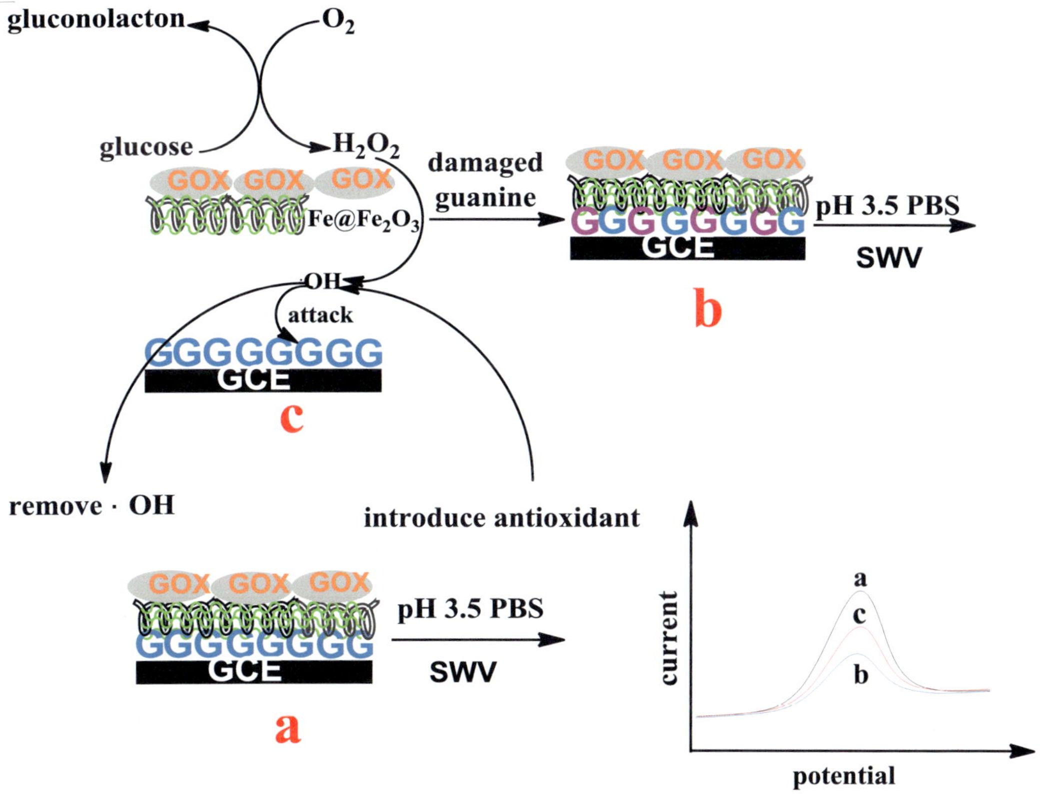

Fig. (4). Mechanism of the biosensor and the detection by square wave voltammetry (SWV). (Figures reproduced from reference [35] by permission of ESG (www.electrochemsci.org) and under the terms and conditions of the Creative Commons Attribution license (http://creativecommons.org/licenses/by/4.0/)).

In line with modern trends of including nanomaterials in biosensors, a more complex biosensor for quercetin was developed using laccase immobilized in a matrix of β-cyclodextrin / epichlorohydrin matrix. Cyclodextrin was used due to its capacity of forming inclusion complexes through hydroxyl groups. Moreover, the electrode was further modified with Ag nanoparticles (AgNPs) in ionic liquid (IL) of 1-butyl-3-methylimidazolium bis(trifluoromethylsulfonyl) imide (BMI.Tf_2N), due to the high ionic conductivity of IL and catalytic properties of AgNPs. The mechanism of the biosensor is quite simple: laccase catalyses the oxidation of quercetin to the corresponding quinone, and the electrochemical reduction back to quercetin by SWV generates a cathodic current, whose magnitude is proportional with the concentration of quercetin in the sample. Following optimization studies, the biosensor presented high selectivity towards quercetin, the responses decreasing in the order quercetin>rutin> myricetin>ascorbic acid>caffeic acid>catechin. No response was observed for apigenin, kaempferol, naringin and morin [8].

Very recently, another laccase, nanomaterial-based biosensor was described for analysis of caffeic acid, chlorogenic acid and epicatechin in wines [9]. The biosensor incorporates a novel nanobiocomposite from molybdenum disulphide (MoS_2) and graphene quantum dots (GQDs), which improved the conductivity of screen-printed electrodes and provided a compatible matrix for immobilization of laccase, with the result of achieving good sensitivities and wide linear ranges for the 3 analytes. The detection limits were 0.32 μM, 0.19 μM and 2.04 μM for caffeic acid, chlorogenic acid and epicatechin respectively and the results of the analysis for several red wines were correlated with the classic Folin-Ciocalteu test for total polyphenols.

Biosensors for Polyphenols Based on Optical Transducers

Most often, the detection of phenols with optical biosensors targets the assessment of total phenolic content. In one low- cost, paper biosensor for the total phenolic content [11] a round filter paper was used as support for the deposition of tyrosinase and 3-methyl-2-benzothiazolinone hydrazone hydrochloride (MBTH). The detection relied on the formation of a coloured adduct by reaction between MBTH and quinones generated from phenols in a tyrosinase-catalysed reaction. For calibration, after the reaction of the biosensor with L-DOPA, a digital image was taken and analysed using the ImageJ software, attributing a numerical value for the intensity of the colour (related to concentration of L-DOPA). Wine samples were examined and compared with two other different techniques: Winder-Harris enzymatic spectrophotometrically assay in solution and the Folin-Ciocalteu assay. The results were in good agreement, especially with Winder-Harris assay.

An important parameter when it comes to red wines is astringency, associated with the capacity of phenols to bind to salivary proteins. Recently, a LSPR (Localized Surface Plasmon Resonance) biosensor was described, based on covalent binding of α-amylase (AMY) on Au nanodisks deposited on glass substrates and modified with carboxyl–ended thiols [13]. Besides covalent binding of AMY, the thiol had a second role, to reduce the non-specific interactions. The interaction of AMY with pentagalloyl glucose (PGG) was studied with plasmonic enhanced circular dichroism spectroscopy, as this technique can reveal very sensitively conformational changes. Upon binding PGG, the conformation of the protein changed and it was estimated that ˜12 PGG molecules bind to AMY. Moreover, various concentrations of dextran were also used to study the non-specific interactions with the biosensor, which turned out to be negligible. The biosensor was applied for determining astringency levels and polyphenol concentration expressed in terms of PGG equivalents for real wine samples.

A surface plasmon resonance (SPR) biosensor was developed using tyrosinase immobilized in polyacrylamide gel [14]. The fibre was constructed from plastic clad silica and the enzyme was entrapped over the fibre surface. Different phenolic compounds were tested measuring the change in transmittance with concentration and the sensitivities were in the following order:

$$S_{\text{catechol}} > S_{\text{m-cresol}} > S_{\text{4-chlorophenol}} > S_{\text{phenol}}$$

In general, detection limits achieved with optical biosensors for polyphenols were higher than 500 nM. After the electrochemical biosensors, the optical ones were the most researched for polyphenol detection in wines.

Resveratrol was detected with mVenus protein fragments fused at either end of the estrogen receptor α (ERα) binding domain and using the bimolecular fluorescence complementation (BiFC) technique. After transfection of the HeLa cells (where the replication of the plasmid-fusion gene can take place properly) a conformational change occurs and generation of Venus fluorophore can be correlated with the concentration of the added ligand. This biosensor was designed for the detection of estrogenic compounds. Resveratrol, being a phytoestrogenic compound, acts as a proper ligand for the complementation of protein fragments. This biosensing method was not tried in wine samples, but can present a great potential in identifying resveratrol, or other phenolic compounds that can exert estrogenic activity [44].

Fisetin, a flavonol, was detected with a double stranded DNA containing an abasic site [40]. Abasic sites result from the loss of a DNA base, when a deoxyribose residue remains in the strand. By embedding an abasic site of specific design in a double-stranded DNA, a binding pocket was formed where fisetin binds with high selectivity over other phenolic compounds such as morin, rutin, apigenin, kaempferol, myricetin, quercetin, luteolin, baicalin, naringenin, genistein, chrysin, and galangin [15]. Upon binding to the abasic site, the fluorescence of fisetin is greatly enhanced due to intramolecular transfer of proton in excited state. The intensity of the fluorescence emission spectra correlates with the concentration of fisetin. The detection limit (LOD) of this biosensor ($5x10^{-8}$ M) is one of the lowest encountered in biosensors for polyphenolic compounds. For comparison, detection limits typically achieved with enzyme biosensors for phenolic compounds, are in the micromolar range, as it can be observed in Table **1**.

Biosensors for Phenolic Compounds Based on Magnetic Transducers

Magnetic biosensors are quite rare and exploit the magnetoelastic effect, which refers to the change in magnetic susceptibility due to mechanical stress. A magnetoelastic biosensor was developed for the detection of tannic acid [45]. The authors coated the magnetoelastic ribbon with polyurethane to provide the sensors with amino groups for covalent binding of bovine serum albumin (BSA). The principle of the biosensor relies on the change in resonance frequency associated with the formation of the insoluble tannin-BSA complex (mass increase will decrease the frequency). A linear range between $6x10^{-4}$ M and $1.08x10^{-3}$ M, was reported with a limit of detection of $1x10^{-4}$ M and a sensitivity of -433.3 Hz M^{-3}. Additionally, the biosensor was used to follow the kinetics for the formation of the complex tannin-BSA.

Biosensors for Phenolic Compounds Based on Piezoelectric Transducers

Biosensors with detection by quartz crystal microbalance (QCM) were rarely used to analyse phenolic compounds from wine. Lipid coated and respectively gelatine-coated QCM chips were used to determine astringency (tannins) and polyphenols and the results when applied to real samples corresponded with the literature [46].

Coupling QCM with electrochemistry (EQCM) opened the gate to more complex investigations using biosensors. Although it is not a biological recognition element *per se*, but is considered a biomimetic compound [47], lutetium bisphthalocyanine ($LuPc_2$), was used to catalyse the oxidation of four antioxidants (caffeic acid, gallic acid, ferrulic acid and vanillic acid) on the surface of Pt/quartz electrode, in a simulated wine matrix. The mass sensitive quartz oscillators were used as working electrodes to register simultaneously cyclic voltammograms and massograms. Mass changes were observed for all the oxidants, during the catalytic oxidation of phenols: mass increasing in the case of vanillic acid (due to the electro-polymerization) and in the case of ferrulic acid, due to the high tendency to compensate solvated ions expulsion. On the other hand, mass decreasing is observed for ortho-diphenols, due to the incapacity to compensate solvated ions expulsion [48].

Detection of Biogenic Amines

Biogenic amines (melatonin, histamine, cadaverine, putrescine, tyramine, spermine, spermidine, phenylethylamine, agmatine, tryptamine, *etc.*, see structure in Fig. (**5**) are formed in fermented foods and beverages by the microbial decarboxylation of the corresponding free amino acids. At low concentrations they are essential for normal metabolism, but present a health risk at high concentrations [49], *e.g.* excess histamine can induce hypertension, headache and abdominal cramps, putrescine and agmatine are considered carcinogenic amines, and tyramine triggers headaches [50].

The European Food Safety Authority (EFSA) indicated histamine and tyramine as the most toxic biogenic amines [51] and has accepted health claims for melatonin concerning its effectiveness at reducing sleep onset latency [52]. Among the biogenic amines, melatonin is the most prevalent in wine because it is contained in grapes and synthesized during both the alcoholic and malolactic fermentations [53]. Numerous analytical procedures based on (bio) sensors have been developed as an alternative to the chromatographic analysis methods.

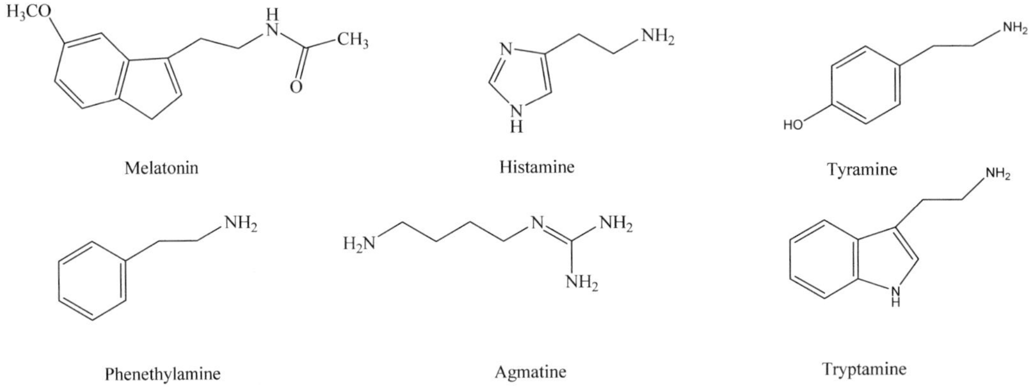

Fig. (5). Chemical structure of some biogenic amines.

Enzymatic biosensors employed for the detection of biogenic amines use either enzymes able to detect multiple amines simultaneously such as diamine oxidase (DAO, E. C. 1.4.3.22) and monoamine oxidase (MAO, E. C. 1.4.3.4) or enzymes that are selective for a specific substrate such as putrescine oxidase (PAO, E. C. 1.4.3.10). These enzymes catalyse the oxidation of (di)amines in presence of oxygen to aldehyde, ammonia and hydrogen peroxide according to the general reaction (1):

$$R\text{-}NH_2 + H_2O + O_2 \rightarrow R\text{-}CHO + NH_3 + H_2O_2 \quad \mathbf{(1)}$$

The principle of enzymatic biosensors for biogenic amines is based on the specific recognition of the substrate by the enzyme from all the chemicals found in wine coupled with the monitoring of a common participant in the enzymatic reaction (oxygen or hydrogen peroxide) using selective electrodes. One has to take into consideration the overall kinetic process: the enzymes catalyse the oxidation of all substrates in parallel and have different preferences for each amine in function of its respective relative affinity constant (K_M). The reaction rate obtained for a common participant in the enzymatic reaction with multiple substrates can be written based on Michaelis-Menten kinetics [54], as given in equation (2) in the case of two enzymatic substrates

$$-\frac{d\,O_2}{dt} = \frac{d[H_2O_2]}{dt} = k_2^1[ES^1] + k_2^2[ES^2] = \frac{V_{max}^1[S^1]K_M^2 + V_{max}^2[S^2]K_M^1}{K_M^1K_M^2 + K_M^2[S^1] + K_M^1[S^2]} \qquad \textbf{(2)}$$

where V_{max} represents the reaction rate, K_M the Michaelis-Menten constant, $[S^1]$ is the concentration of substrate 1, $[S^2]$ is the concentration of substrate 2, $[ES^1]$, the concentration of enzyme-substrate 1 complex and $[ES^2]$ the concentration of enzyme-substrate 2 complex.

The total content of biogenic amines is expressed as equivalent concentration of a single biogenic amine, but this approach may lead to underestimated results. For example, a DAO-biosensor was used for detection of biogenic amines in alcoholic beverages and the response was given as equivalent putrescine (the biogenic amine presenting maximum sensitivity). The results obtained with DAO-biosensor were compared with analyses carried out by GS-MS and the results were similar for white and red wines, but underestimated up to 50% for beer samples [55]. A comparison of biosensors based on different enzymes reveals that (i) DAO allows the estimation of the total biogenic amine content, (ii) MAO is useful for determination of tyramine, tryptamine and phenylethylamine and (iii) PAO is, as expected, selective towards putrescine [56].

Exploiting these differences in selectivity, an interesting analytical application described the combination of the responses obtained from different enzyme biosensors, and identification of different amines by pattern recognition using artificial neural network signal interpretation strategies. Thus, the combined response obtained from a MAO-biosensor and two biosensors based on DAOs with different origins allowed the development of an array for the simultaneous determination of the three biogenic amines (histamine, tyramine and putrescine) in 20 minutes that provided results with satisfactory correlation with conventional LC analyses [57]. Nevertheless, biosensor arrays are difficult to extend to a high number of substrates, cannot achieve the resolution of chromatographic techniques and provide unreliable results in sample with unknown amines. The main utility of biosensors is for the purpose of alarm systems, for screening a large number of samples to detect those that contain high concentrations of biogenic amines rather than for identification of each individual amine.

Besides enzymatic biosensors used for evaluation of total biogenic amine, affinity based biosensors were developed for specific detection of single biogenic amines, based on optical or electrochemical detection. A detection limit of 0.5 pg/mL was reported for a competitive electrochemical immunosensor for histamine with detection by Square Wave Voltammetry. While graphene was used as electrode modifier, the horseradish peroxidase (HRP)-catalysed deposition of insulating PDB film ensured effective and ultrasensitive signal transduction [58]. In a different work, sensitive determination of histamine content in fermented vegetable juices was achieved with an Optical Waveguide Lightmode Spectroscopy (OWLS)-based immunosensor, where a conjugate of histamine with bovine serum albumin was attached on the sensor surface with glutaraldehyde. The results compared well with an HPLC method. Still, immunosensor applications for biogenic amines in wine remain to be developed [59].

Artificial recognition elements have also been developed for the analysis of biogenic amines. The molecularly imprinted polymers (MIPs) have several advantages over antibodies such as: higher stability, inter-lot reproducibility and production without animal use. Computational models based on simulated cavity docking procedures and theoretical prediction of the polymer affinity for the analytes in function of the monomers and solvents are used in order reduce the experimental workload required for MIPs development and optimization [60]. A potentiometric sensor able to detect histamine at micromolar concentration in wine was developed by incorporation of MIPs in a membrane used for the construction of a ion-selective electrode [61]. Picomolar concentrations of melatonin were detected using amperometric detection based on a MIP from poly(ethylene-co-vinyl alcohol) that was deposed by evaporation of the solvent on the working electrode of an electrochemical sensing chip [62]. The MIPs are not only useful for sensor development, but also for analyte extraction and sample clean-up by solid-phase extraction prior to chromatographic analysis, a typical example being the histamine detection in wine samples [63].

Detection of Mycotoxins

The grape mycobiota present in vineyards depends on the geographic location and climate, but mycotoxigenic species such as *Aspergillus, Alternaria, Cladosporium and Penicillium* may be found [64]. Among the mycotoxins that could be found in wines are: ochratoxin A, aflatoxins (B1, B2, G1, G2), fumonisins (B1, B2, B3) and zearalenone (Fig. **6**) [65]. Dozens of mycotoxins in wines can be detected, usually at low concentrations that do not present a health risk, using highly sensitive analytical techniques based on liquid chromatography–tandem mass spectrometry [65]. Biosensors have also been developed for rapid sample screening.

The biosensors used for mycotoxins analysis are based on analyte recognition by antibodies, aptamers or MIPs with signal transduction and quantification by different electrochemical, optical or mass sensitive techniques. Paramagnetic microbeads modified with a monoclonal antibody specific for ochratoxin A were mixed with wine sample and subsequently separated using a magnetic field removing thus the sample matrix. Detection of ochratoxin A was achieved by deposition of the magnetic beads on a screen-printed electrode using a magnet placed underneath the electrode, followed by a competition between the ochratoxin A in the sample and peroxidase-conjugated ochratoxin [66]. An alternative to the antibodies is represented by the aptamers (oligonucleotide short chains able bind to the analyte, obtained using systematic evolution of ligands by exponential enrichment SELEX technique). The main advantages of the aptamers are the easy and reproducible production (once the exact structure was found) and their good operational stability. The use of aptamers in analytical chemistry is based on similar protocols as with immunosensors. For example, a direct competitive format in which the analyte ochratoxin competed with alkaline phosphatase-labelled ochratoxin for the binding to the DNA aptamer immobilized on magnetic beads was used for wine analysis using disposable screen-printed electrodes [67]. Another possibility of detection is based on quantification of conformational changes in aptamer structure after its interaction with the analyte. Thus, an aptamer specific to ochratoxin A was marked with alkaline phosphatase and immobilized on gold electrodes using a short complementary nucleotide with a thiol moiety. The interaction of the ochratoxin A with the enzymatically-labelled aptamer leads to a reduced enzymatic activity. Differential pulse voltammetry shows a decrease of the substrate conversion (4-nitrophenyl phosphate to 4-nitrophenol) that is electrochemically detected at 225 mV. Interestingly, the biosensor can be regenerated by a short potential pulse (2 minutes at -400 mV) before reuse [68].

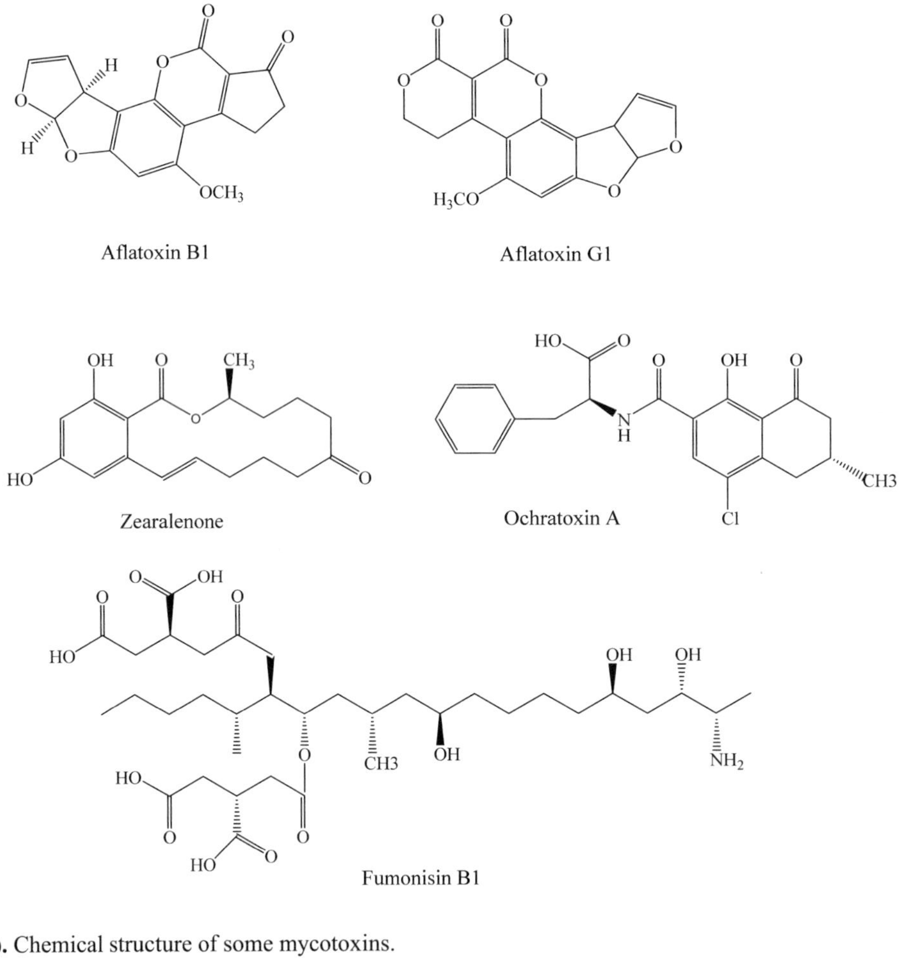

Fig. (6). Chemical structure of some mycotoxins.

The need of labelled analytes or antibodies/aptamers is a major disadvantage of the molecular recognition assays due to high costs and increased complexity/working time. An alternative is to use label-free detection methods such as electrochemical impedance spectroscopy or surface plasmon resonance. An aminated aptamer selective to ochratoxin was immobilized on screen-printed carbon electrode surface modified with an electropolymerized film of polythionine and iridium oxide nanoparticles. Electrochemical impedance spectroscopy allows the detection of analyte immobilised on the electrode surface by quantification of the modifications in Nyquist spectra recorded in standard solutions that contain $[Fe(CN)_6]^{-3/-4}$ redox probe [69].

There are numerous immobilization strategies, and nanoparticles may be used for an improved limit of detection. A thiolated aptamer specific to ochratoxin was immobilized on gold nanoparticles using Au-S bonding. The aptamer-ochratoxin interaction induces a conformational change of the aptamer that restricts the transfer of the $[Fe(CN)_6]^{-3/-4}$ redox probe and leads to an increase of the charge tra-nsfer resistance measured with electrochemical impedance spectroscopy (Fig. **7**) [70].

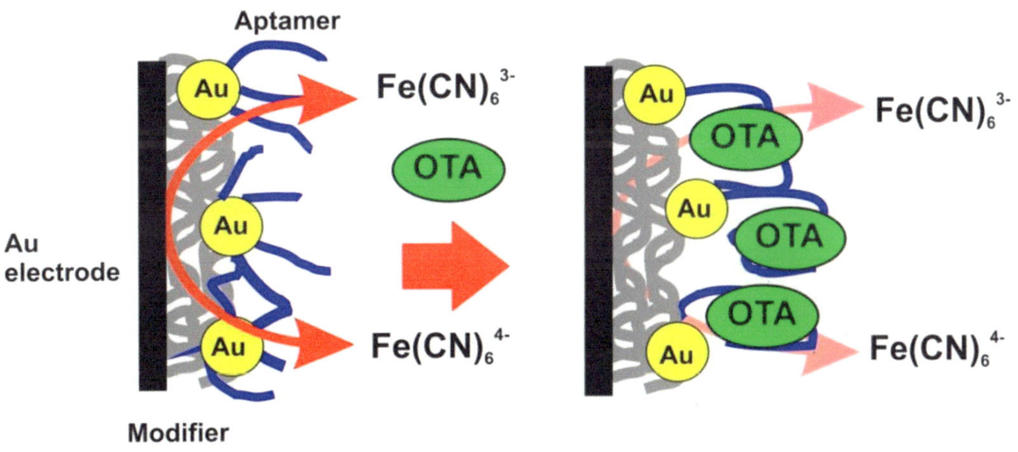

Fig. (7). Detection principle of the impedimetric aptasensor: ochratoxin induces a modification of the aptamer structure that restrict the $[Fe(CN)_6]^{-3/-4}$ redox probe interaction with the electrode surface. OTA= ochratoxin A. [Reproduced from reference [70], under the terms and conditions of the Creative Commons Attribution license (http://creativecommons.org/licenses/by/4.0)].

Another label-free detection strategy is to extract only the analyte from the wine sample and oxidize it at high potential in standard solutions avoiding the electrochemical interferences from the sample matrix. Thus, a glassy carbon electrode was modified with multiwalled carbon nanotubes and a polypyrrole-based molecularly imprinted polymer for ochratoxin A. The modified electrodes were incubated in wine during 15 min and, subsequently, transferred into a voltammetric cell containing phosphate buffer solution where the differential pulse voltammetry of the captured ochratoxin A was performed from +0.45 V to +1.1 V [71].

Optical measurement strategies were also developed for ochratoxin A using aptamers as biological recognition elements. An aptamer modified with carboxyfluorescein was mixed with the wine together with single-walled carbon nanohorns as fluorescence quencher. In the absence of ochratoxin the aptamer is adsorbed on carbon nanohorns and the fluorescence is quenched, while in the presence of analyte carbon nanohorns will desorb and the fluorescence is restored [72]. Other techniques reported for ochratoxin detection are based on surface plasmon resonance using aptamers for analyte binding on the chip [73] and optical waveguide lightmode spectroscopy based on antibodies [74].

Rapid visual assays were developed for medical analysis (*e.g.* pregnancy home test) and subsequently adapted for food applications using adapted antibodies. These rapid tests have the advantage of simplicity, being easy to use by everyone, and do not require equipment. Different formats have been developed (such as: lateral flow or flow-through) and the visualization reaction is based on various strategies such as: colloidal gold or enzymes (peroxidase). For intensely coloured samples (such as red wine) the direct analysis using visual test is not possible and clean-up columns are used in tandem with the immunoassay as depicted in Fig. (**8**) [75].

One major advantage of the biosensors is the simplification of the analytical procedures and reduced workload in comparison with classical methodologies based on chromatographic methods. Thus, while numerous sample clean up and extraction steps (liquid-liquid extraction, solid phase purification, head space separations, *etc.*) are necessary prior to the chromatographic analysis, the biosensors analyse the sample either directly [68] or after a simple dilution with the working buffer to adjust the pH according to the enzyme requirements [76]. Sample preparation is a bottleneck and some biosensors studies still required liquid-liquid extraction of analyte prior to analysis (*e.g.* the overnight extraction with toluene of ochratoxin A followed by another clean-up with a buffer prior to actual analysis [73]. Such procedures are too long for a practical biosensor assay, which tends to be developed for rapid, on site and cost-effective analysis.

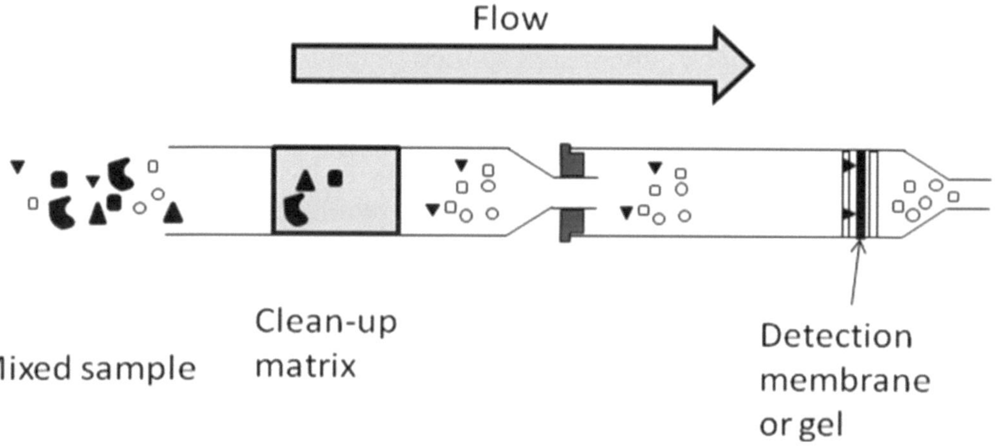

Fig. (8). The working principle of a clean-up column in combination with a lateral flow detection. [Reproduced from reference [75] under the terms and conditions of the Creative Commons Attribution license (http://creativecommons.org/licenses/by/3.0/)].

Detection of Allergens

Food allergens are substances that cause adverse reactions in the human body, involving IgE-type antibodies. In wine, exogenous allergens such as casein from milk, egg proteins (*e.g.* ovalbumin, lysozyme) and sulphites can cause allergic reactions in hypersensitive individuals. Their presence is related to treatments applied along the wine-making process to improve the stability of wine and to protect it from microbial spoilage. For example, casein from milk and various products based on egg-proteins are used for fining, to improve the clarity of musts and wine and remove colloidal particles in suspension. Lysozyme is added also to control the malolactic fermentation, and as an antimicrobial agent to partially replace sulphite. Sulphite is used as an antibacterial agent and to prevent oxidation ("browning") of white wines. Several comprehensive reviews have been recently published on allergenic proteins in oenology [77], biosensing strategies for food allergens [78] and recent advances in the field of food allergen detection using biosensors [79, 80]. There are a variety of biosensors for allergen detection developed so far, based on antibodies, aptamers or MIPs, coupled in general with electrochemical or optical detection. Recent years have witnessed the development of novel detection strategies with exquisite sensitivity obtained by incorporating nanomaterials at all levels in the biosensors: the basic transducer, the recognition element and the amplification step. Despite these impressive achievements, the number of biosensors applied to food and particularly to wine analysis remains quite small.

Both electrochemical and optical biosensors have been developed for the detection of lysozyme from egg-white [81, 82] while an immunosensor was described for ovalbumin [83]. Detection of sulphite was achieved with various enzymatic biosensors [84 - 88]. For casein, immunosensors with detection by SPR have been applied to the analysis of milk samples but remain to this date unproven in wines [89]. Table **2** summarizes data on representative biosensors for allergen detection in wines and some of these biosensing strategies are discussed in more detail below.

Table 2. Example of biosensors for allergen detection in wines.

Bioactive compound	Biorecognition element	Details of detection/method	Figures of merit	Reference
Lysozyme	aptamer	Direct detection/EIS	**APT 1:** LR: 0.025 - 1.75μM LOD: 725nM **APT2:** LR: 0.025 - 1 μM LOD: 31.7nM	[81]
	aptamer	Sandwich aptamer/lysozyme/ALP-Ab/ Electrochemical Detection of 1-naphtol by DPV	LR: 5 fM - 5 nM LOD: 4.3 fM	[90]
	aptamer	Direct detection/SPR	LR: 0.05–1mg/mL LOD:0.035 μg/mL	[82]

(Table 2) contd.....

Bioactive compound	Biorecognition element	Details of detection/method	Figures of merit	Reference
Ovalbumin	polyclonal anti-OVA antibody/SPR	Direct detection/SPR	LR:0.03 -0.2 µg/mL	[83]
Sulphites	Sulphite oxidase and cytc/PASA LBL assembly	Au electrode/ Amperometric detection of cyt c oxidation at 0.1 V	LR: 1 -60 µM	[91]
Sulphites	Sulphite oxidase cross- linked with glutaraldehyde	Oxidation of H_2O_2 produced in enzymatic reaction, mediated by TTF, amperometric detection at +0.2 V	LR: 9.9–82.6 µM	[84]
	Several layers of Sulphite oxidase	SPCE with Os polymer wired sulfite oxidase/ Amperometry at −0.1 V in a flow system	LR: 0.5-100 µM LOD: 0.5 µM	[87]
	Starkeya novella sulfite dehydrogenase (SDH), horse heart cyt *c*	SAM of 11-mercaptoundecanol, gold electrode /Amperometric, Oxidation of cyt c	LR: 1-6 µM LOD:44 pM	[88]
	PPO obtained from crude extract of sweet potato (*Ipomoea batatas* immobilized on AuNps	Inhibition of PPO by sulfite; GCE /MWCNT/ Au NPs/ PAH electrode/ detection of catechol by DPV	LR: 0.5 - 22 µmol L^{-1} LOD:0.4 µmol L^{-1}	[86]

ALP: alkaline phosphatase; ALP-Ab: labelled antibody.SPE: screen-printed electrode; SAM: self-assembled monolayer. NSA: non-specific adsorption;BSA: bovine serum albumin. PPO: polyphenol oxidase. poly(allylamine hydrochloride) (PAH). MWCNT: multi-wall carbon nanotubes. LBL: layer-by-layer; cyt c: cytochrome c; PASA: polyaniline sulfonic acid

Lysozyme is a small protein of 14 kDa, also called "muramidase" as it is an enzyme catalysing the hydrolysis of glycosidic bonds between N-acetyl muramic acid and N-acetyl D-glucosamine in peptidoglycan in the cellular walls of Gram-positive bacteria. In wine industry, lysozyme from egg white is used as an antimicrobial agent, to partially replace sulphite and in egg-protein based products used for fining. The recommendation of the International Organization of Vine and Wine (OIV) specifies a maximum amount of lysozyme allowed for use for these purposes of 500 mg/L wine. Lysozyme was rarely detected in bottled wines on the market and even in those cases its content was quite small, in the low ppm range [77]. However, the potential to induce allergic effects in sensitive individuals is still considered to be important and labelling rules are in place in the European Union [92]. Any wine where lysozyme was detected using a method characterized by a detection limit of 0.25 mg/L and a limit of quantification of 0.5

mg/L or lower must include the allergen warning on the label. The official methods for lysozyme detection in wines are based on chromatography with UV or fluorescent detection and on capillary electrophoresis [93], in the same time analysis by ELISA is largely preferred due to availability of several commercial kits, fast and sensitive analysis, convenience for routine testing and less sophisticated analytical equipment.

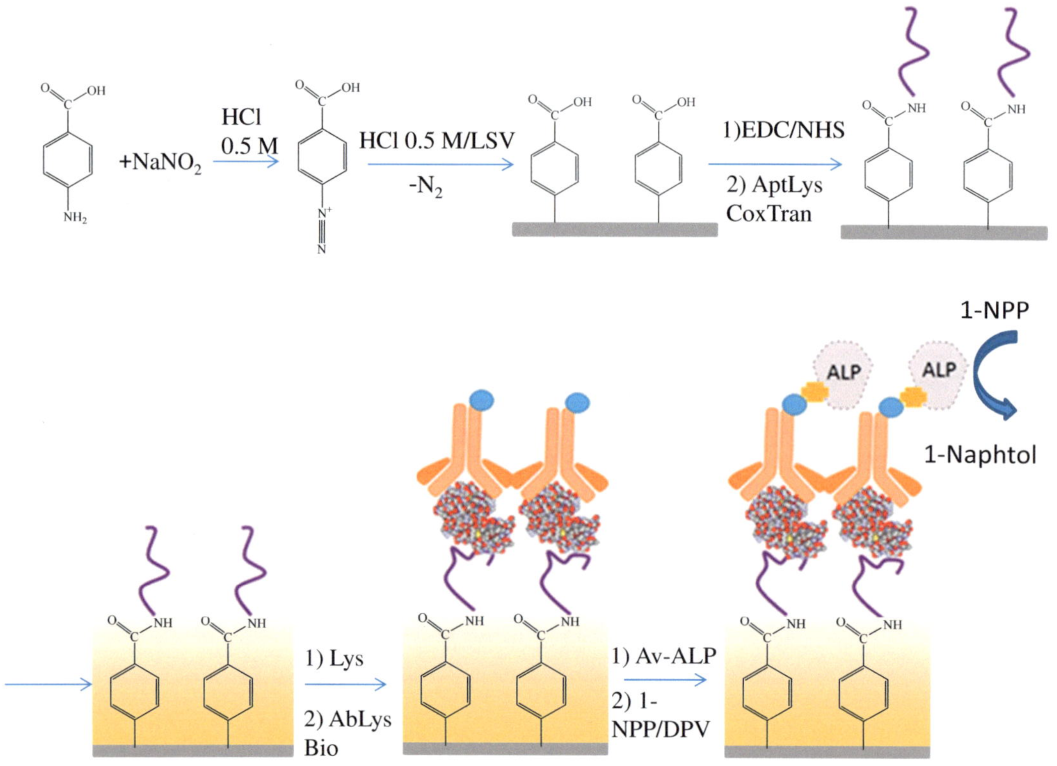

Fig. (9). Schematic illustration of a sandwich-type assay using an electrochemical aptasensor for lysozyme and an enzyme-labelled antibody. Reprinted from [90], by permission of the Royal Society of Chemistry, 2015.

Lysozyme was one of the preferred target molecules in biosensors and various analytical approaches have been described in the literature based on specific recognition elements such as antibodies, aptamers and MIPs. Despite the variety of biosensors developed for lysozyme, only a very small number of reports included practical applications in food and beverages. One paper compared the performances of two aptamers obtained by two different selection methods, - SELEX and capillary electrophoresis, respectively - when used for the development of label-free impedimetric aptasensors [81]. In a second work by the same research group, a sandwich-based assay was shown to lead to very sensitive detection of lysozyme, reaching a detection limit of 5 fM [90]. The lysozyme aptamer was immobilised on the electrode. After binding lysozyme, an antibody was introduced which binds also to lysozyme, making thus a sandwich (Fig. **9**). The antibody was labelled with alkaline phosphatase and detection was achieved by DPV, based on the electrochemical oxidation of 1-naphtol, the product of enzymatic reaction.

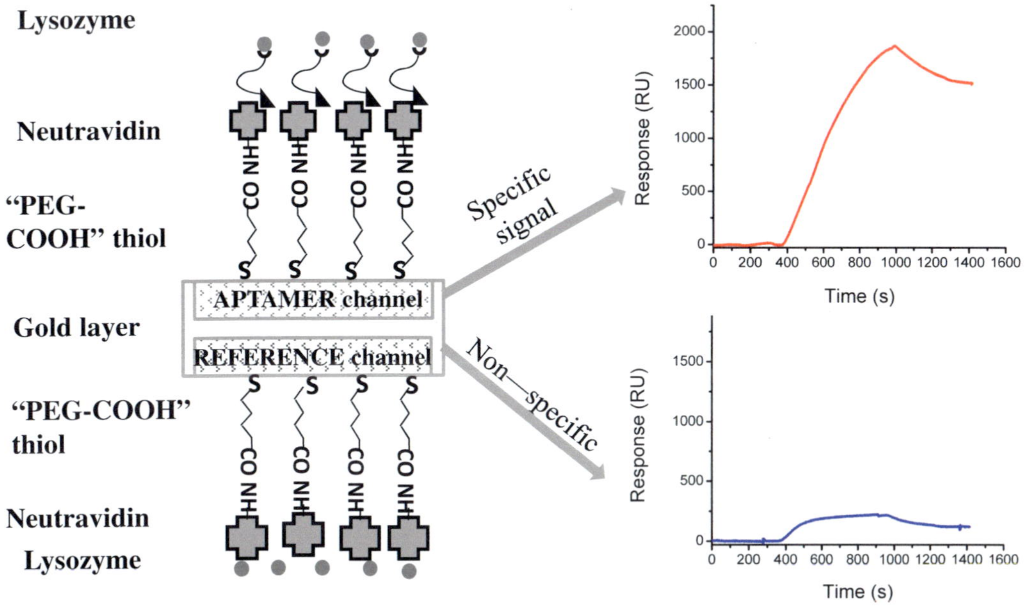

Fig. (10). Design of the aptasensor for lysozyme detection in wines. [Reprinted from [82] with permission from Elsevier, 2015].

Besides these electrochemical biosensors, optical detection by Surface Plasmon Resonance was also used with an aptasensor for lysozyme detection in wines [86].

A gold SPR chip was covered with a self-assembled monolayer of carboxyl-ended thiols, neutravidin was covalently attached by amine coupling to the SAM and a biotinylated aptamer was immobilised by affinity binding (Fig. **10**).

Non-specific adsorption effects were mitigated by the use of a thiol including six ethylene glycol groups, while a simple sample pre-treatment strategy allowed to minimize interferences due to the complex wine matrix. The results obtained for several wines, in which lysozyme was intentionally added were compared to those provided by a chromatographic method. The biosensor can be used not only for detection of allergen lysozyme but also to study its binding to various phenolic compounds in wine.

A direct assay for ovalbumin in wines has been proposed recently based on an immunosensor that uses a polyclonal anti-OVA antibody as biorecognition element, with detection by SPR. The presence of allergenic ovalbumin in fined wines can be detected at levels between 0.03 and 0.2 μg/mL and the results provided by the biosensor were correlated to those obtained by a liquid chromatography mass spectrometry (LC-MS) method [83].

Sulphites

Sulphur dioxide is an additive widely used in winemaking as an antibacterial agent and to prevent oxidation of wines. However, sulphur dioxide may cause allergic reactions, which include nausea, diarrhea, gastric irritation or asthmatic attacks. For this reason labels on wine bottles must include the appropriate allergen warning pertaining to sulphites. The maximum allowed level of sulphur dioxide in wine range from 160 mg. L^{-1} in red wines in Europe to 350 mg L^{-1} in the US. Currently, the standard method of the AOAC relies on distillation of sulphur dioxide and iodometric titration [94]. Other methods for the determination of sulfite have been developed, relying on spectrophotometry [95] capillary electrophoresis [96] or chromatography [97], these methods still require significant time per analysis, skilled personnel or expensive equipment. Consequently, the development of biosensors for sulphites detection is of great interest allowing a fast, selective and accurate determination without significant sample preparation or expensive equipment.

Biosensors for sulphites are based on enzymes as specific recognition elements with electrochemical detection. Various approaches have been described in literature, ranging from direct detection using sulphite oxidase (EC 1.8.3.1) (i) or sulphite dehydrogenase (EC 1.8.2.1) (ii) to, alternatively detection based on the principle of the enzymatic inhibition of polyphenol oxidase (EC 1.14.18.1) by sulphite (iii).

Most biosensors for sulphites are based on sulphite oxidase immobilised on a transducer and catalysing the oxidation of sulphite to sulphate:

$$2SO_3^{2-} + 2H_3O^+ + 2O_2 = 2SO_4^{2-} + H_2O_2 \quad \textbf{(3)}$$

Sulphite oxidase is a homodimeric metalloprotein, which contains two cofactors, molybdenum and a cytochrome b_5-type heme, located in two different domains of the protein. Oxidation of sulfite takes place at the molybdenum site, simultaneously with reduction of Mo (VI) to Mo(IV). Next, electrons are transferred intramolecularly from Mo(IV) to the heme b5 situated in another domain of the protein. Thus, Mo(VI) is regenerated.

Either the consumed oxygen, the produced hydrogen peroxide or the heat generated by the enzymatic reaction can be measured to detect sulphite. As summarized by Pundir *et al.* [85] enzymatic biosensors for sulphite have response time between 2 and 900 s, can reach detection limits of 0.2-200 µM and display best analytical performances at pH 6.5-9.0 and temperatures between 25 and 40 °C. Moreover, these sulfite biosensors are appropriate for measurements in food and beverages, can be reused many times and some are stable up to 240 days.

One of the preferred approaches for sulphite measurement relies on the amperometric detection of H_2O_2, typically by direct electrochemical oxidation at potentials around 0.6 V:

$$H_2O_2 = O_2 + 2H^+ + 2e^-. \quad \textbf{(4)}$$

The electrical potential required for H_2O_2 oxidation being rather high, many electroactive species in the samples might be oxidized as well and these electrochemical interferences could drastically affect the accuracy of the measurements. To minimize interferences, various electron transfer mediators have been proposed such as tetrathiafulvalene [98], Os redox polymer [87] *etc.* Additionally, direct electron transfer communication between the enzyme and the electrode is expected to be more efficient. For example, sulphite oxidase was co-immobilized together with cytochrome c by the layer-by-layer technique using an anionic polyelectrolyte. Besides providing stability of the multilayer coating by electrostatic interactions, the polyelectrolyte also facilitated the electron transfer [91].

General Considerations-allergen Detection in Wines

As for any real sample applications, biosensors devoted to wine analysis face challenges related to non-specific adsorption and interferences due to complex sample matrix. The challenges are particularly important in the case of red wines.

Several strategies have been reported for interference minimization: the use of permselective membranes such as Nafion [99], measurement in parallel at +600 mV with two electrodes (one without enzyme for evaluation of interference and the biosensor for analytical signal detection [55]) and use of mediators to decrease the working overpotential, such as tetrathiafulvalene able to quantify hydrogen peroxide at +250 mV [100]. Non-specific adsorption was typically blocked with bovine serum albumin (BSA). For Au-based sensor surfaces, a common approach was to coat the transducer with a protective layer with anti-fouling properties: *e.g.* a thiol containing ethylene-glycol groups [101].

Red wines are known to have high contents of polyphenolic compounds, which might affect biosensor performance in several ways. For example, in enzymatic biosensors, polyphenols inhibit sulphite oxidase and have antioxidant activity. Direct electrochemical oxidation of polyphenols cause large electrochemical interferences, affecting the analytical signal for sulphite. In aptamer-based assays, large differences in the analytical signal recorded by SPR were observed for lysozyme in wine, where is bound to polyphenols as compared to standard solutions of lysozyme. Minimized interferences in all biosensors for allergens were granted by combined optimisation of sample pre-treatment and of the operational conditions. For example, sample pre-treatment using salt and detergent [101] was used to break the bonds between lysozyme and phenolic compounds in wine. Performing an appropriate sample dilution is a simple and effective way of mitigating interferences, *e.g* a 500 fold sample dilution was used when analysing sulphites in wines with an amperometric, enzymatic biosensor. Electrochemical interferences have been avoided by applying low potentials close to 0 V. To increase the effectiveness of electron transfer and improve selectivity of the biosensing method, electron transfer mediators or wiring redox enzymes to electrode surfaces have been envisaged.

Sensory Evaluation

The sensory evaluation of the wine is typically carried out by one or a group of experts, but this is expensive and irreproducible. In consequence, multiparametric analytical systems (electronic nose, tongue or eyes) could be useful complementary tools to sensory human panels. The sensory evaluation is carried out with a combination of different sensors, each one with different sensitivity towards the wine components, and the brute analytical results are assessed using numerous chemometric tools based on multivariate calibration (such as principal component analysis PCA) or supervised pattern recognition (such as artificial neural networks ANNs) for pattern recognition [102 - 105]. The overall process is depicted in Fig. (**11**) for an e-nose and consists in different steps: recording of the analytical signals form the sensors, a computational stage to transform the analytical signals into mathematical data (spreadsheets, clusters, vectors, *etc.*) that can be further used for prediction, generally classification, regression or clustering [104].

Most systems are based on chemical sensors for voltammetric, conductometric or impedimetric measurements, but also biosensors have also been included to obtain a more selective signal (*e.g.* a group of biosensors each one specific for one or a group of analytes can be regarded as a bioelectronic tongue) [103].

The most recent review gathering biosensors for taste sensors, including recent biosensors with application in wine, describes several potentiometric, voltammetric, amperometric and impedance biosensors [105]. One complex analytical platform used to determine either tannin, anthocyanin or total phenolic content, consists of 4 coupled sensors, capable to register, in the same time and from the same sample, different phenolic indexes (without any dilution of the wine) and afterwards, the Artificial Neural Networks (ANNs) are used to correlate the results and to determine the plot score of the different components. The biological elements for the two biosensors were: laccase and tyrosinase, and the other two sensors were: an unmodified and a copper nanoparticle modified electrode.

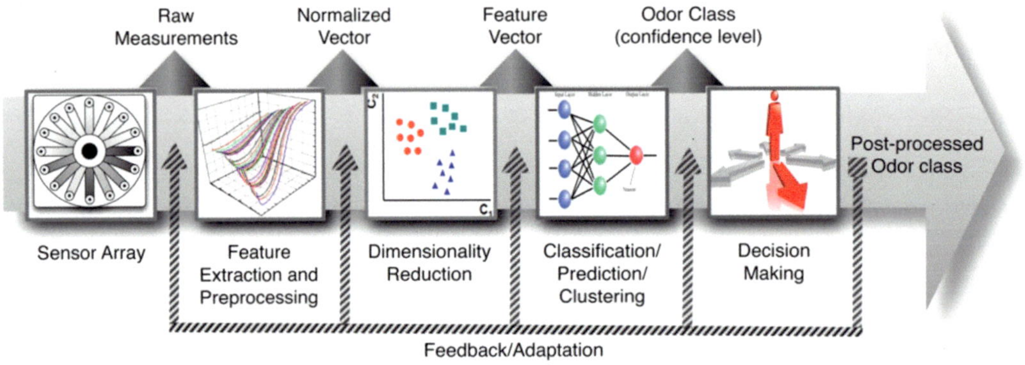

Fig. (11). Typical steps involved in signal processing of an electronic nose for wine sensory evaluation. [Reproduced from reference [104] under the terms and conditions of the Creative Commons Attribution license (http://creativecommons.org/licenses/by/4.0)].

An array composed of histidine-containing, metal-binding short peptides (WEEHEE, WAHEDEFF, and FHFPHHF), divalent metals ions (Cu^{2+} and Ni^{2+}) and colorimetric pH indicators (Chrome Azurol S, Bromopyrogallol Red, and Pyrocatechol Violet) was used for the pattern-based sensory discrimination of Cabernet Sauvignon wines. The displacement of the indicator by the wine tannins causes a spectral change measured by spectroscopy. The peptide-based receptors are sensitive to differences of tannin content of the Cabernet Sauvignon wines made from grapes harvested at different post-ripening times [106].

Even biosensors based on whole microbial cells were developed for the selective measurement of ethanol and glucose in mixtures: amperometric sensors with *Gluconobacter oxidants* were sensitive to both substrates, while the biosensors based on *Pichia methanolica* oxidized only ethanol [107].

An important advantage of the electronic noses is their ability to remotely monitor the fermentation in each fermentation tanks because the winemaking process may proceed differently due to microbiological and chemical factors. Such a demonstrative control system was developed for two sensors (pH and temperature) to test the communication and management hardware, but the equipment is easily scalable to multiple detectors (Fig. **12**) [108].

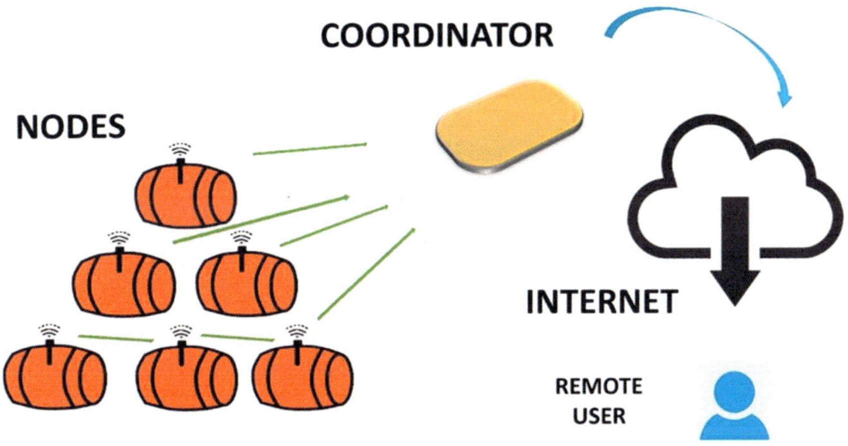

Fig. (12). Remote control system for wireless control of sensors in different fermentation tanks. [Reproduced from reference [108]; under the terms and conditions of the Creative Commons Attribution license (http://creativecommons.org/licenses/by/4.0/)].

CONCLUSION

Biosensors have appropriate analytical performances for wine analysis and the scientific literature has proven that most technical issues have been successfully solved. With the increased exploration of nanomaterials the limits of sensitive detection with biosensors will continue to be pushed in the coming years. Some biosensing approaches need further tuning of experimental conditions and sample pre-treatment, to be adequate for analysis of red wines. Biosensors developed for bioactive compounds in wines do not intend to eliminate classic chromatographic methods (especially with regards for safety control of toxins), but represent a viable alternative suitable for low-cost monitoring of wine production, safety screening for allergens and toxins and are valuable for a better management of wineries. The major hurdle in form of commercial application remains a wider market acceptance essential to gain access to a sufficient number of customers in order to be able to decrease the cost per analysis to competitive levels and reach economic viability.

CONFLICT OF INTEREST

The authors confirm that they have no conflict of interest to declare for this publication.

ACKNOWLEDGEMENTS

This work was supported by grants of the Romanian National Authority for Scientific Research, CNDI-UEFISCDI, PN-II-PT-PCCA-2013-4-0297. (BB), PN-II-ID-PCE-2011-3-1076, (ISH) and PN-II-PT-PCCA-2011-3.1-1809 (AV).

REFERENCES

[1] Cancer Institute Dictionary of Cancer Terms 2015. Available from: http://www.cancer.gov/publications/dictionaries/cancer-terms?CdrID=703278.

[2] Guaadaoui A, Benaicha S, Elmajdoub N, Bellaoui M, Hamal A. What is a bioactive compound? a combined definition for a preliminary consensus. Int J Nut Food Sci 2014; 3(3): 174-9.

[3] Timur S, Pazarlioğlu N, Pilloton R, Telefoncu A. Detection of phenolic compounds by thick film sensors based on Pseudomonas putida. Talanta 2003; 61(2): 87-93. [http://dx.doi.org/10.1016/S0039-9140(03)00237-6] [PMID: 18969166]

[4] Kim SK, Kwen HD, Choi SH. Fabrication of a microbial biosensor based on QD-MWNT supports by a one-step radiation reaction and detection of phenolic compounds in red wines. Sensors (Basel) 2011; 11(2): 2001-12. [http://dx.doi.org/10.3390/s110202001] [PMID: 22319395]

[5] Litescu SC, Eremia SA, Bertoli A, Pistelli L, Radu GL. Laccase-nafion based biosensor for the determination of polyphenolic secondary metabolites. Anal Lett 2010; 43(7-8): 1089-99. [http://dx.doi.org/10.1080/00032710903518518]

[6] Gomes S, Rebelo MJ. A new laccase biosensor for polyphenols determination. Sensors (Basel Switzerland) 2003; 3(6): 166-75. [http://dx.doi.org/10.3390/s30600166]

[7] Singh S, Jain DV, Singla ML. Sol-gel based composite of gold nanoparticles as matix for tyrosinase for amperometric catechol biosensor. Sens Actuators B Chem 2013; 182: 161-9. [http://dx.doi.org/10.1016/j.snb.2013.02.111]

[8] Franzoi AC, Vieira IC, Scheeren CW, Dupont J. Development of quercetin biosensor through immobilizing laccase in a modified beta-cyclodextrin matrix containing Ag nanoparticles in ionic liquid. Electroanalysis 2010;22(12);1376-8535. Wang XL, Chen R, Sun L, Yu ZY. Study on the antioxidant capacities of four antioxidants based on oxidizing guanine in a composite membrane. Int J Electrochem Sci 2014; 9(12): 6834-42.

[9] Vasilescu I, Eremia SA, Kusko M, Radoi A, Vasile E, Radu GL. Molybdenum disulphide and graphene quantum dots as electrode modifiers for laccase biosensor. Biosens Bioelectron 2016; 75: 232-7.
[http://dx.doi.org/10.1016/j.bios.2015.08.051] [PMID: 26319166]

[10] Granero AM, Fernandez H, Agostini E, Zon MA. An amperometric biosensor for trans-resveratrol determination in aqueous solutions by means of carbon paste electrodes modified with peroxidase basic isoenzymes from Brassica napus. Electroanalysis 2008; 20(8): 858-64.
[http://dx.doi.org/10.1002/elan.200704109]

[11] Arciuli M, Palazzo G, Gallone A, Mallardi A. Bioactive paper platform for colorimetric phenols detection. Sens Actuators B Chem 2013; 186: 557-62.
[http://dx.doi.org/10.1016/j.snb.2013.06.042]

[12] Sanz J, de Marcos S, Galbán J. Autoindicating optical properties of laccase as the base of an optical biosensor film for phenol determination. Anal Bioanal Chem 2012; 404(2): 351-9.
[http://dx.doi.org/10.1007/s00216-012-6061-0] [PMID: 22562544]

[13] Guerreiro RL, De Freitas V, Sutherland DS, Goreti M, Salesa F. SPR based studies for pentagalloyl glucose binding to α-amylase. Procedia Eng 2012; 47: 498-501.
[http://dx.doi.org/10.1016/j.proeng.2012.09.193]

[14] Singh S, Mishra SK, Gupta BD. SPR based fibre optic biosensor for phenolic compounds using immobilization of tyrosinase in polyacrylamide gel. Sens Actuators B Chem 2013; 186: 388-95.
[http://dx.doi.org/10.1016/j.snb.2013.06.034]

[15] Xu SJ, Shao Y, Ma K, *et al.* DNA abasic site-based aptamer for selective fluorescence light-up detection of fisetin by excited-state intramolecular proton transfer. Sens Actuators B Chem 2012; 171: 666-71.
[http://dx.doi.org/10.1016/j.snb.2012.05.051]

[16] Gutowski M, Kowalczyk S. A study of free radical chemistry: their role and pathophysiological significance. Acta Biochim Pol 2013; 60(1): 1-16.
[PMID: 23513192]

[17] Tetens I. EFSA Panel on Dietetic Products, Nutrition and Allergies (NDA); Scientific Opinion Part I on the substantiation of health claims related to various food (s)/food constituent (s) not supported by pertinent human data (ID 411, 559, 1174, 1184, 1197, 1380, 1409, 1656, 1667, 1670, 1763, 1767, 1806, 1884, 1908, 1997, 2141, 2159, 2243, 2244, 2325, 2331, 2333, 2336, 2652, 2717, 2727, 2752, 2788, 2861, 2870, 2885, 2894, 3077, 3101, 3516, 3595, 3726, 4252, 4288, 4290, 4406, 4509, 4709) pursuant to Article 13 (1) of Regulation (EC) No 1924/2006. European Food Safety Authority 2011.

[18] USDA. Oxygen Radical Absorbance Capacity (ORAC) of Selected Foods, Release 2. 2010 [updated 16th May 2012; cited 2015 10th October] Available from: http://www.ars.usda.gov/Services/docs.htm?docid=15866

[19] Pandey KB, Mishra N, Rizvi SI. Myricetin may provide protection against oxidative stress in type 2 diabetic erythrocytes. Z Naturforsch, C, J Biosci 2009; 64(9-10): 626-30. [http://dx.doi.org/10.1515/znc-2009-9-1004] [PMID: 19957428]

[20] Carrizzo A, Forte M, Damato A, *et al.* Antioxidant effects of resveratrol in cardiovascular, cerebral and metabolic diseases. Food Chem Toxicol 2013; 61: 215-26. [http://dx.doi.org/10.1016/j.fct.2013.07.021] [PMID: 23872128]

[21] Levantesi G, Marfisi R, Mozaffarian D, *et al.* Wine consumption and risk of cardiovascular events after myocardial infarction: results from the GISSI-Prevenzione trial. Int J Cardiol 2013; 163(3): 282-7. [http://dx.doi.org/10.1016/j.ijcard.2011.06.053] [PMID: 21737162]

[22] Arranz S, Chiva-Blanch G, Lamuela-Raventos RM, Estruch R. Wine polyphenols in the management of cardiovascular risk factors. In: Preedy VR, Zibadi S, Eds. Polyphenols in human health and disease. Watson, RR: Academic Press 2013; pp. 993-1003.

[23] Sun L, Fan H, Yang L, Shi L, Liu Y. Tyrosol prevents ischemia/reperfusion-induced cardiac injury in H9c2 cells: involvement of ROS, Hsp70, JNK and ERK, and apoptosis. Molecules 2015; 20(3): 3758-75. [http://dx.doi.org/10.3390/molecules20033758] [PMID: 25723850]

[24] Schneider Y, Vincent F, Duranton B, *et al.* Anti-proliferative effect of resveratrol, a natural component of grapes and wine, on human colonic cancer cells. Cancer Lett 2000; 158(1): 85-91. [http://dx.doi.org/10.1016/S0304-3835(00)00511-5] [PMID: 10940513]

[25] Vang O, Ahmad N, Baile CA, *et al.* What is new for an old molecule? Systematic review and recommendations on the use of resveratrol. PLoS One 2011; 6(6): e19881. [http://dx.doi.org/10.1371/journal.pone.0019881] [PMID: 21698226]

[26] Jeong JH, Jung H, Lee SR, *et al.* Anti-oxidant, anti-proliferative and anti-inflammatory activities of the extracts from black raspberry fruits and wine. Food Chem 2010; 123(2): 338-44. [http://dx.doi.org/10.1016/j.foodchem.2010.04.040]

[27] Bijak M, KoLODziejczyk-Czepas J, Ponczek MB, Saluk J, Nowak P. Protective effects of grape seed extract against oxidative and nitrative damage of plasma proteins. Int J Biol Macromol 2012; 51(3): 183-7. [http://dx.doi.org/10.1016/j.ijbiomac.2012.05.009] [PMID: 22584076]

[28] Habib S, Ahmad S, Dixit K, Moinuddin , Ali A. Peroxynitrite modified DNA may be an antigenic trigger for antibodies in various cancers of gynecologic origin. Hum Immunol 2013; 74(10): 1239-43. [http://dx.doi.org/10.1016/j.humimm.2013.07.015] [PMID: 23911359]

[29] Olas B, Nowak P, KoLODziejczyk J, Ponczek M, Wachowicz B. Protective effects of resveratrol against oxidative/nitrative modifications of plasma proteins and lipids exposed to peroxynitrite. J Nutr Biochem 2006; 17(2): 96-102. [http://dx.doi.org/10.1016/j.jnutbio.2005.05.010] [PMID: 16111878]

[30] Minussi RC, Rossi M, Bologna L, *et al.* Phenolic compounds and total antioxidant potential of commercial wines. Food Chem 2003; 82(3): 409-16. [http://dx.doi.org/10.1016/S0308-8146(02)00590-3]

[31] Howitz KT, Bitterman KJ, Cohen HY, *et al.* Small molecule activators of sirtuins extend Saccharomyces cerevisiae lifespan. Nature 2003; 425(6954): 191-6. [http://dx.doi.org/10.1038/nature01960] [PMID: 12939617]

[32] Valenzano DR, Terzibasi E, Genade T, Cattaneo A, Domenici L, Cellerino A. Resveratrol prolongs lifespan and retards the onset of age-related markers in a short-lived vertebrate. Curr Biol 2006; 16(3): 296-300. [http://dx.doi.org/10.1016/j.cub.2005.12.038] [PMID: 16461283]

[33] Baur JA, Pearson KJ, Price NL, *et al.* Resveratrol improves health and survival of mice on a high-calorie diet. Nature 2006; 444(7117): 337-42. [http://dx.doi.org/10.1038/nature05354] [PMID: 17086191]

[34] Udenigwe CC, Ramprasath VR, Aluko RE, Jones PJ. Potential of resveratrol in anticancer and anti-inflammatory therapy. Nutr Rev 2008; 66(8): 445-54. [http://dx.doi.org/10.1111/j.1753-4887.2008.00076.x] [PMID: 18667005]

[35] Stervbo U, Vang O, Bonnesen C. A review of the content of the putative chemopreventive phytoalexin resveratrol in red wine. Food Chem 2007; 101(2): 449-57. [http://dx.doi.org/10.1016/j.foodchem.2006.01.047]

[36] Quintieri AM, Baldino N, Filice E, *et al.* Malvidin, a red wine polyphenol, modulates mammalian myocardial and coronary performance and protects the heart against ischemia/reperfusion injury. J Nutr Biochem 2013; 24(7): 1221-31. [http://dx.doi.org/10.1016/j.jnutbio.2012.09.006] [PMID: 23266283]

[37] Guerreiro JR, Frederiksen M, Bochenkov VE, De Freitas V, Sales MG, Sutherland DS. Multifunctional biosensor based on localized surface plasmon resonance for monitoring small molecule-protein interaction. ACS Nano 2014; 8(8): 7958-67. [http://dx.doi.org/10.1021/nn501962y] [PMID: 25003494]

[38] Litescu SC, Eremia S, Radu GL. Biosensors for the Determination of Phenolic Metabolites. In: Giardi MT, Rea G, Berra B, Eds. Bio-Farms for Nutraceuticals: Functional Food and Safety Control by Biosensors. Berlin: Springer-Verlag Berlin 2010; pp. 234-40. [http://dx.doi.org/10.1007/978-1-4419-7347-4_17]

[39] Karim F, Fakhruddin AN. Recent advances in the development of biosensor for phenol: a review. Rev Environ Sci Biotechnol 2012; 11(3): 261-74. [http://dx.doi.org/10.1007/s11157-012-9268-9]

[40] Sochor J, Dobes J, Krystofova O, *et al.* Electrochemistry as a tool for studying antioxidant properties. Int J Electrochem Sci 2013; 8(6): 8464-89.

[41] Cortina-Puig M, Prieto-Simon B, Campas M, Calas-Blanchard C, Marty JL. Determination of the antioxidants' ability to scavenge free radicals using biosensors. In: Giardi MT, Rea G, Berra B, Eds. Bio-Farms for Nutraceuticals: Functional Food and Safety Control by Biosensors. Berlin: Springer-Verlag Berlin 2010; pp. 222-33. [http://dx.doi.org/10.1007/978-1-4419-7347-4_16]

[42] Vasilescu A, Sharpe E, Andreescu S. Nanoparticle-based technologies for the detection of food antioxidants. Curr Anal Chem 2012; 8(4): 495-505.

[http://dx.doi.org/10.2174/157341112803216780]

[43] Barthelmebs L, Calas-Blanchard C, Istamboulie G, Marty JL, Noguer T. Biosensors as Analytical Tools in Food Fermentation Industry. In: Giardi MT, Rea G, Berra B, Eds. Bio-Farms for Nutraceuticals: Functional Food and Safety Control by Biosensors. Berlin: Springer-Verlag Berlin 2010; pp. 293-307.
[http://dx.doi.org/10.1007/978-1-4419-7347-4_22]

[44] McLachlan MJ, Katzenellenbogen JA, Zhao H. A new fluorescence complementation biosensor for detection of estrogenic compounds. Biotechnol Bioeng 2011; 108(12): 2794-803.
[http://dx.doi.org/10.1002/bit.23254] [PMID: 21732327]

[45] Gao X, Ge S, Cai Q, Zeng K, Grimes CA. Kinetic study on the interaction between tannin and bovine serum albumin with a wireless magnetoelastic biosensor. Sens Actuators B Chem 2008; 129(2): 929-33.
[http://dx.doi.org/10.1016/j.snb.2007.10.005]

[46] Kaneda H, Watari J, Takashio M, Okahata Y. Measuring astringency of beverages using a quartz-crystal microbalance. J Am Soc Brew Chem 2003; 61(3): 119-24.

[47] Sehlotho N. Phthalocyanines: Photochemical, electrochemical and biomimetic catalytic behaviour. Rhodes University 2007.

[48] Gay-Martin M, Diez-Arevalo E, Rodriguez-Mendez ML, Saez JA. Electrochemical quartz crystal microbalance analysis of the oxidation reaction of phenols found in wines at lutetium bisphthalocyanine electrodes. Sens Actuators B Chem 2013; 185: 24-31.
[http://dx.doi.org/10.1016/j.snb.2013.04.115]

[49] Moreno-Arribas MV, Polo MC. Amino acids and biogenic amines. In: Moreno-Arribas MV, Polo MC, Eds. Wine chemistry and biochemistry. New York, NY: Springer New York 2009; pp. 163-89.
[http://dx.doi.org/10.1007/978-0-387-74118-5_9]

[50] Smit AY, du Toit WJ, du Toit M. Biogenic amines in wine: understanding the headache. S Afr J Enol Vitic 2008; 29(2): 109-27.

[51] Scientific opinion on risk based control of biogenic amine formation in fermented foods. EFSA J 2011; 9(10): 2393.
[http://dx.doi.org/10.2903/j.efsa.2011.2393]

[52] Scientific Opinion on the substantiation of a health claim related to melatonin and reduction of sleep onset latency. EFSA Journal 2011; 9(6): 2241.

[53] Rodriguez-Naranjo MI, Ordóñez JL, Callejón RM, Cantos-Villar E, Garcia-Parrilla MC. Melatonin is formed during winemaking at safe levels of biogenic amines. Food Chem Toxicol 2013; 57: 140-6.
[http://dx.doi.org/10.1016/j.fct.2013.03.014] [PMID: 23531627]

[54] Bucur B, Radu GL, Toader CN. Analysis of methanol–ethanol mixtures from falsified beverages using a dual biosensors amperometric system based on alcohol dehydrogenase and alcohol oxidase. Eur Food Res Technol 2008; 226(6): 1335-42.
[http://dx.doi.org/10.1007/s00217-007-0662-4]

[55] Di Fusco M, Federico R, Boffi A, Macone A, Favero G, Mazzei F. Characterization and application of a diamine oxidase from *Lathyrus sativus* as component of an electrochemical biosensor for the

determination of biogenic amines in wine and beer. Anal Bioanal Chem 2011; 401(2): 707-16. [http://dx.doi.org/10.1007/s00216-011-5131-z] [PMID: 21644017]

[56] Boka B, Adanyi N, Virag D, Sebela M, Kiss A. Spoilage detection with biogenic amine biosensors, comparison of different enzyme electrodes. Electroanalysis 2012; 24(1): 181-6. [http://dx.doi.org/10.1002/elan.201100419]

[57] Lange J, Wittmann C. Enzyme sensor array for the determination of biogenic amines in food samples. Anal Bioanal Chem 2002; 372(2): 276-83. [http://dx.doi.org/10.1007/s00216-001-1130-9] [PMID: 11936099]

[58] Yang M, Zhang J, Chen X. Competitive electrochemical immunosensor for the detection of histamine based on horseradish peroxidase initiated deposition of insulating film. J Electroanal Chem 2015; 736: 88-92. [http://dx.doi.org/10.1016/j.jelechem.2014.11.002]

[59] Adányi N, Székács I, Szendro I, Székács A. Determination of histamine content in vegetable juices by using direct and competitive immunosensors. Food Agric Immunol 2014; 25(1): 20-33. [http://dx.doi.org/10.1080/09540105.2012.731686]

[60] Zołek T, Luliński P, Maciejewska D. A computational model for selectivity evaluation of 2-(3,--dimethoxyphenyl)ethylamine (homoveratrylamine) imprinted polymers towards biogenic compounds. Anal Chim Acta 2011; 693(1-2): 121-9. [http://dx.doi.org/10.1016/j.aca.2011.03.009] [PMID: 21504819]

[61] Basozabal I, Guerreiro A, Gomez-Caballero A, Aranzazu Goicolea M, Barrio RJ. Direct potentiometric quantification of histamine using solid-phase imprinted nanoparticles as recognition elements. Biosens Bioelectron 2014; 58: 138-44. [http://dx.doi.org/10.1016/j.bios.2014.02.054] [PMID: 24632140]

[62] Lee M-H, O'Hare D, Chen Y-L, *et al.* Molecularly imprinted electrochemical sensing of urinary melatonin in a microfluidic system. Biomicrofluidics 2014; 8(5): 054115. [http://dx.doi.org/10.1063/1.4898152] [PMID: 25584113]

[63] Basozabal I, Gomez-Caballero A, Diaz-Diaz G, *et al.* Rational design and chromatographic evaluation of histamine imprinted polymers optimised for solid-phase extraction of wine samples. J Chromatogr A 2013; 1308: 45-51. [http://dx.doi.org/10.1016/j.chroma.2013.08.002] [PMID: 23953615]

[64] García-Cela E, Crespo-Sempere A, Gil-Serna J, Porqueres A, Marin S. Fungal diversity, incidence and mycotoxin contamination in grapes from two agro-climatic Spanish regions with emphasis on *Aspergillus* species. J Sci Food Agric 2015; 95(8): 1716-29. [http://dx.doi.org/10.1002/jsfa.6876] [PMID: 25131651]

[65] Pizzutti IR, de Kok A, Scholten J, *et al.* Development, optimization and validation of a multimethod for the determination of 36 mycotoxins in wines by liquid chromatography-tandem mass spectrometry. Talanta 2014; 129: 352-63. [http://dx.doi.org/10.1016/j.talanta.2014.05.017] [PMID: 25127606]

[66] Vidal JC, Bonel L, Ezquerra A, Duato P, Castillo JR. An electrochemical immunosensor for ochratoxin A determination in wines based on a monoclonal antibody and paramagnetic microbeads. Anal Bioanal Chem 2012; 403(6): 1585-93.

[http://dx.doi.org/10.1007/s00216-012-5951-5] [PMID: 22466259]

[67] Barthelmebs L, Hayat A, Limiadi AW, Marty J-L, Noguer T. Electrochemical DNA aptamer-based biosensor for OTA detection, using superparamagnetic nanoparticles. Sens Actuators B Chem 2011; 156(2): 932-7.
[http://dx.doi.org/10.1016/j.snb.2011.03.008]

[68] Prieto-Simón B, Samitier J. "Signal off" aptasensor based on enzyme inhibition induced by conformational switch. Anal Chem 2014; 86(3): 1437-44.
[http://dx.doi.org/10.1021/ac402258x] [PMID: 24377312]

[69] Rivas L, Mayorga-Martinez CC, Quesada-González D, Zamora-Gálvez A, de la Escosura-Muñiz A, Merkoçi A. Label-free impedimetric aptasensor for ochratoxin-A detection using iridium oxide nanoparticles. Anal Chem 2015; 87(10): 5167-72.
[http://dx.doi.org/10.1021/acs.analchem.5b00890] [PMID: 25901535]

[70] Evtugyn G, Porfireva A, Stepanova V, *et al.* Impedimetric aptasensor for ochratoxin A determination based on Au nanoparticles stabilized with hyper-branched polymer. Sensors (Basel) 2013; 13(12): 16129-45.
[http://dx.doi.org/10.3390/s131216129] [PMID: 24287535]

[71] Pacheco JG, Castro M, Machado S, *et al.* Molecularly imprinted electrochemical sensor for ochratoxin A detection in food samples. Sens Actuators B Chem 2015; 215: 107-12.
[http://dx.doi.org/10.1016/j.snb.2015.03.046]

[72] Lv L, Cui C, Liang C, *et al.* Aptamer-based single-walled carbon nanohorn sensors for ochratoxin A detection. Food Contr 2016; 60: 296-301.
[http://dx.doi.org/10.1016/j.foodcont.2015.08.002]

[73] Zhu Z, Feng M, Zuo L, *et al.* An aptamer based surface plasmon resonance biosensor for the detection of ochratoxin A in wine and peanut oil. Biosens Bioelectron 2014; 65C: 320-6.
[PMID: 25461176]

[74] Adányi N, Levkovets IA, Rodriguez-Gil S, Ronald A, Váradi M, Szendro I. Development of immunosensor based on OWLS technique for determining Aflatoxin B1 and Ochratoxin A. Biosens Bioelectron 2007; 22(6): 797-802.
[http://dx.doi.org/10.1016/j.bios.2006.02.015] [PMID: 16600588]

[75] Bazin I, Nabais E, Lopez-Ferber M. Rapid visual tests: fast and reliable detection of ochratoxin A. Toxins (Basel) 2010; 2(9): 2230-41.
[http://dx.doi.org/10.3390/toxins2092230] [PMID: 22069682]

[76] Bucur B, Mallat E, Gurban A-M, *et al.* Strategies to develop malic acid biosensors based on malate quinone oxidoreductase (MQO). Biosens Bioelectron 2006; 21(12): 2290-7.
[http://dx.doi.org/10.1016/j.bios.2005.10.022] [PMID: 16318918]

[77] Peñas E, di Lorenzo C, Uberti F, Restani P. Allergenic proteins in enology: a review on technological applications and safety aspects. Molecules 2015; 20(7): 13144-64.
[http://dx.doi.org/10.3390/molecules200713144] [PMID: 26197307]

[78] Alves RC, Barroso MF, González-García MB, Oliveira MBP, Delerue-Matos MC. New trends in food allergens detection: Towards biosensing strategies. Crit Rev Food Sci Nutr 2016; 56(14): 2304-19.

[79] Pilolli R, Monaci L, Visconti A. Advances in biosensor development based on integrating nanotechnology and applied to food-allergen management. Trac-Trends Anal Chem 2013; 47: 12-26. [http://dx.doi.org/10.1016/j.trac.2013.02.005]

[80] Vasilescu A, Vezeanu A, Badea M. Electrochemical impedance spectroscopy investigations focused on food allergens. In: Kalcher K, Metelka R, Švancara I, Vytřas K, Eds. Sensing in Electroanalysis 2013/2014. Pardubice: University Press Centre 2014; pp. 59-83.

[81] Ocaña C, Hayat A, Mishra RK, Vasilescu A, Del Valle M, Marty JL. Label free aptasensor for Lysozyme detection: A comparison of the analytical performance of two aptamers. Bioelectrochemistry 2015; 105: 72-7. [http://dx.doi.org/10.1016/j.bioelechem.2015.05.009] [PMID: 26011131]

[82] Mihai L, Vezeanu A, Polonschii C, *et al.* Label-free detection of lysozyme in wines using an aptamer based biosensor and SPR detection. Sens Actuators B Chem 2015; 206: 198-204. [http://dx.doi.org/10.1016/j.snb.2014.09.050]

[83] Pilolli R, Visconti A, Monaci L. Rapid and label-free detection of egg allergen traces in wines by surface plasmon resonance biosensor. Anal Bioanal Chem 2015; 407(13): 3787-97. [http://dx.doi.org/10.1007/s00216-015-8607-4] [PMID: 25796525]

[84] Molinero-Abad B, Alonso-Lomillo MA, Domínguez-Renedo O, Arcos-Martínez MJ. Sulfite oxidase biosensors based on tetrathiafulvalene modified screen-printed carbon electrodes for sulfite determination in wine. Anal Chim Acta 2014; 812: 41-4. [http://dx.doi.org/10.1016/j.aca.2013.12.041] [PMID: 24491762]

[85] Pundir CS, Rawal R. Determination of sulfite with emphasis on biosensing methods: a review. Anal Bioanal Chem 2013; 405(10): 3049-62. [http://dx.doi.org/10.1007/s00216-013-6753-0] [PMID: 23392406]

[86] Sartori ER, Vicentini FC, Fatibello-Filho O. Indirect determination of sulfite using a polyphenol oxidase biosensor based on a glassy carbon electrode modified with multi-walled carbon nanotubes and gold nanoparticles within a poly(allylamine hydrochloride) film. Talanta 2011; 87: 235-42. [http://dx.doi.org/10.1016/j.talanta.2011.10.003] [PMID: 22099673]

[87] Spricigo R, Richter C, Leimkuhler S, *et al.* Sulfite biosensor based on osmium redox polymer wired sulfite oxidase. Colloids Surf A Physicochem Eng Asp 2010; 354(1-3): 314-9. [http://dx.doi.org/10.1016/j.colsurfa.2009.09.001]

[88] Kalimuthu P, Tkac J, Kappler U, Davis JJ, Bernhardt PV. Highly sensitive and stable electrochemical sulfite biosensor incorporating a bacterial sulfite dehydrogenase. Anal Chem 2010; 82(17): 7374-9. [http://dx.doi.org/10.1021/ac101493y] [PMID: 20698497]

[89] Muller-Renaud S, Dupont D, Dulieu P. Quantification of β casein in milk and cheese using an optical immunosensor. J Agric Food Chem 2004; 52(4): 659-64. [http://dx.doi.org/10.1021/jf034723o] [PMID: 14969512]

[90] Ocaña C, Hayat A, Mishra RK, Vasilescu A, Del Valle M, Marty JL. Label free aptasensor for Lysozyme detection: A comparison of the analytical performance of two aptamers. Bioelectrochemistry 2015; 105: 72-7. [http://dx.doi.org/10.1016/j.bioelechem.2015.05.009] [PMID: 26011131]

[91] Spricigo R, Dronov R, Lisdat F, Leimkühler S, Scheller FW, Wollenberger U. Electrocatalytic sulfite biosensor with human sulfite oxidase co-immobilized with cytochrome c in a polyelectrolyte-containing multilayer. Anal Bioanal Chem 2009; 393(1): 225-33. [http://dx.doi.org/10.1007/s00216-008-2432-y] [PMID: 18932024]

[92] No 579/2012 of 29 June 2012 amending Regulation (EC) No 607/2009 laying down certain detailed rules for the implementation of Council Regulation (EC) No 479/2008 as regards protected designations of origin and geographical indications, traditional terms, labeling and presentation of certain wine sector products. Off J. Eur Union 2012; L171: 4-7.

[93] Compendium of International Methods of Analysis of Wines and Musts. Paris: International Organisation of Vine and Wine 2015.

[94] Official methods of analysis of AOAC international. 16th ed., USA: AOAC International 1995.

[95] Fatibello-Filho O, da Cruz Vieira I. Flow injection spectrophotometric determination of sulfite using a crude extract of sweet potato root (*Ipomoea batatas* (L.) Lam.) as a source of polyphenol oxidase. Anal Chim Acta 1997; 354(1–3): 51-7. [http://dx.doi.org/10.1016/S0003-2670(97)00427-3]

[96] Trenerry VC. The determination of the sulphite content of some foods and beverages by capillary electrophoresis. Food Chem 1996; 55(3): 299-303. [http://dx.doi.org/10.1016/0308-8146(95)00117-4]

[97] Koch M, Köppen R, Siegel D, Witt A, Nehls I. Determination of total sulfite in wine by ion chromatography after in-sample oxidation. J Agric Food Chem 2010; 58(17): 9463-7. [http://dx.doi.org/10.1021/jf102086x] [PMID: 20690603]

[98] Molinero-Abad B, Alonso-Lomillo MA, Dominguez-Renedo O, Arcos-Martinez MJ. Malate quinone oxidoreductase biosensors based on tetrathiafulvalene and gold nanoparticles modified screen-printed carbon electrodes for malic acid determination in wine. Sens Actuators B Chem 2014; 202: 971-5. [http://dx.doi.org/10.1016/j.snb.2014.06.057]

[99] Telsnig D, Kalcher K, Leitner A, Ortner A. Design of an amperometric biosensor for the determination of biogenic amines using screen printed carbon working electrodes. Electroanalysis 2013; 25(1): 47-50. [http://dx.doi.org/10.1002/elan.201200378]

[100] Henao-Escobar W, Domínguez-Renedo O, Asunción Alonso-Lomillo M, Julia Arcos-Martínez M. Simultaneous determination of cadaverine and putrescine using a disposable monoamine oxidase based biosensor. Talanta 2013; 117: 405-11. [http://dx.doi.org/10.1016/j.talanta.2013.09.043] [PMID: 24209360]

[101] Mihai I, Vezeanu A, Polonschii C, *et al.* Low-fouling SPR detection of lysozyme and its aggregates. Anal Methods 2014; 6(19): 7646-54. [http://dx.doi.org/10.1039/C4AY01237B]

[102] Fernandes JR, Pereira L, Jorge P, *et al.* Wine fingerprinting using a bio-geochemical approach. 2015; 5: 02021. [http://dx.doi.org/10.1051/bioconf/20150502021]

[103] Zeravik J, Hlavacek A, Lacina K, Skl Ã. ¡dal P. State of the art in the field of electronic and

bioelectronic tongues – towards the analysis of wines. Electroanalysis 2009; 21(23): 2509-20. [http://dx.doi.org/10.1002/elan.200900285]

[104] Aguilera T, Lozano J, Paredes JA, Álvarez FJ, Suárez JI. Electronic nose based on independent component analysis combined with partial least squares and artificial neural networks for wine prediction. Sensors (Basel) 2012; 12(6): 8055-72. [http://dx.doi.org/10.3390/s120608055] [PMID: 22969387]

[105] Ha D, Sun Q, Su K, *et al.* Recent achievements in electronic tongue and bioelectronic tongue as taste sensors. 2015; 207: 1136-46. [http://dx.doi.org/10.1016/j.snb.2014.09.077]

[106] Umali AP, Ghanem E, Hopfer H, *et al.* Grape and wine sensory attributes correlate with pattern-based discrimination of Cabernet Sauvignon wines by a peptidic sensor array. Tetrahedron 2015; 71(20): 3095-9. [http://dx.doi.org/10.1016/j.tet.2014.09.062]

[107] Lobanov AV, Borisov IA, Gordon SH, Greene RV, Leathers TD, Reshetilov AN. Analysis of ethanol-glucose mixtures by two microbial sensors: application of chemometrics and artificial neural networks for data processing. Biosens Bioelectron 2001; 16(9-12): 1001-7. [http://dx.doi.org/10.1016/S0956-5663(01)00246-9] [PMID: 11679281]

[108] Di Gennaro SF, Matese A, Mancin M, Primicerio J, Palliotti A. An open-source and low-cost monitoring system for precision enology. Sensors (Basel) 2014; 14(12): 23388-97. [http://dx.doi.org/10.3390/s141223388] [PMID: 25490596]

[109] Choi SH, Chung DJ, Kwen HD. Fabrication of biosensors using vinyl polymer-grafted carbon nanotubes. New perspectives in biosensors technology and applications. Prof. Pier Andrea Serra (Ed.), ISBN: 978-953-307-448-1, InTech, 2011. [http://www.intechopen.com/books/newperspectives-i--biosensors-technology-and-applications/fabrication-of-biosensors-using-viny--polymer-graftedcarbon-nanotubes]

[110] Vasilescu A, Vezeanu A, Liu Y, Hosu I S, Worden R M, Peteu S F. Meat freshness: Peroxynitrite's oxidative role, its natural scavengers, and new measuring tools. Instrumental methods for the analysis of bioactive molecules, ACS symposium series. 2014; 1185: pp. 303-32.

CHAPTER 6

Electronic Sensory Systems for Characterization of Bioactive Compounds

Jesús Lozano[1], Daniel Matatagui[2], José Pedro Santos[3] and M. Carmen Horrillo[3,*]

[1] *Escuela de Ingenierías Industriales, Universidad de Extremadura, Badajoz, Spain*

[2] *Centro de Ciencias Aplicadas y Desarrollo Tecnológico, UNAM, México D.F.*

[3] *GRIDSEN, Instituto de Tecnologías Físicas y de la Información (ITEFI-CSIC), Madrid, Spain*

Abstract: This chapter provides a review of the most recent works in electronic sensory system for characterization of Bioactive Compounds. Focus is placed on the food applications and those compounds that are good for health. This chapter reviews the main electronic sensory techniques, from electronic noses to sensors and biosensors devices used for Bioactive Compound Characterization. The chapter demonstrates that although Bioactive Compounds are difficult to detect with sensor systems due to their high molecular weight and low volatility of the compounds and therefore few references are found in the literature, it is important to stand out that these systems could be a promising field for research and developing of industrial instruments applied to real contexts.

Keywords: Bioactive compound characterization, Biosensors, Electronic noses, Electronic sensory systems, Nanoparticles, Optical sensors, Sensors, Surface-enhanced raman spectroscopy, Surface plasmon resonance.

INTRODUCTION

Nutrition research has shown repeatedly and consistently that diets, the best adapted to the goals of healthy diet, are those based mainly on food consumption

* **Corresponding author M Carmen Horrillo:** GRIDSEN: Instituto de Tecnologías Físicas y de la Información (ITEFI-CSIC). C/ Serrano, 144. 28006 Madrid, Spain; E-mail: carmen.horrillo.guemes@csic.es

Constantin Apetrei (Ed.)

of plant sources (fruits, vegetables, whole grains, nuts and legumes), using sparingly foods of animal origin. In fact, low intake of fruits and vegetables, along with hypertension, hypercholesterolemia, smoking, physical inactivity and overweight and obesity are among the main risk factors of total mortality in the European population. Today we know that the benefits of a prudent diet are not limited to content of nutrients, must also provide other protective factors against stress, carcinogenesis and oxidative contents especially in plant and foods, called "bioactive components".

Today, nutritional currents are influencing on the concept that we have above health. New dietary strategies are being carried out to see the effect of diet on regulation of the genetic and molecular levels and therefore improving the human health. Hereby, the bioactive compounds (extracts of plants or single compounds therefore) need to be identified and developed to complement a balanced diet. Therefore it is very important to evaluate the efficacy and safety of these compounds. The development of characterization techniques presents a great interest for the scientist community, as well as the nanotechnology, which is also being directed to the development of delivery systems for bioactive compounds.

What are Bioactive Substances?

Bioactive substances are food components that influence the cellular activity and the physiological mechanisms with beneficial effects on the health [1]. These substances have an important role [2 - 4] in the prevention of many diseases (CVD, coronary heart disease, stroke, hypertension, various cancers [GI, prostate, breast, *etc.*], neurodegenerative, inflammatory diseases, ocular ones [macular degeneration, -DMAE- cataracts], obesity, diabetes, osteoporosis, longevity [lifespan essential], *etc.*).

Besides their important antioxidant action (through mechanisms such as sequestration of free radicals, inhibition of the production of hydrogen peroxide, activation of endogenous defense mechanisms [catalase, superoxide dismutase], metal chelation, *etc.*), many other biologically plausible mechanisms may be responsible for their protective effect.

Bioactive Compound Classification

The main bioactive components and the foods where they can be found are listed in Table **1** [5]. They can be classified into three groups: terpenoids (carotenoids and sterols); phenolic compounds (such as phytoestrogens and flavonoids quercetin, the most common dietary flavonoid); and sulfur compounds.

SENSORS AND ELECTRONIC SENSORY SYSTEMS FOR CHARACTERIZATION OF BIOACTIVE COMPOUNDS

There are several electronic sensory systems, including optical, nanotechnology and gas sensors based systems for the characterization of bioactive compounds and some experiments have been described in the bibliography. Table **2** summarizes the sensors and sensory systems applications for detection and characterization of different bioactive compounds, mainly in the food industry.

As a general rule, the sensing of Bioactive Compounds is not an easy task due to the molecular weight of these substances. Their low volatility makes it difficult to detect them with gas detection techniques. However, the sensing of these compounds is very important because of high value of these compounds for the health. Main compounds and their sensing techniques can be found in Table **2** and some of them described in the following subsections. Electrochemical methods for wine bioactive compounds have been widely discussed elsewhere in the book thus few references to these methods will be given in this chapter.

Table 1. Bioactive compounds classification.

Bioactive compound	Food
Terpenoids (several thousand): - **Carotenoids**: • Carotene: - α-carotene, beta-carotene (precursor of vitamin A). - Lycopene. • Xanthophylls: - ß-cryptoxanthin (provitamin A). - Lutein, zeaxanthin.	ß-carotene vegetables and orange fruit (*e.g.*, carrots, mango, apricot, melon, peach, passion fruit, plum); dark green leafy vegetables (such as spinach); tomato and derivatives. Lycopene: tomatoes, watermelon, red pepper, pink grapefruit. ß-cryptoxanthin: oranges, papaya. Lutein and zeaxanthin: green leafy vegetables (*e.g.*, beets, spinach, lettuce, celery), oranges, potatoes, tomatoes, red peppers, corn, avocado, melon.

(Table 1) contd.....

Bioactive compound	Food
Phytosterols (> 250): • Sterols and stanols: ß-sitosterol, stigmasterol, campesterol, sitostanol, campestanol.	Vegetable oils (corn, sunflower, soybean, olive), cereals, legumes, nuts, vegetables. Fortified foods.
Phenolic compounds (> 8000) **Phenolic acids and simple alcohols:** (tyrosol, hydroxytyrosol, hydroxybenzoic and hydroxycinnamic acids [ellagic, gallic, vanilínico, capsaicin, coumaric, caffeic, ferulic, chlorogenic, *etc.*]).	Citrus, olives, olive oil, other fruits, vegetables, oats, soybeans, nuts, wine, beer, tea, *etc.*
Polyphenols: • Flavonoids (> 5000): - Flavonol (quercetin, kamferol, myricetin, rutin, *etc.*). - Flavanols (flavan-3-ols): - Catechins (catechin, epicatechin, *etc.*). - Pro (anto) cyanidins or condensed tannins. - Flavanones (naringenin, hesperitin, naringin, hesperidin, *etc.*). - Anthocyanins (sugar at position 3) and anthocyanidins (anthocyanins sugarless) (cyanidin, *etc.*). - Flavones (apigenin, luteolin, *etc.*). - Isoflavones (genistein, daidzein, *etc.* [phytoestrogens]). • Stilbenes (resveratrol). • Curcuminoids (curcumin). • Lignans (main source of phytoestrogens in the West).	Quercetin, kamferol, myricetin and anthocyanins: onions, leeks, lettuce, broccoli, tomatoes, grapes, oranges (pulp), apples, cherries, blackberries, raspberries, blueberries, olives, red wine, tea, oregano and other herbs. Catechins and proanthocyanidins: apples, pears, cherries, grapes, apricot, peach, nuts, beans, cocoa, black chocolate, wine, cider, beer, tea. Hesperidin, naringenin: citrus, grape juice Apigenin, luteolin: parsley, celery, pepper, thyme, olives. Resveratrol: grape skins, wine, grape juice, blueberries. Phytoestrogens: isoflavones (genistein, daidzein) and lignans: soy products, other legumes, whole grains, nuts, berries, broccoli, garlic, carrots, *etc.*
Sulphur compounds - From allium (alliin, allicin, ajoene, dialilsulfuro, *etc.*). - Glucosinolates (> 120) (isothiocyanate, sulforaphane, indole-3-carbinol).	Alliin, dialilsulfuro: onion, chives, scallions, leeks, garlic. Isothiocyanate, sulforaphane, I3C: cabbage, cauliflower, broccoli, cabbage, Brussels sprouts, cabbage, garlic, onions, turnips, mustard.

Table 2. Bioactive compounds and sensing techniques.

Compound	Technique	References
Lycopene in olive oil and Ginseng bioactive compounds	Electronic nose	[6, 7]
Alcaloids from plants (*Chelidonium majus* L.)	Surface Plasmon Resonance Spectroscopy	[8, 9]
Theophilyne (drug used in therapy for respiratory diseases)	Surface-enhanced Raman spectroscopy	[10]
Total polyphenol content in tea	Diffuse reflectance spectroscopy	[11]

(Table 2) contd.....

Compound	Technique	References
Polyphenols	Fluorescence spectroscopy	[12]
Olive oil polyphenols Polyphenols Phenolic compounds Tannins	Colorometric-absorption spectroscopy	[13 - 17]
Crocin (carotenoid)	SPE-HPLC	[18]

GAS SENSORS COMBINED WITH PATTERN RECOGNITION TECHNIQUES (ELECTRONIC NOSES)

Introduction

The term electronic nose (e-nose) refers to an instrument that tries to mimick the functionality and structure of the biological sense of smell for being used in applications related with the sensory analysis of a product. The way of working of an e-nose tries to be similar to the human nose: the first step is the interaction between smells or volatile organic compounds (VOCs), usually a complex mixture, and the appropriate receptors: the biological nose uses olfactory receptors and the e-nose uses an array of chemical sensor. In both cases, one odorant receptor (olfactory receptor or gas sensor) responds to multiple odorants and one odorant can be detected by multiple odorant receptors. The next step in both systems is the storage of the signal generated by the receptors in the brain or in a database (learning stage). Once learned, the identification of one of the measured odour can be achieved (classification stage).

An accepted definition for the scientific community of an e-nose was given by Gardner in 1994: "an electronic nose is an instrument which comprises an array of electronic chemical sensors with partial specificity and an appropriate pattern recognition system, capable of recognising simple or complex odours" [19]. This definition restricts the term e-nose to those types of sensory systems based on the combination of gas sensors and pattern recognition methods, which are specifically used to sense odorous molecules in a similar way to the biological nose. However, the architecture of an e-nose has much in common with sensory systems designed for the discrimination and quantification of individual components in a simple carrier gas or in a vapour mixture [20]. An electronic nose is usually composed by a sampling method, a gas sensor array, a control and instrumentation system and a pattern recognition method. A simple diagram of the typical structure of an e-nose and the signals involved is shown in Fig. (**1**) [21].

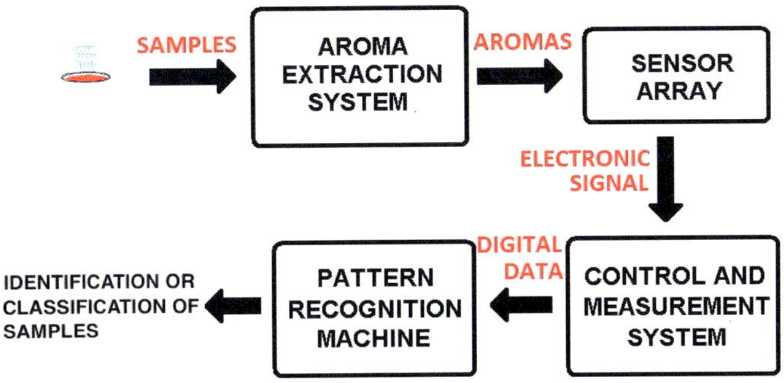

Fig. (1). Block diagram of an electronic nose system.

The sampling methods extract the aromatic (volatile) compounds of the liquid or solid samples and carry the aromatic (volatile) compounds from the samples to the sensor cell. Several sampling techniques are adapted from other analythical techniques and are available for electronic noses [22]. The most common are: static headspace (HS), purge & trap (P&T) and solid-phase micro-extraction (SPME).

E-noses use chemical sensors as sensory receptors. They can be defined as a device capable of converting a chemical change in the environment into an electric signal or data, which is function of the concentration of specific particles (atoms, molecules, or ions) in gases or liquids [23]. Chemical sensors can be classified according to their principle of working: electrical, mass, thermal or optical. Examples of gas sensors used in e-noses can be found in [24]: conducting polymers (CP) [25], Quartz Cristal Microbalances (QCM) [26], surface acoustic wave devices (SAW) and [27] metal oxide and semiconductor sensors. E-noses have the following advantages: high portability, easy to use, quick measurements, low costs compared with other methods of analysis and good reliability. They are appropriate for making *in situ* and on-line measurements

The third part of an e-nose is the control and measurement system, that consists of diverse electronic circuitry used for the measurements of sensors signals, such as interface circuits, drivers, signal conditioning circuits, communication processors, and A/D and D/A converters.

The main objective of an electronic nose is to identify an odorant or chemical component in a sample and try to estimate its concentration. It involves several tasks related to signal and data processing usually made with the appropriate pattern recognition system. However, those two tasks may be subdivided into the following four steps: preprocessing, feature extraction, classification, and decision-making [28].

But first, a database of expected odorants and concentrations must be compiled, and a collection of samples must be measured with the e-nose. Preprocessing methods try to compensate for sensor drift, to compress the transient response of the sensor array into a vector of responses, and reduces variations from one sample to another. Typical preprocessing techniques include: manipulation of the baseline of each sensor; normalization of sensor response and compression of sensor transients.

Feature extraction has different purposes: to reduce the dimension and reduce the redundancy of the measurement data, and to extract information relevant for further analysis (pattern recognition). This extraction is generally performed with well-known methods like principal component analysis (PCA) and linear discriminant analysis (LDA), based in linear transformations applied to the data. PCA is a statistical method used by many researchers to reduce the dimensionality of the measurement space leading to a better visualization of data clustering that uses an orthogonal transformation to convert a set of measurements of possibly correlated variables into a new set of values of linearly uncorrelated variables called principal components, sometimes referred to vector decomposition and thus helps to display multivariate data in plots of two or three axis. PCA finds projection of maximum variance and is the most widely used linear feature-extraction technique [28]. But PCA, is not commonly used for classification since it ignores the category or class of the examples in the database. LDA, by the contrary, looks at the identity of each example in order to find that projections that maximize the distance among measurements from different classes to minimize the distance among measurements of the same class [29, 30].

Finally, the classification task is usually performed by artificial intelligence methods, like Fuzzy Logic Classifiers (FL) and Artificial Neural Networks (ANN). FL are based algorithms that assign a class label to an unknown measurement, based on the application of several previously defined rules expressed in linguistic values. ANNs are information processing systems that try to mimick biological neural networks. It allows the e-nose to function as the biological sense of smell in the interpretation of responses from olfactory sensors. A typical ANN consists of two or more layers (each one composed of a variable number of neurons) connected with synaptic weights (number that connect the outputs of the previous layer neurons to the inputs of the next layer neurons) [29 - 31]. During training, the ANN optimizes the synaptic weights to learn the patterns of the different measurements in order to obtain a minimum error. After training, when presented with an unknown odorant, the ANN feeds its data through the different layers of neurons and calculates the class label that provides the minimum error or maximum response. The classifier makes a prediction of the class for an unknown measurement along with the estimation of the confidence placed on the class assignment. A final decision-making stage may be used for specific applications such as confidence thresholds calculation or risk associated with different errors in classification. The decision-making module may modify the classifier assignment and even include new categories to classify the unknown sample if it does not belong to any of the odorants in the database [32]. Validation techniques are also used to estimate the performance of the classifiers.

Electronic Noses for Sensing Application

The practical application of biological nose and sensory panel as a smell instrument for assessment in detecting odorants is limited by the fact that the sense of smell has a great subjective component, saturates easily, and is therefore difficult to train. In this way, there is a great need for an artificial instrument that could work as the sense of smell of mammals and could be use in industrial applications replacing sensory panels of routine and tedious tasks. To promote this technology to industrial applications, chemical sensors based on semiconductor thin films became good candidates in areas like environment, automotive, food industry and medicine for different tasks like pollution control, prediction and monitoring of indoor and outdoor air quality, industrial plants control, detection of illnesses, *etc*. Scientific community worldwide are investigating chemical sensors and materials giving importance to the various aspects of gas/odour sensing properties.

E-nose instruments are attractive for the industry due to a number of significant features: a fast assessment of the headspace of a sample, a quantitative repres-entation or signature of a gas or mixture, and low-cost and miniature sensors, which can be easily integrated in current production processes. A consi-derable number of applications of e-noses have been reported for sensing application similar to biological nose. A broad list of e-nose reviews can be found in the literature: e-noses based on mass spectrometry techniques [33], health care and biomedical applications [34], forestry and agriculture applications [35], microbial quality control of food products [36] and food industry [37], pharmaceutical industry applications [38], for developing chemical sensor arrays [39].

Electronic Noses for the Analysis of Bioactive Compounds

A portable electronic nose was developed and designed for the detection of lycopene in olive oil samples [6]. The designed electronic nose (Fig. **2a**) is equipped with a wireless communication processor in order to connect to a host or forming a network of similar devices for distributed monitoring [40] (Fig. **2b**). It has been designed to work with conductive sensors, static or headspace with effluent transfer as sampling method and a portable instrumentation and measurement system, it includes 9V rechargeable batteries, touch color screen and WI-FI transceiver for wireless communication (Fig. **3**).

It consists of two gas inlet pipes that are switched through a three way electro valve connected to the sensors chamber that contains the sensor array. One of the gas inlets has a carbon filter in order to provide clean and humidity free air as a reference. Downstream are located the temperature and humidity sensor and a micropump to carry the gas samples to the sensors cell. The whole system is controlled by a digital signal controller (Fig. **3**). It also includes several analog to digital converters (A/D) for measurements of the different signals involved in the experiments and several pulse width modulation (PWM) outputs for generating the signal for sensor heating. Main measurement parameters can be configured using the touchscreen. It allows the change of several measurement parameters as pump and heaters power as well as electrovalve switching. Rechargeable batteries give more than 8 hours of autonomy to the device. Wireless communications are provided by a WI-FI transceiver. A network could be established with a host computer for remote monitoring. The sensor chamber and board are designed for different microsensors using different packages and it is easily adaptable to other packages and sensors. The device can measure up to 4 resistive sensors and provides independent heating for each sensor. The device is controlled by a home-made program developed in Testpoint. The program controls the measurement parameters and generates the response database in the computer. Algorithms for both on-line and off-line pattern recognition techniques has been developed in Matlab and integrated in the program through mathcripts. External classification using a web server can also be performed [41].

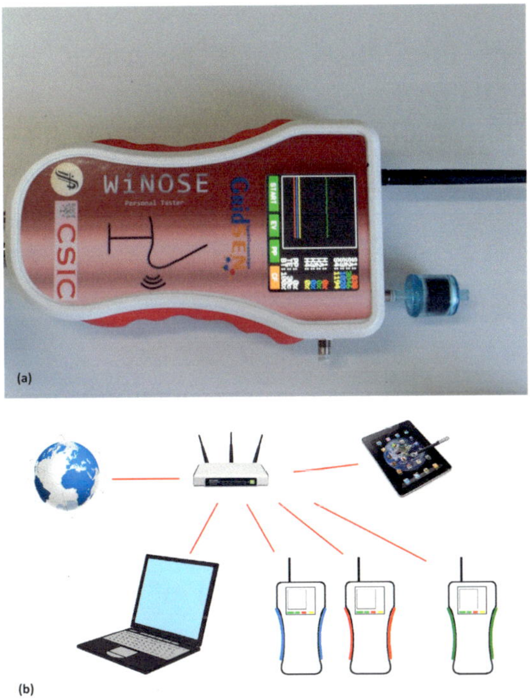

Fig. (2). (a) Portable electronic nose; **(b)** Configuration of the e-nose in the network.

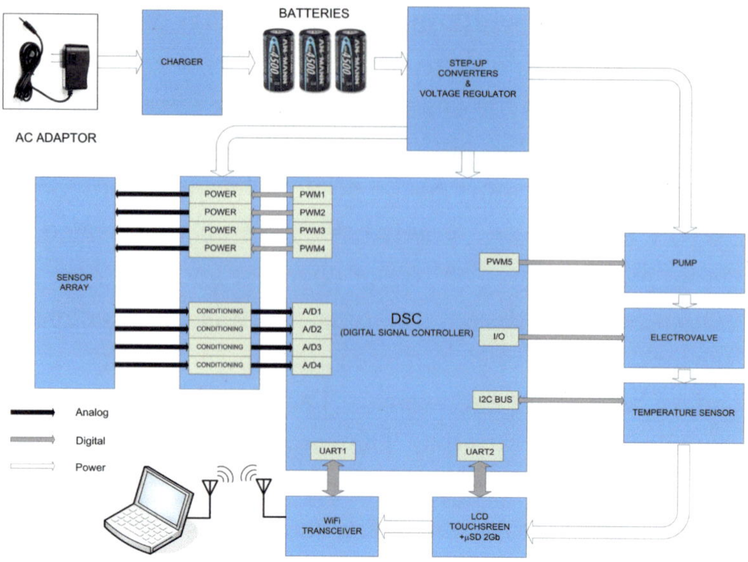

Fig. (3). Block diagram of the main components of the developed portable e-nose.

The electronic nose has been used on a consistent application, in the differentiation of olive oils with and without added lycopene. The company "Tomates del Guadiana Soc. Coop." produces olive oil enriched with Lycopene from an extraction technique of Lycopene by direct solubilization in the oil without using organic solvents [42]. Lycopene is a powerful antioxidant found in tomatoes and other vegetables [43]. It is a matter of great interest for their beneficial health effects as it prevents cancer and cardiovascular diseases. Until now Lycopene was extracted with organic solvents, a method that has big problems due to the remaining traces of solvent with the consequent health risk [44]. Lycopene and liposoluble VOCs are extracted by the process developed in the aforementioned patent [42], which gives the oil a defined tomato flavor. In this section, oil samples containing 250 ppm of Lycopene have been used.

Measurements have been taken every second of the following parameters: resistance of gas sensors, relative humidity, temperature, flow and battery voltage. In this application, SILSENS commercial sensors comprising 4 single tin oxide sensors have been used, one of them doped with platinum. Temperature sensors were between 350 and 450°C. Oil samples (10 mL in 22 mL vials) were kept at a constant temperature of 28°C using a thermostated bath. Measuring cycles of 10 minutes were done: 1 minute of air through the sample and 9 minutes of reference air for cleaning the sensors. These times allowed fast and reproducible responses for the measurements. The flow rate was held constant at 150 mL / min in all cases. 40 measures have been taken for each sample to have a significant number when applying statistical and pattern recognition techniques.

Fig. (**4**) shows a radial diagram average sensor response against oil samples with and without lycopene. It can be noticed that the response profile is different for the two types of samples analyzed.

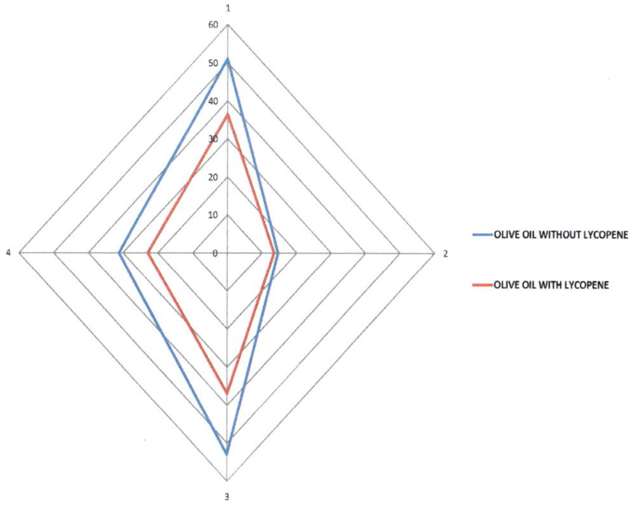

Fig. (4). Diagram of the average radial response of the sensors to the two oils.

Once the measurements have been made and responses of the sensors to the samples were stored and preprocessed, the feature extraction (relative response baseline manipulation and sensor normalization) was performed, a calibration for reducing sensors drift was applied, PCA was performed to the response vectors in order to reduce the dimension of the data space in order to show it in a 2D plot. The PCA plot is shown in Fig. (**5**).

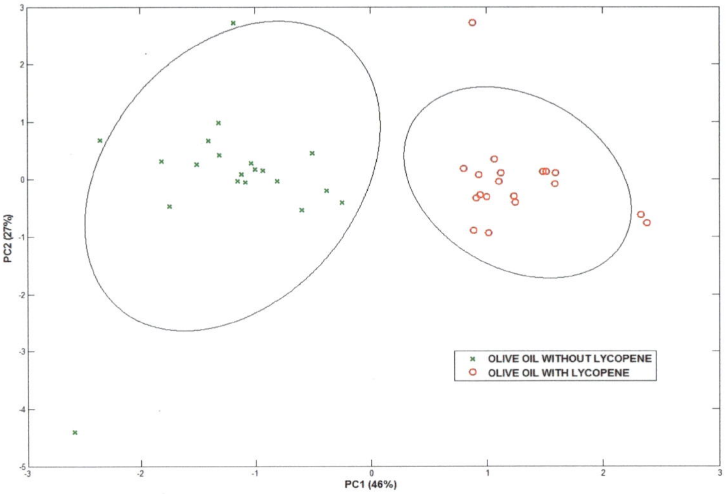

Fig. (5). PCA score plot of the measurements of olive oil with and without added lycopene.

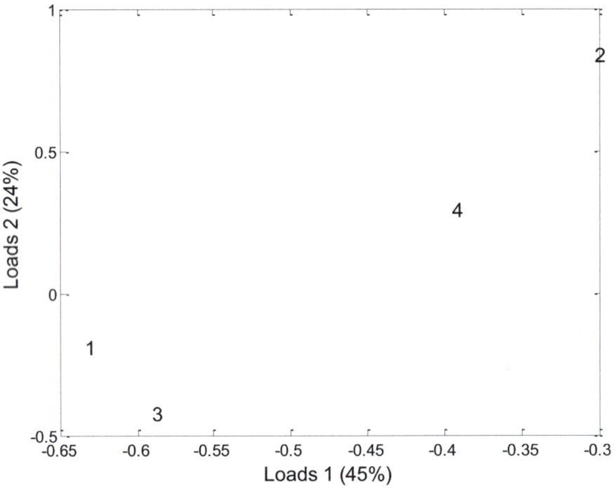

Fig. (6). Loading plot of the PCA.

A complete separation between the two classes can be observed. The variance of each of the components is shown in brackets on each axis. Fig. (**6**) shows the load diagram, in which the contribution of each of the four sensors to each one of the principal components is explained. It can be seen that the sensor 2 is the one that has greater contribution to the PCA.

Further analysis are done to the data. Different classifiers such as with Radial Basis Functions based Probabilistic Neural Network (RBF) and Backpropagation networks (BP) were used. Leave One Out [30] was used as validation method for performance estimation. This method was used due to the low quantity of measurements. The success rate (percentage of cases correctly classified) obtained in validation was 90% for RBF and 92.5% for BP classifier. The confusion matrix obtained in validation with both neural nets is shown in Tables **3** and **4**.

Table 3. Confusion matrix of RBF network obtained in Leave One Out validation.

	Without added Lycopene	**With added Lycopene**
Without added Lycopene	37	3
With added Lycopene	3	37

Further version of this prototype includes a web-based application for remote classification. This tool integrates an Artificial Neural Network to provide the classification service and enhances the memory and calculation power of the system [41].

Table 4. Confusion matrix of BP network obtained in Leave One Out validation.

	Without added Lycopene	**With added Lycopene**
Without added Lycopene	36	4
With added Lycopene	4	36

SURFACE PLASMON RESONANCE SPECTROSCOPY

Introduction

Between 1902 and 1912 R.M. Wood observed patterns of unusual dark and bright bands in the reflected light from a metal backed diffraction grating. It was not until the late sixties that Kretschmann and Otto reported the optical excitation of surface plasmons through the method of attenuated total reflection [45, 46]. Surface plasmons have been investigated over the years, obtaining very significant results and establishing their properties [47]. It is only during the eighties that surface plasmon resonance (SPR) was proposed as a technique to study biomolecular interactions [48]. Nowadays, SPR is recognized as one of the best available technologies to detect and characterize biomolecules and their interactions [49, 50], and is especially useful for identifying bioactive compounds

[8, 9].

Surface Plasmon Resonance for Sensing Application

SPR is a charge density fluctuation that occurs at the interface between a metal and a dielectric. This phenomenon happens when a photon strikes the metal surface. The oscillation caused in SPR produces an evanescent field, which extends into the dielectric around 500 nm from the metal surface.

The prism-based system is the most common and simple configuration to produce excitation of surface plasma wave, coupling a single layer nanometric metal-coated prism (isosceles triangle or semi-cylindrical) and incident light under conditions of attenuated total reflection [45]. Typically, in prism-based configuration, the laser is reflected in the metal-coated surface and perpendicular to the inlet and outlet surfaces of the prism. Under conditions of a constant light source wavelength the plasmon resonance occurs for a specific angle, which is determined by the refractive index of the material on the metal surface. Consequently, changes in the refractive index on the metal surface are detected by analysing of the reflected light intensity for different angles. Other versions of SPR are obtained by replacing the prism for different optical configurations, such as grating coupled SPR [51], waveguide based SPR [52] and optic fiber SPR [53], which are among the most common.

Surface Plasmon Resonance Sensors for the Analysis of Bioactive Compounds

Some examples of biomolecular interactions, which have been successfully studied using SPR can be, DNA or RNA detecting protein, protein detecting cell, protein detecting virus or phage, carbohydrate detecting protein, carbohydrate detecting cell, liposome detecting protein, drugs detecting protein, *etc*. A particular case of successful implementation of SPR is DNA-based biosensor to detect bioactive compounds in plants. María Minunni *et al.* [8] applied this sensor to the analysis of bioagents contained in an extract of Chelidonium majus L. (great celandine) obtained by chromatographic separation. The great celandine has different pharmaceutical effects, such as anti-inflammatory and anti-microbial activity, inhibition of SH-enzymes and microtubule assembly. The main reason for the pharmaceutical effects is related to alkaloids, from DNA interaction [8, 9]. The results of this study show that DNA-based SPR sensors possess a great ability to detect bioactive constituents from plants, which could be used for medical purposed.

SURFACE-ENHANCED RAMAN SPECTROSCOPY

Introduction

In 1974 Martin Fleischmann, Patrick J. Hendra and A. James McQuillan took the first step in discovering the surface enhanced Raman scattering effect (SERS) [54]. When they tried to carry out Raman on a silver electrode with high surface area; due to the roughened metal, an amplified Raman signal was observed, which was attributed to enhanced surface area for adsorption. The debate over the exact mechanism of signal enhancement began in 1977, the moment in which it was observed, by two different groups, that the intensity of the Raman signal is enhanced due to rough silver electrode. Therefore, Jeanmarie and Van Duyne proposed an electromagnetic effect [55], while Albrecht and Creighton suggested a charge-transfer effect [56]. The mechanism of SERS is a matter that is still being debated. In the late nineties, the possibility to detect single-molecule by SERS was demonstrated using silver and gold nanoparticles [57, 58]. Recently, experimental and theoretical studies have demonstrated that enhancement Raman factors are higher than 10^{11} [59]. Nowadays, due to its sensitivity and selectivity, SERS is one of the most promising technologies for biochemical and biomedical sensing applications [60, 61] and, more specifically, in bioactive compounds analysis [10].

Surface-enhanced Raman Spectroscopy for Sensing Application

There are different configurations for SERS biosensing applications [61, 62]. The most common configuration of this technique is locating the analyte on the nanostructured surfaces, acquiring the enhanced signal. To obtain a high specificity in analyte detection, Raman difference spectroscopy can also be employed, in which Raman signals before and after analyte conjugation are compared. Additionally, analyte detection can be achieved using indirect Raman spectroscopy; in this case, a molecule with a well-known spectrum is used as a label to which the analyte is associated.

Tip-Enhanced Raman Spectroscopy (TERS) is one of the latest SERS techniques that emerged from the combination of scanning probe microscope (SPM) by using metal-coated tip or metal-wire probe to generate SERS effect. The main advantage of this technique is thar analyte does not need to be immobilized on the substrate [58].

Surface Enhanced Raman Scattering Effect for the Analysis of Bioactive Compounds

An example of SERS to detect a bioactive compound is that of theophylline. Theophylline is used to treat respiratory diseases; however, it can cause side effects in the neurological and cardiovascular systems. Consequently, when theophylline is used with therapeutic purposes it requires specific concentrations, between 10 and 20 $mg \cdot L^{-1}$ in serum, to avoid adverse or lethal effects. This compound is naturally found in foods and beverages, as is the case of chocolate tea, and coffee, so there is an urgent interest for food safety in developing quick methods of early detection of high theophylline concentrations. Ping Liu *et al.* developed a SERS-based sensor to detect theophylline and the system was tested in real drink samples. The sensor was based on the molecular imprinting polymer (MIP) on the surface of silver nanoparticles, detecting less than micromolar concentration of theophylline [10]. High selectivity is an additional advantage of the developed sensor that was tested for caffeine, which have a similar structure.

DIFFUSE REFLECTANCE SPECTROSCOPY

Introduction

In 1666 Sir Isaac Newton introduced “spectrum” term, describing the component colours by the dispersion of sunlight. Diffuse reflectance spectroscopy (DRS) is a technique based on analysis of scattered light emitted back through a determinate medium. Since the early sixties DRS has been a well-established technique to determine chemical compounds [63, 64]. At the end of the seventies, the first biological chromophores, such as haemoglobin, were studied under DRS technique [65]. Nowadays, DRS is a tool that is of great interest for biochemical and biomedical sensing applications [66], and for the detection of bioactive compounds [11]. The fact that DRS is easy to use and has extensive availability makes in an attractive technique attractive for biological sensing application, and specifically for bioagents analysis purposes.

Diffuse Reflectance Spectroscopy for Sensing Application

Reflectance spectroscopy systems are composed of a light source, a fiber optic probe, and a spectrometer.

The light source for reflectance spectroscopy can be provided by a coherent light source or a white-light lamp. The advantage of the coherent light source is its high power, resulting in a reflected light that is much higher than noise. However, it sends out light at only one wavelength and due to the high heating power, the sample can be damaged. On the other hand, lamps send out low power light for each of the wavelength of a wide spectrum, including the ultraviolet and infrared regions, thought this may present spectral peaks and infrared regions can heat the sample. The ideal light source would be that which has high power for the most informative wavelength in the particular application. The light from the source to the sample is transmitted by optical fiber.

The reflected light collection from the sample is also transmitted to the spectrometer by optical fiber. Then, the spectrometer uses a monochromator to split the light in an individual wavelength, measuring their intensities by means of a CCD chip or a multi-channel detector.

Diffuse Reflectance Spectroscopy for the Analysis of Bioactive Compounds

It is well known that some of the most relevant therapeutic actions of the polyphenols are: anti-carcinogenic, antimicrobial, antidiabetic, anti-arteriosclerotic and anti-oxidative. On the other hand, tea has high polyphenonl content and, next to water, tea is the second most consumed drink in the world that. Both of these considerations create a need for a system of measuring tea quality *in situ* and real-time. For this purpose, Shuai Qi *et al.* have developed a low-cost and portable system based on near infrared reflectance (NIR) spectroscopy to analyse tea quality [11]. The system consists of a light source

(10-W halogen lamp), a backscattering fiber-probe, a grating system (with a slit 25 μm), and a detector (CCD). The system is controlled by a computer, which incorporates data acquisition and statistical processing software.

A data treatment based on partial least squares (PLS) algorithm to predict polyphenols concentration in tea was implemented. In the first step, the spectra intervals were decided by using synergy interval PLS algorithm. Then, the variables were selected from decided intervals *via* competitive adaptive reweighted sampling and genetic algorithms. To verify if the applied method of analysis generated reliable results, the models were generated on the basis of 30 samples, and then, the correct prediction of the model was tested with 20 samples. The results showed that the presented system combined with PLS algorithm is adequate to quantify the polyphenols concentration in tea.

FLUORESCENCE SPECTROSCOPY

Introduction

In 1852, G. G. Stokes published an article entitled "On the Change of Refrangibility of Light" [67], describing the fluorescence phenomenon. Thus, he explained that certain materials exposed to an invisible part of the spectrum, ultraviolet light, glowed with a blue light. As a conclusion to the experiments, Stocks explained that the emission light always result in longer wavelengths than the excitation light. Then, in the next paper "On the Change of Refrangibility of Light. No II" [68], he defined this phenomenon with the word fluorescence. Chemical compounds that cause the fluorescence phenomenon, are called fluorophores or fluorescent dyes. Fluoresce mechanism occurs in different steps. First, excitation is produced, when a photon with a determined energy is absorbed and generates an excited electronic singlet state, whose lifetime is between 1 to 10 nanoseconds. Following this excited state, the molecule returns to its ground state by losing a photon with a determinate energy, which is lower than the energy of the excitation photon, having a larger wavelength. This difference between excitation photon energy and emitted photon energy is known as the Stokes shift.

Nowadays, fluorescence spectroscopy is a technique widely used in biosensing [69, 70] and in bioactive compounds analysis [12].

Fluorescence Spectroscopy for Sensing Application

The most detailed fluorescence studies are carried out using a fluorescence spectrometer. This instrument consists of a light source, which can be used a coherent light source or white-light lamp, one or more monochromator and the detector. In the case of a white-light lamp as a source and with a monochromator after this, the fluorescence spectrometer may be used to determinate the emission spectra, the detection wavelength is scanned for a specific excitation wavelength, or determinated excitation spectra, detection wavelength is fixed and the excitation wavelength is tuned.

Fluorescence Spectroscopy for the Analysis of Bioactive Compounds

The fact that polyphenols have versatile activity *in vivo*, and their quantification is very much important for a healthy diet, motivated Uchangi Satyaprasad Akshath *et al.* to develop fluorescence method for ultrasensitive detection of polyphenols by using laccase enzyme-immobilized on CdTe quantum dots [12]. Therefore, the sensing method proportioned linear quenching of QDs fluorescence with respect to the concentration of three different polyphenols: catechin, epicatechin and epigallocatechin-gallate, in the range of 100 mg to 1 ng/mL. In addition, differential quenching response provided "finger print" for individual polyphenol detection.

COLORIMETRIC-ABSORPTION SPECTROSCOPY

Introduction

In 1802, William Hyde Wollaston observed the dark lines in the sunlight spectrum, and he deduced that this was separation among colours. Joseph von Fraunhofer found 574 dark lines in the solar spectrum, which, in his honour, are called Fraunhofer lines. However, their origin was not deduced until 1859, when Gustav Kirchhoff and Robert Bunsen explained that the dark lines are caused by sunlight when crossing the atmosphere, due to absorption by vapours in the atmosphere [71], thus leading to differentiate between samples through absorption spectrum.

Colorimetric sensors are promising systems to carry out specific studies in detecting and characterizing bioagent compounds in food [13, 14, 17].

Colorimetric-absorption Spectroscopy for Sensing Application

In sensor application, a reaction between analyte and the substrate is carried out. The result of the reaction is a product that has a specific colour and, in many cases, it is enough to be visually detected. The analyte concentration could also be determined by the intensity of the colour over the exposure time. Papers or tubes are the most common platform for colorimetric sensors. Because of their low cost and easy operation, the colorimetric sensors paper is one of the more widespread techniques for biological detection. In detection process, a drop is absorbed in a paper, in which dyes and a pH indicator were previously deposited. Thus, due to the reaction of the different elements, the paper colour changes. On the other hand, colorimetric sensor tubes are based on sorbent material that contains the reagent solution. Detection is known due to the change of the colour when the target reacts with the tube sorbent material.

In the case that visual detection can not be carried out, an absorption spectrometer can be used. The absorption spectrometer detects the intensity of the radiation transmitted through the sample and it is compared with the intensity of the radiation passing through a reference. Therefore, a lamp provides a spectrum in the range required that is separated in different wavelengths by a monochromator and then a detector quantify the intensity of the light transmitted.

Colorimetric-absorption Spectroscopy for the Analysis of Bioactive Compounds

Colorimetric technique has been successfully used in polyphenol detection. Therefore Alona P. Umali *et al.* developed a peptide based array composed of a combinatorial library of colorimetric indicators, in which a shift of the indicator by the wine tannins produces a spectral change, which is monitored using absorbance spectroscopy [17]. The variability among the ensembles shows that differential sensing is an effective way for fingerprinting complex mixtures using competitive binding between the analytes and the sensing array. Classification of wine samples was carried out by linear discriminant analysis (LDA), and correlation of wine sensory attributes from receptor responses was executed by partial least squares regression (PLSR).

Polyphenol antioxidant activity is another example of the importance of detecting bioactive compounds. According to Vilela *et al.* the gold nanoparticles formation driven by polyphenols standards and extract can be fitted with a sigmoidal curve, the peak of absorbance being a function of the polyphenol concentration [15]. Therefore, Della Pelle *et al.* used gold nanoparticles formation driven by the polyphenols in olive oil as a based colorimetric assay in order to assess the antioxidant activity in olive oils [14].

CONCLUSION AND FUTURE TRENDS

The present chapter attempts to provide information about the recent and latest advanced in electronic sensory systems for detection and characterization of Bioactive Compounds. The main problems encountered in Bioactive Compounds applications are related with their high molecular weight and low volatility, which makes them difficult to detect with gas sensors and another techniques. However, the detection of these compounds is very important because of their high value for the health. There are several electronic sensory systems, including optical ones and nanotechnology and gas sensors based systems for the characterization of bioactive compounds and some experiments have been described in this chapter and bibliography. The main applications found in the literature have also reviewed (Table **2**). It is important to stand out that these systems could be a promising field for research and developing of industrial instruments applied to real contexts.

The concept 'e-nose' is very attractive since researchers have attempted the mimicking of mammalian sense of olfaction without the subjective component implicit on it. In any case, there are not much e-noses in Bioactive Compounds field, although they can be used for the characterization of these compounds in food applications through the differences in the aroma due to the addition of them. But in other hand, some of the well-known problems (robustness, selectivity and reproducibility) of e-noses must be solved before its use in industrial applications. Nanotechnology will decisively influence miniaturization and cheapening of e-noses, integrating gas sensors in smart devices and creating large sensor networks combined with advanced instrumentation and electronics.

An improving of the signal and of the data processing system could help to enhance the features of the detection system and therefore the robustness of these systems. There is also a great need to compensate the inherent drift of the sensors, the humidity effect and other interferents. Moreover, these electronic sensory systems require to recognize specific patterns representing different foods or substances with Bioactive Compounds. All these improvements are necessary to incorporate these devices into real industrial processes.

Novel and amazing applications of Electronic Sensory Systems are expected for the characterization of Bioactive Compounds for the coming years that benefit both food industry and users. It will be necessary to develop robust systems for long time measurements that can be easily operated and maintained, and allow the handling by non expert technicians. This is the main challenge for the scientific groups that develops Electronic Sensory Systems.

CONFLICT OF INTEREST

The authors confirm that they have no conflict of interest to declare for this publication.

ACKNOWLEDGEMENTS

Declarad none.

REFERENCES

[1] Biesalski HK, Dragsted LO, Elmadfa I, *et al.* Bioactive compounds: definition and assessment of activity. Nutrition 2009; 25(11-12): 1202-5. [http://dx.doi.org/10.1016/j.nut.2009.04.023] [PMID: 19695833]

[2] Denny A, Buttriss J. Plant foods and health: focus on plant bioactives. London, UK.: British Nutrition Foundation, 2007.

[3] Muñoz S, Olza J, Gómez C. Compuestos bioactivos de los alimentos de origen vegetal En: Tratado de Nutrición Tomo II. Madrid: Panamericana 2010.

[4] Holst B, Williamson G. Nutrients and phytochemicals: from bioavailability to bioefficacy beyond antioxidants. Curr Opin Biotechnol 2008; 19(2): 73-82. [http://dx.doi.org/10.1016/j.copbio.2008.03.003] [PMID: 18406129]

[5] Available from: www.kelloggs.es/content/dam/newton/media/manual_de_nutricion_new/Manual_

Nutricion_Kelloggs_Capitulo_02.2.pdf 2012; pp. 32-36.

[6] Santos JP, Elizalde J, Lozano J, Fernández MJ, Fontecha JL, Horrillo MC. Desarrollo de una nariz electrónica inalámbrica portátil. Aplicación al campo agroalimentario. Actas del congreso del Seminario Anual de Automática. Badajoz, Spain: Electrónica Industrial e Instrumentación SAAEI 2011; pp. 445-50.

[7] Cui S, Wang J, Yang L, Wu J, Wang X. Qualitative and quantitative analysis on aroma characteristics of ginseng at different ages using E-nose and GC-MS combined with chemometrics. J Pharm Biomed Anal 2015; 102: 64-77.
[http://dx.doi.org/10.1016/j.jpba.2014.08.030] [PMID: 25244512]

[8] Minunni M, Tombelli S, Mascini M, Bilia A, Bergonzi MC, Vincieri FF. An optical DNA-based biosensor for the analysis of bioactive constituents with application in drug and herbal drug screening. Talanta 2005; 65(2): 578-85.
[http://dx.doi.org/10.1016/j.talanta.2004.07.020] [PMID: 18969838]

[9] Minunni M, Bilia AR. SPR in drug discovery: searching bioactive compounds in plant extracts. Methods Mol Biol 2009; 572: 203-18.
[http://dx.doi.org/10.1007/978-1-60761-244-5_13] [PMID: 20694694]

[10] Liu P, Liu R, Guan G, Jiang C, Wang S, Zhang Z. Surface-enhanced Raman scattering sensor for theophylline determination by molecular imprinting on silver nanoparticles. Analyst (Lond) 2011; 136(20): 4152-8.
[http://dx.doi.org/10.1039/c1an15318h] [PMID: 21853172]

[11] Qi S, Ouyang Q, Chen Q, Zhao J. Real-time monitoring of total polyphenols content in tea using a developed optical sensors system. J Pharm Biomed Anal 2014; 97: 116-22.
[http://dx.doi.org/10.1016/j.jpba.2014.04.034] [PMID: 24867295]

[12] Akshath US, Shubha LR, Bhatt P, Thakur MS. Quantum dots as optical labels for ultrasensitive detection of polyphenols. Biosens Bioelectron 2014; 57: 317-23.
[http://dx.doi.org/10.1016/j.bios.2014.01.038] [PMID: 24607583]

[13] Sharpe E, Andreescu S. Portable nanoparticle based sensors for antioxidant analysis. Methods Mol Biol 2015; 1208: 221-31.
[http://dx.doi.org/10.1007/978-1-4939-1441-8_16] [PMID: 25323510]

[14] Della Pelle F, Vilela D, González MC, *et al.* Antioxidant capacity index based on gold nanoparticles formation. Application to extra virgin olive oil samples. Food Chem 2015; 178: 70-5.
[http://dx.doi.org/10.1016/j.foodchem.2015.01.045] [PMID: 25704685]

[15] Vilela D, González MC, Escarpa A. Nanoparticles as analytical tools for *in vitro* antioxidant-capacity assessment and beyond. Trends Analyt Chem 2015; 64: 1-16.
[http://dx.doi.org/10.1016/j.trac.2014.07.017]

[16] Arciuli M, Palazzo G, Gallone A, Mallardi A. Bioactive paper platform for colorimetric phenols detection. Sens Actuators B Chem 2013; 186: 557-62.
[http://dx.doi.org/10.1016/j.snb.2013.06.042]

[17] Umali AP, Ghanem E, Hopfer H, *et al.* Grape and wine sensory attributes correlate with pattern-based discrimination of cabernet sauvignon wines by a peptidic sensor array. Tetrahedron 2015; 71(20):

3095-9.
[http://dx.doi.org/10.1016/j.tet.2014.09.062]

[18] Zhang XL, Yu HJ. Sensitive determination of bioactive crocin analogues in gardenia jasminoides ellis using SPE prior to HPLC analysis. Appl Mech Mater 2013; 303-306: 654-8.
[http://dx.doi.org/10.4028/www.scientific.net/AMM.303-306.654]

[19] Gardner JW, Bartlett PN. A brief history of electronic noses. Sens Actuators B Chem 1994; 18(1): 210-1.
[http://dx.doi.org/10.1016/0925-4005(94)87085-3]

[20] Gardner JW, Bartlett PN. Electronic noses: principles and applications. New York: Oxford University Press 1999; Vol. 233: p. 245.

[21] Lozano-Rogado J, Teixeira da Silva J A. New technology in sensing odours: from human to artificial noses. Floriculture, ornamental and plant biotechnology 2006; 4: 152-61.

[22] Lozano J, Santos JP, Gutiérrez J, Horrillo MC. Comparative study of sampling systems combined with gas sensors for wine discrimination. Sens Actuators B Chem 2007; 126(2): 616-23.
[http://dx.doi.org/10.1016/j.snb.2007.04.018]

[23] Pearce TC, Schiffman SS, Nagle HT, Gardner JW, Eds. Handbook of machine olfaction: electronic nose technology. USA: John Wiley & Sons 2006.

[24] Guadarrama A, Fernandez JA, Iniguez M, Souto J, De Saja JA. Array of conducting polymer sensors for the characterisation of wines. Anal Chim Acta 2000; 411(1): 193-200.
[http://dx.doi.org/10.1016/S0003-2670(00)00769-8]

[25] Sharma P, Ghosh A, Tudu B, *et al.* Monitoring the fermentation process of black tea using QCM sensor based electronic nose. Sens Actuators B Chem 2015; 219: 146-57.
[http://dx.doi.org/10.1016/j.snb.2015.05.013]

[26] Lozano J, Fernández MJ, Fontecha JL, *et al.* Wine classification with a zinc oxide SAW sensor array. Sens Actuators B Chem 2006; 120(1): 166-71.
[http://dx.doi.org/10.1016/j.snb.2006.02.014]

[27] Di Natale C, Macagnano A, Nardis S, *et al.* Comparison and integration of arrays of quartz resonators and metal-oxide semiconductor chemoresistors in the quality evaluation of olive oils. Sens Actuators B Chem 2001; 78(1): 303-9.
[http://dx.doi.org/10.1016/S0925-4005(01)00830-9]

[28] Gutierrez-Osuna R. Pattern analysis for machine olfaction: a review. Sens J IEEE 2002; 2(3): 189-202.
[http://dx.doi.org/10.1109/JSEN.2002.800688]

[29] Waldemark J, Roppel T, Wilson D, Dunman K, Padgett ML, Lindblad T. Neural networks and PCA for determining region of interest in sensory data pre-processing. Proc SPIE 1998.
[http://dx.doi.org/10.1117/12.343057]

[30] Duda RO, Hart PE, Stork DG, Eds. Pattern classification. 2nd ed., New York: Wiley 2001.

[31] Bishop CM. Neural networks for pattern recognition. Oxford: Oxford University Press 1999.

[32] Nagle H T, Gutierrez-Osuna R, Schiffman S S. The how and why of electronic noses. Spectrum, IEEE 1998; 35(9): 21-31.

[http://dx.doi.org/10.1109/6.715180]

[33] Peris M, Escuder-Gilabert L. A 21st century technique for food control: electronic noses. Anal Chim Acta 2009; 638(1): 1-15. [http://dx.doi.org/10.1016/j.aca.2009.02.009] [PMID: 19298873]

[34] Wilson AD, Baietto M. Advances in electronic-nose technologies developed for biomedical applications. Sensors (Basel) 2011; 11(1): 1105-76. [http://dx.doi.org/10.3390/s110101105] [PMID: 22346620]

[35] Wilson AD. Diverse applications of electronic-nose technologies in agriculture and forestry. Sensors (Basel) 2013; 13(2): 2295-348. [http://dx.doi.org/10.3390/s130202295] [PMID: 23396191]

[36] Falasconi M, Concina I, Gobbi E, Sberveglieri V, Pulvirenti A, Sberveglieri G. Electronic nose for microbiological quality control of food products. Int J Electrochem 2012; 2012: 1-12. [http://dx.doi.org/10.1155/2012/715763]

[37] Loutfi A, Coradeschi S, Mani GK, Shankar P, Rayappan JB. Electronic noses for food quality: A review. J Food Eng 2015; 144: 103-11. [http://dx.doi.org/10.1016/j.jfoodeng.2014.07.019]

[38] Alam H, Saeed SH, Engg C. Electronic nose in food and healh applications: a review. Int J Comput Corp Res 2012; 2: 1-17.

[39] James D, Scott SM, Ali Z, O'Hare WT. Chemical sensors for electronic nose systems. Mikrochim Acta 2005; 149: 1-17. [http://dx.doi.org/10.1007/s00604-004-0291-6]

[40] Santos JP, Lozano J, Suárez JI, Aleixandre M. Electronic Nose Network For Detection Of TATP Precursors. Proceedings of ISOEN 2013. Daegu (South Korea). 2013.

[41] Lozano J, Santos JP, Suárez JI, Arroyo P, Herrero JL, Martín A. Detection of Pollutants in Water Samples with a Wireless Hand-held E-nose. Procedia Eng 2014; 87: 556-9. [http://dx.doi.org/10.1016/j.proeng.2014.11.548]

[42] Eduardo SR. Method of prepairing lycopene - enriched formulations that are free of organic solvent, formulations thus ostained, composition composing said formulaions and use of same. Patent. WO 2006/111591; 2006

[43] Burton GW. Antioxidant action of carotenoids. J Nutr 1989; 119(1): 109-11. [PMID: 2643692]

[44] E. Sabio, "Method of preparing lycopene-enriched formulations that are free of organic solvents, formulations thus obtained, compositions comprising said formulations and use of same". Patent. WO 96/13178

[45] Kretschmann E, Raether H. Radiative decay of nonradiative surface plasmons excited by light. Z Naturforsch B 1968; 23: 2135-6.

[46] Otto A. Excitation of nonradiative surface plasma waves in silver by the method of frustrated total reflection. Z Phys 1968; 216(4): 398-410. [http://dx.doi.org/10.1007/BF01391532]

[47] Raether H. Surface plasmons on smooth and rough surfaces and on gratings tracts in modren physics. Geemany: Springer-Verlag, Berlin, 1988. 1988; (b) Boardman, A. D. Electromagnetic surface modes USA: Wiley. 1982.

[48] Flanagan MT, Pantell RH. Surface plasmon resonance and immunosensors. Electron Lett 1984; 20(23): 968-70.
[http://dx.doi.org/10.1049/el:19840660]

[49] Kim M, Park K, Jeong E-J, Shin Y-B, Chung BH. Surface plasmon resonance imaging analysis of protein-protein interactions using on-chip-expressed capture protein. Anal Biochem 2006; 351(2): 298-304.
[http://dx.doi.org/10.1016/j.ab.2006.01.042] [PMID: 16510110]

[50] Majka J, Speck C. Analysis of protein–DNA interactions using surface plasmon resonance. In: Seitz H, Ed. Analytics of Protein–DNA Interactions. Springer Berlin Heidelberg 2007; Vol. 104: pp. 13-36.
[http://dx.doi.org/10.1007/10_026]

[51] Heuberger K, Lukosz W. Embossing technique for fabricating surface relief gratings on hard oxide waveguides. Appl Opt 1986; 25(9): 1499-504.
[http://dx.doi.org/10.1364/AO.25.001499] [PMID: 18231367]

[52] Dostálek J, Čtyroký J, Homola J, *et al.* Surface plasmon resonance biosensor based on integrated optical waveguide. Sens Actuators B Chem 2001; 76(1–3): 8-12.
[http://dx.doi.org/10.1016/S0925-4005(01)00559-7]

[53] Slavík R, Homola J, Čtyroký J, Brynda E. Novel spectral fiber optic sensor based on surface plasmon resonance. Sens Actuators B Chem 2001; 74(1–3): 106-11.
[http://dx.doi.org/10.1016/S0925-4005(00)00718-8]

[54] Fleischmann M, Hendra PJ, McQuillan AJ. Raman spectra of pyridine adsorbed at a silver electrode. Chem Phys Lett 1974; 26(2): 163-6.
[http://dx.doi.org/10.1016/0009-2614(74)85388-1]

[55] Van Duyne RP. Laser excitation of raman scattering from adsorbed molecules on electrode surfaces. In: Moore CB, Ed. Chemical and Biochemical Applications of Lasers. USA: Academic Press 1979; pp. 101-85.
[http://dx.doi.org/10.1016/B978-0-12-505404-1.50009-X]

[56] Albrecht MG, Creighton JA. Anomalously intense Raman spectra of pyridine at a silver electrode. J Am Chem Soc 1977; 99(15): 5215-7.
[http://dx.doi.org/10.1021/ja00457a071]

[57] Nie S, Emory SR. Probing single molecules and single nanoparticles by surface-enhanced Raman scattering. Science 1997; 275(5303): 1102-6.
[http://dx.doi.org/10.1126/science.275.5303.1102] [PMID: 9027306]

[58] Kneipp K, Wang Y, Kneipp H, *et al.* Single molecule detection using surface-enhanced raman scattering (SERS). Phys Rev Lett 1997; 78(9): 1667-70.
[http://dx.doi.org/10.1103/PhysRevLett.78.1667]

[59] Pettinger B. Single-molecule surface- and tip-enhanced Raman spectroscopy. Mol Phys 2010; 108(16): 2039-59.

[http://dx.doi.org/10.1080/00268976.2010.506891]

[60] Lyandres O, Yuen JM, Shah NC, VanDuyne RP, Walsh JT, Glucksberg MR. Progress toward an *in vivo* surface-enhanced Raman spectroscopy glucose sensor. Diabetes Technol Ther 2008; 10(4): 257-65.
[http://dx.doi.org/10.1089/dia.2007.0288] [PMID: 18715199]

[61] Bantz KC, Meyer AF, Wittenberg NJ, *et al.* Recent progress in SERS biosensing. Phys Chem Chem Phys 2011; 13(24): 11551-67.
[http://dx.doi.org/10.1039/c0cp01841d] [PMID: 21509385]

[62] Xie W, Qiu P, Mao C. Bio-imaging, detection and analysis by using nanostructures as SERS substrates. J Mater Chem 2011; 21(14): 5190-202.
[http://dx.doi.org/10.1039/c0jm03301d] [PMID: 21625344]

[63] Companion AL. The diffuse reflectance spectra of zinc oxide and zinc peroxide. J Phys Chem Solids 1962; 23(12): 1685-8.
[http://dx.doi.org/10.1016/0022-3697(62)90205-6]

[64] Wendlandt WW, Franke PH, Smith JP. High temperature diffuse reflectance spectroscopy. Anal Chem 1963; 35(1): 105-7.
[http://dx.doi.org/10.1021/ac60194a032]

[65] Reynolds L, Johnson C, Ishimaru A. Diffuse reflectance from a finite blood medium: applications to the modeling of fiber optic catheters. Appl Opt 1976; 15(9): 2059-67.
[http://dx.doi.org/10.1364/AO.15.002059] [PMID: 20165338]

[66] Rahman S, Lipert RJ, Porter MD. Rapid screening of pathogenic bacteria using solid phase concentration and diffuse reflectance spectroscopy. Anal Chim Acta 2006; 569(1-2): 83-90.
[http://dx.doi.org/10.1016/j.aca.2006.03.083]

[67] Stokes GG. On the change of refrangibility of light. Philos Trans R Soc Lond 1852; 142: 463-562.
[http://dx.doi.org/10.1098/rstl.1852.0022]

[68] Stokes GG. On the Change of Refrangibility of Light. No. II. Philos Trans R Soc Lond 1853; 143: 385-96.
[http://dx.doi.org/10.1098/rstl.1853.0016]

[69] Ding C, Tian Y. Gold nanocluster-based fluorescence biosensor for targeted imaging in cancer cells and ratiometric determination of intracellular pH. Biosens Bioelectron 2014; 65C: 183-90.
[PMID: 25461156]

[70] Sun J, Ji J, Sun Y, Abdalhai MH, Zhang Y, Sun X. DNA biosensor-based on fluorescence detection of E. coli O157:H7 by Au@Ag nanorods. Biosens Bioelectron 2015; 70: 239-45.
[http://dx.doi.org/10.1016/j.bios.2015.03.009] [PMID: 25829221]

[71] James J. A brief history of spectroscopy Spectrograph design fundamentals. UK: Cambridge University Press 2007.
[http://dx.doi.org/10.1017/CBO9780511534799]

CHAPTER 7

Bioactive Biomolecules Interacting with Cellular Membranes: Modelling with Langmuir Monolayers

Felippe J. Pavinatto[1,*], **Thatyane M. Nobre**[1], **Luciano Caseli**[2] and **Osvaldo N. Oliveira, Jr.**[1]

[1] *University of Washington, Clean Energy Institute, Seattle - WA, United States*

[2] *Institute of Environmental, Chemical, and Pharmaceutical Sciences, Federal University of São Paulo, Diadema, São Paulo, Brasil*

Abstract: Cell membranes are essential for life, since they compartmentalize cells, the basic unit of living matter, and mediate their symbiosis with the environment. In many events in which cells participate, interaction with biomolecules such as polysaccharides, proteins and peptides, is crucial. In spite of the developments in physics, chemistry and biology in the last century, the study of cells and bioactive biomolecules interacting "*in vivo*" is still elusive. Therefore, simplified models are employed to obtain fundamental knowledge from these systems. In this chapter, the use of Langmuir monolayers as cell membrane models made to interact with biomolecules is described. Firstly, a brief introduction on the nature of nanostructured Langmuir monolayers is provided. In particular, we describe technical aspects related to the experimental setup and procedures to guide novice researchers in the field. We also describe the fundamentals and possibilities of the most useful techniques applied to film characterization. In the final section, examples are given of Langmuir monolayers successfully employed to obtain relevant molecular-level information from systems involving biomolecules. Trends in the field are discussed as motivation for innovative research to bridge the gap between biology (and medicine) and the physicochemical framework used in cell membrane modeling.

* **Corresponding author Felippe J. Pavinatto:** University of Washington, Clean Energy Institute, Seattle - WA, United States; E-mail: felippe@uw.edu

Constantin Apetrei (Ed.)

Keywords: Biomembrane Models, Biomolecules, Cell Membrane Interactions, Langmuir Monolayers, Peptides and Polysaccharides, Proteins.

1. INTRODUCTION

Cells are the basic unit of living matter, and in every living system (*e.g.* human body) they function to perform specific tasks and keep the balance of life. This very specialized duty takes place in constant interaction with the environment, and is subject to the action of external agents, naturally present or artificially administered to the medium. In this context, the membrane "represents" the cell in the outside environment, for it faces interacting molecules and mediates their action on the cell. The cell membrane also controls the influx and efflux of molecules, thus determining some intracellular processes and ultimately cell composition. Needless to say, it is virtually impossible to characterize this complex, dynamic system "on the fly", even using state-of-the-art techniques. The characterization of molecular-level interactions, in particular, is hard to attain for such elaborated structure. Hence, simplified models are used to mimic the membrane and study its interaction with bioactive compounds in a controlled fashion. In this chapter, we will describe and exemplify the use of Langmuir films for this goal.

The topic covered here has been reviewed and discussed a few times, and most of the relevant literature in the field is listed in the references. We conceived this chapter with a different focus, to contribute with something unique to the literature. Approximately half of the chapter is devoted to discussing systems in which Langmuir films are used as models to study interaction of cell membranes with biomolecules (section 4). In this part, our chapter distinguishes itself by highlighting studies where (in our opinion) more compelling and close-to-application cases are studied, and contributions from the models are really relevant. In addition, in section 2 we describe Langmuir films and why they are useful as membrane models. In section 3, we introduce the many characterization techniques used for these films. We opted to concentrate on practical aspects rather than on elaborated theoretical description, since we aim at offering "shortcuts" for the effective, precise learning of the methods by scientists interested in the field and in its possibilities. In other words, we aim at providing an accurate and robust material for a textbook.

2. LANGMUIR MONOLAYERS AND THEIR USE AS MEMBRANE MODELS

Langmuir films are named after Irving Langmuir, a prominent American scientist that studied liquid surfaces in the first decades of the 20th century at the General Electric Laboratories, New York [1]. In his seminal paper, Langmuir described with great detail the action of surface forces as well as the nature of interfacial ultrathin (monomolecular) floating films formed on several liquid surfaces. Among other contributions, this work rendered him the Nobel Prize in Chemistry in 1932. Since then, knowledge has evolved and the (nano)structure of the films is nowadays known in detail. If deposited in the right amount on the water surface, amphiphilic molecules can spread as a monolayer, with their hydrophilic portion facing the water and the hydrophobic part oriented towards the air. A scheme of the Langmuir film architecture is shown in Fig. (**1**).

The resemblance of the film structure with one of the leaflets of a cell membrane is clear. Although in some cases, as in gram-negative bacteria, the cell membrane has a multilayered structure, in the vast majority of the cells of either prokaryotes or eukaryotes the membrane is a bilayer of lipids in which proteins are inserted. The drawing in Fig. (**1**) illustrates a film of a single component (*e.g.* phospholipid) with two tails attached to a single head. However, the Langmuir film can be made of more than one component, have transmembrane or peripheral proteins incorporated to it, or even be prepared from a membrane extract with a composition very similar to the one found in the actual membrane.

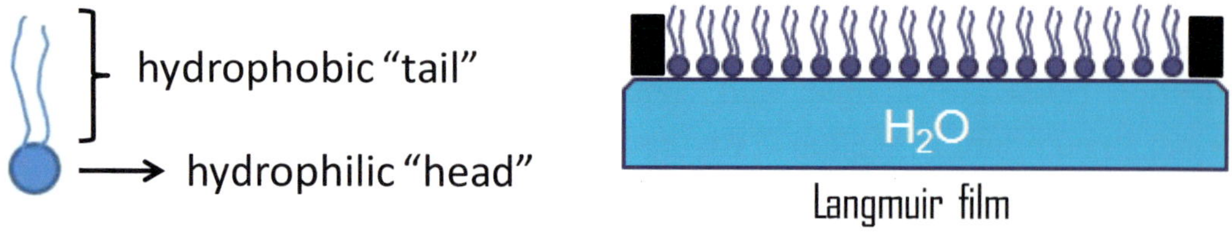

Fig. (1). Left: representation of an amphiphilic molecule with its hydrophilic and hydrophobic parts. Right: schematics of a Langmuir film, where amphiphilic molecules are assembled as a monolayer at the air-water interface. [Adapted with permission from [2]. Copyright 2010, American Chemical Society].

The similarity and equivalence between a Langmuir film made of phospholipids and half of a cell membrane are well accepted in the scientific community [3]. The only point that is still confusing even to some experienced scientists is the correspondence between the concepts of film lateral pressure and packing state of the membrane. A Langmuir film compressed to a high surface pressure value (see definition below) has a packing state similar to the one of a real biomembrane. This is to say, the molecules in the film have average orientation and area density similar to the one displayed by real membrane molecules. However, the major difference for a real biomembrane is that there is no external mechanical constraint to the area phospholipids can occupy. Thus, it does not make sense to speak about "pressure of a real biomembrane". In the real system, instead, what holds the molecule packed together is the so-called "hydrophobic effect" [4]. This effect arises from the hydrophobic chains of the phospholipids avoiding the aqueous environment, thus resulting in the chains assembling as a bilayer structure.

As for the modeling of cell membranes, the main advantage of using Langmuir films is the precise control over interfacial film composition and packing state [5]. Some disadvantages are its non-suitability for permeability studies, the planarity of the surface that differs from the real curvature of a cell membrane, and the need for a clean environment (usually a clean room) to get reliable results. In the next sub-section, the apparatus and a few experimental features in Langmuir film preparation are explained, with special emphasis on scientific-technical issues for "hands on" research in the field.

2.1. Technical Aspects

Langmuir films are formed in dedicated equipment referred to as a Langmuir trough picture in Fig. (**2**). Essentially, the trough is made of Teflon filled with the liquid (usually water) that acts as the subphase for the spreading of amphiphilic molecules. Spreading is carried out by dispensing small droplets of a solution of the film material in a volatile and water immiscible solvent (*e.g.* chloroform). Another important component of the apparatus is a pair of movable barriers that allow for the control of the area occupied by the interfacial film. In the picture in Fig. (**2**), a surface pressure sensor and a dipper device (for transferring Langmuir-

Blodgett films) are also shown, in the most typical configuration of the system.

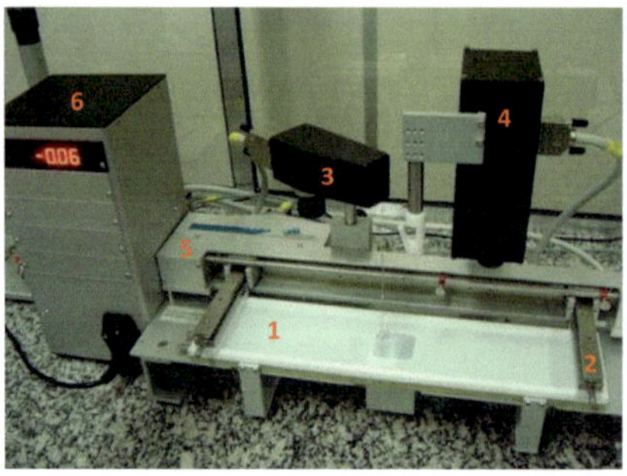

Fig. (2). Picture of a Langmuir trough with its most typical configuration. 1 - Teflon trough. 2 - movable barriers. 3 - surface pressure sensor. 4 - "dipper" device. 5 - barriers driver unit. 6 - computer interface and control unit.

The aqueous subphase for Langmuir films can mimic the extracellular or intracellular medium, with controlled pH, ionic strength and temperature. Interacting materials can either be co-spread with the lipids at the air-water interface or diluted in the aqueous subphase, depending on their solubility. Actually, this is the first feature of the biomolecules that must be known in order to plan the experiments. Most biological molecules are water soluble as almost all biological processes happen in this solvent. However, transmembrane proteins and some long-chain peptides with a significant number of hydrophobic residues can be highly lipophilic and completely insoluble. In the next sub-sections, practical aspects related to each of these two possibilities are discussed for performing accurate and artifact-free experiments.

2.1.1. Co-spreading

Phospholipids are the main constituent of cell membranes, and the models usually contain a high proportion of these molecules to properly reflect the chemistry of the biomembranes. In this regard, the phospholipids mostly studied in Langmuir films are DPPC (dipalmitoyl-phosphatidyl-choline) and DPPG (dipalmitoyl-phosphatidyl-glycerol). Although these lipids possess saturated palmitoyl chains that are not so abundant in real biomembranes, they form stable monolayers and bear choline and glycerol headgroups, which are prevalent in cell surfaces. Together with the availability of high-purity samples of these lipids, this justifies their historical preference. Pure films of these phospholipids are studied as the simplest model of half a cell membrane. However, when interactions with biomolecules are to be evaluated, an extra "guest" material is usually mixed with this "matrix" film. There are two main reasons why additional insoluble molecules are spread together with phospholipids. Firstly, lipids of different composition (*e.g.* cholesterol, unsaturated lipids, *etc.*) may be introduced at the interface for the film composition better match that of a real membrane. Secondly, the interacting molecule can be itself insoluble, thus having to be placed at the interface along with the phospholipid.

Co-spreading of insoluble molecules is advantageous from the point of view that only in this case film composition is precisely controlled. However, the way the materials are placed together at the interface, and the way molecular area is calculated for the isotherms can have profound effects over the results. Insoluble materials can be co-spread in three different ways:

i) Co-dissolution: materials are mixed in the powder form, and then solvent is added to attain the desired concentration for the mixture. In this strategy, the proportion between the mixed materials is precisely controlled, but irreversible aggregation of the compounds can be induced by the physical contact of the materials in the dry from. This option must therefore be used with caution, especially when macromolecules are involved, since they are more susceptible to aggregation;

ii) Co-spreading of individual solutions: the materials to be mixed can be separately dissolved in the same solvent and then both solutions be individually spread on the subphase surface. This will likely avoid aggregation before dissolution, but can lead to another drawback, which is an induced, undesirable inhomogeneous distribution of the materials at the surface. Materials can concentrate preferentially at the spot where the droplets of solution are placed, and this initial inhomogeneous distribution can induce specific interactions that could differ from those the molecules would spontaneously experience in a mixture at the thermodynamic equilibrium;

iii) Mixture of Individual Solutions and then Spreading: This is the most recommended procedure to get reproducible results that better reflect an equilibrium situation for a mixture. Materials are dissolved separately (as in *ii*), but the individual solutions are mixed in the right proportion before spreading. Thus, contact between the molecules is made in solution and not in the dry form (as in *i*), and more balanced interactions are attained. Induced aggregation due to droplets at the surface is also avoided, and the thermodynamic of mixing is less disturbed.

Traditionally, results from the most basic surface characterization of Langmuir films (*e.g.* surface pressure and surface potential) are recorded and plotted against the area per molecule, which is the area available for each molecule at the air-water interface. The latter is calculated by the Langmuir trough apparatus by dividing the area of the surface (which varies during compression with the movable barriers) by the number of molecules spread on the surface (estimated from solution concentration, spread volume and molar mass of the compound). For a single component film this is a straightforward procedure, but for a mixed film it can be tricky.

There are generally two approaches adopted for mixed films. The first one is to inform the apparatus that only the most abundant material is being spread (for instance, the phospholipid matrix), and let the system calculate the area per molecule for that material. This is of course not the protocol that best reflect reality, but it is convenient to estimate the effects from the guest molecules on the matrix (not necessarily interactions). In this first approach, we can think of two scenarios: a) only matrix molecules are taken into account to calculate the area axis; b) all the molecules at the surface are counted as matrix molecules in the calculation. For the sake of exemplification, we can analyze a mixed DPPC:cholesterol (Chol) Langmuir film with a 4:1 molar proportion of molecules. In the first scenario (a), we could inform the apparatus that there is only DPPC at the interface (80% of the actual total of molecules), and neglect Chol, and in a second scenario (b), we could inform that 100% of the molecules are DPPC. The surface pressure-area isotherms would then be placed in different positions of the area per molecule axis, as illustrated in Fig. (**3**) by the blue and green curves for procedures (a) and (b), respectively. These isotherms are useful to estimate the effects of Chol over the shape of DPPC curves. With position calculated based on (b), one learns of the area occupied by Chol relative to the one that would be occupied by DPPC (since Chol is treated as DPPC). However, the nature of materials interaction cannot be estimated from any of the curves, for which we would have to use the second approach discussed below.

A second approach to calculate the area axis for Langmuir isotherms is based on the precise accounting of each material in the film, the "size" of the molecules and their proportion in the mixture. Although without real physical meaning, an "averaged molar mass" calculated from the individual molar masses and the respective proportion of each material is informed to the apparatus, together with the actual solution concentration and spread volume. This is done to induce the trough to consider both molecules in the mixture, even if it uses only a single material in its calculations. In this case, the comparison of the position of the isotherms in the area per molecule axis with the position of the curves from the pure materials in the mixture is an indicative of materials interactions. Moreover, deviations from the "ideal" position, also estimated from the area occupied by the individual molecules in the pure films, can be used to calculate the excess Gibbs free energy involved in the interaction. The curve with area axis calculated in this way is also included in Fig. (**3**) for comparison (pink curve). The isotherm lies closer to the isotherm of pure DPPC, but the replacement of some DPPC by Chol makes the film more expanded (shifted to higher areas). An example of the estimative of the interaction energy using mixtures of different compositions and the approach discussed here can be found in [6]. The authors have studied mixtures of the synthetic phospholipid dimiristoyl-phosphatidic acid (DMPA) and Chol, and their results are reproduced in Fig. (**4**). The measured area for specific surface pressures deviates from the thermodynamically "ideal value" calculated from the individual molecular areas and the proportion of each material at the interface, with the interaction between the materials being attractive at low packing, but repulsive at higher (more biologically-relevant) surface pressures.

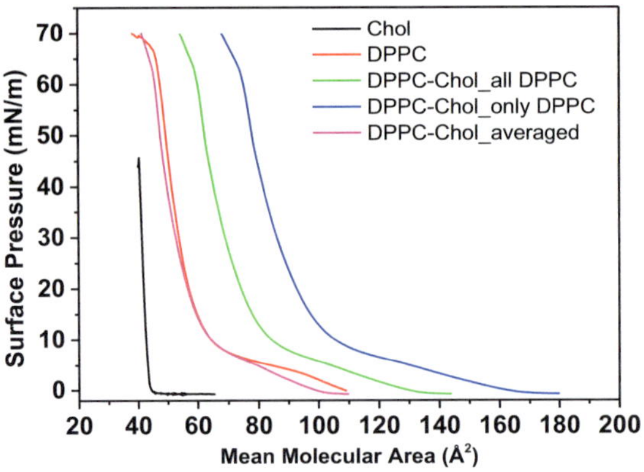

Fig. (3). Surface pressure-area isotherms of a co-spread DPPC:Chol mixed film with a molar proportion of 4:1. Pink, green and blue curves are all from the same measurement and have all the same shape, but their position vary due to the method used to calculate the molecular area axis. For the blue curve, the area axis was calculated taking into account only the host DPPC molecule, and for the green curve, all the molecules at the interface (including cholesterol) were considered as DPPC. The pink curve is plotted considering both molecules and their occupied area, averaged by their molar mass and proportion in the mixture. Curves for pure DPPC and Chol monolayers were added as reference. Courtesy of Celina M. Miyazaki.

2.1.2. Adsorption from the subphase

For water-soluble biomolecules, the design of experiments with Langmuir films as membrane models is more involved than for the case where all molecules are insoluble. Soluble proteins, peptides or polysaccharides can be incorporated into the subphase either by dissolution (previous to film spreading), or injection (after the Langmuir monolayer is already formed). The choice of the method must take into account mainly the biological relevance of the experiment, in view of the application to be mimicked. For instance, one can speculate about a peptide-based drug that is administered to the human body by the oral intake of pills of the compound in the powder form. When the container of the pills burst in our gastrointestinal tract, peptide is released and starts to be absorbed, going to our blood stream and eventually spreading to the whole body. Specifically in the targeted action site, drug concentration will start to rise from zero until it reaches a certain equilibrium value. In the modeling of the described application, peptide injection in the subphase appears as more adequate, since the concentration build up at the surface of biological tissues would be better simulated. In contrast, previous dissolution in the subphase can be more adequate to mimic, for instance, the healing action of chitosan over wounds on skin, a site where the polysaccharide is usually delivered all-at-once by means of patches or membranes.

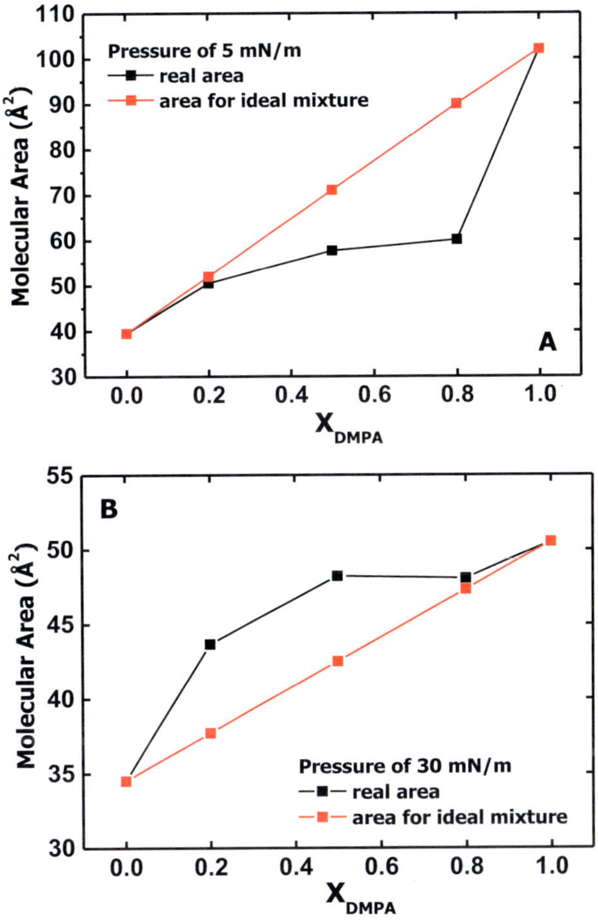

Fig. (4). "Real" and "ideal" molecular area for mixed monolayers of DMPA and Chol, plotted for two surface pressures: (A) 5 mN/m, and (B) 30 mN/m. "Real" area was directly read from the isotherms of the corresponding films, and "ideal" area (A_{12}) was calculated from the equation $A_{12} = X_1A_1 + X_2A_2$, where X_i is the molar proportion of the "i" material in the mixture and A_i is the area occupied by that material in the pure monolayer. [Reprinted with permission from (6). Copyright 2009, American Chemical Society].

Regardless of the methodology used - previous dissolution or injection - the most relevant property to be characterized for soluble biomolecules is surface activity. Water-soluble molecules can, in some cases, inherently present some tendency to accumulate at the air-water interface, being surfactants. In other cases, the presence of a phospholipid film at the interface induces surface activity. Both inherent and induced surface activity can be directly measured in the Langmuir trough by monitoring the evolution of surface pressure (π - decrease in surface tension) with time. As an example, Fig. (**5**) shows kinetic curves for chitosan (a polysaccharide) adsorption on a clean air-water interface and over a Langmuir film of DPPC. Chitosan has no intrinsic surface activity, and π does not change over time for the bare interface. On the other hand, the pressure increases by approximately 7 mN/m in 5 min when there is a phospholipid film at the interface. This induced surface activity is a first indicative of a positive spontaneous interaction of the polysaccharide with DPPC.

The adsorption kinetics measured in a Langmuir trough may lead to apparently contradictory results. For instance, Fig. (**6A**) shows no increase in pressure with the enzyme Tyrosinase (Tyr) in the subphase. In fact, there is even a small decrease caused by instabilities owing to the long duration of the experiment (a little more than 2 h). One could therefore infer that there is no adsorption of Tyr. However, a surface pressure-area isotherm can actually be obtained when the same interface is compressed by the barriers, as shown in Fig. (**6B**). The origin of this unexpected behavior is the large surface area of the Langmuir trough used in the experiment, which is not adequate to detect migration of the biomolecules to the interface. While the curve for chitosan has been measured in a dedicated low-volume/low-surface-area tensiometer, the one for Tyr has been recorded in a trough with a surface area of 488 cm^2. When a small amount of Tyr is adsorbed in this large area, it is insufficient to cause a change in surface tension (similar to a Langmuir film in the gaseous state), thus being undetectable in this setup. In contrast, when the area is reduced by the closing of the barriers, the density of molecules and hence π rise. This clearly serves to demonstrate that extreme care must be taken at recording adsorption kinetics "*in situ*" for soluble biomolecules in a Langmuir trough.

In general, injection is preferred over previous dissolution in experiments with soluble biomolecules. Micro-syringes are employed to cause only minimal disturbances to the floating films when crossing the interface to reach the subphase. Moreover, this simple act of injecting must be planned and performed in a reproducible fashion. Variations like deeper or shallower, single point or multiple injections can have a strong effect on the adsorption of the biomolecules onto the pre-formed Langmuir film, and hence, on the results using the Langmuir model. Once again, the goal of the experiment must direct the choice of the procedure. For the sake of proper and reproducible modeling of a biomembrane, deep, multiple points injection is thought as better. However, in cases in which maximum incorporation to the film is sought (for instance, when immobilizing biomolecules for the fabrication of biosensors), it is fair to foster accumulation of materials at the interface by injecting them right underneath the film in a single central position in the trough. Some of the latest Langmuir trough models already come with "injection ports" to facilitate the experimental work described above.

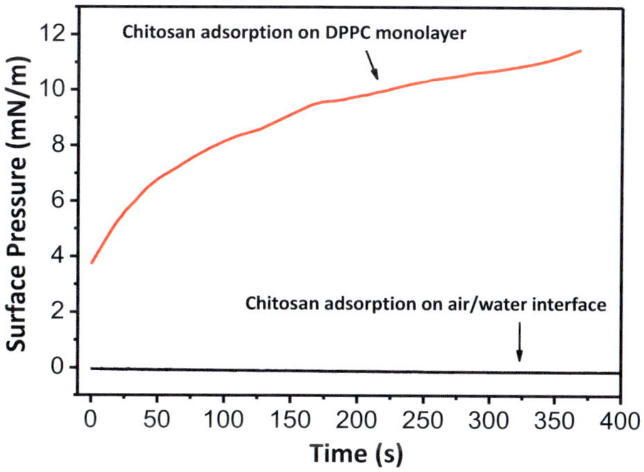

Fig. (5). Adsorption kinetics of the polysaccharide chitosan onto a bare air-water interface (black curve) or to a Langmuir film of DPPC (red curve). [Reprinted with permission from [7]].

Diffusion phenomena, and their dependence on geometrical features of the trough, are also important for the incorporation of soluble biomolecules to interfaces. For macromolecules such as proteins and polysaccharides, in particular, migration to the surface after injection heavily depends on their rate of diffusion in the liquid subphase. It is known that the diffusion coefficient of these biomacromolecules depends on parameters like size, shape of molecule, viscosity and interaction with solvent, as well as their concentration in the medium [9]. In addition to these molecular parameters, the surface area-to-volume ratio of the Langmuir trough may affect the kinetics, either in injection or previous dissolution experiments. It is the surface area-to-volume ratio that determines the "average" proximity of a biomolecule dissolved in the subphase to the interfacial film. Large surface areas might never be significantly populated if the volume of the subphase "repository" is small compared to it.

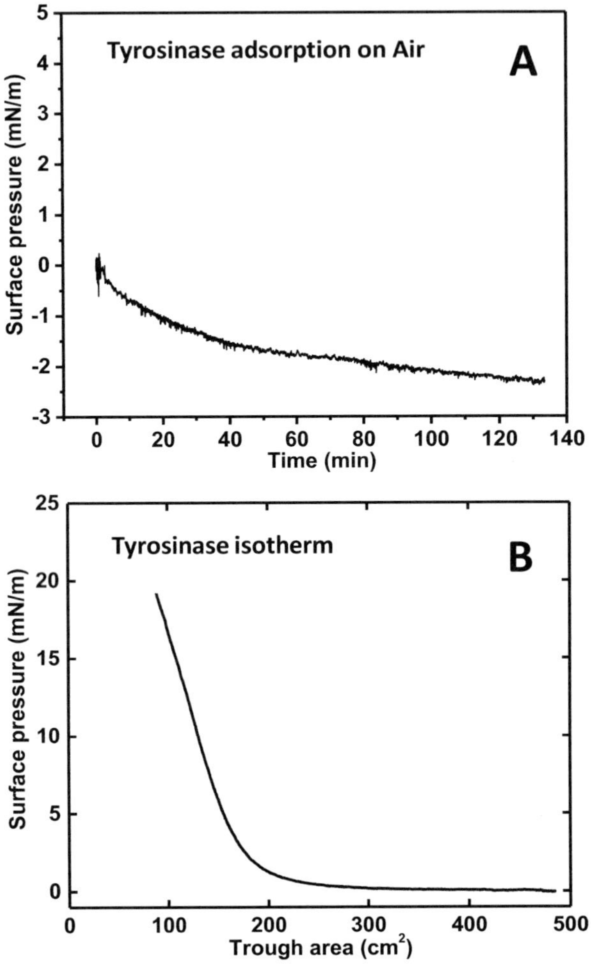

Fig. (6). (A) adsorption kinetics for the enzyme Tyrosinase onto a bare air-water interface. (B) Surface-pressure-area isotherm obtained after the experiment in (A). (B) Reprinted with permission from [8]. Copyright 2011, Royal Society of Chemistry.

To conclude the discussion in this sub-section, we bring in Fig. (7) an interesting example of how dramatically different results can be obtained by using previous dissolution, injection or co-spreading for biomolecules incorporated in Langmuir films. Schmidt *et al.* [10] used different methodologies to prepare mixed films of DPPG and the enzyme Horseradish Peroxidase (HRP). Some of the strategies (injection and previous dissolution) were natural in view of the soluble character of the enzyme. However, the authors also tested co-dissolution with DPPG and co-spreading from chloroform solutions, in an attempt to estimate the impacts over the phospholipid matrix. The changes in the surface pressure isotherms for DPPG varied with the method of enzyme incorporation. From the higher expansion (shift to higher molecular areas) imposed to the films, injection is taken as the more effective method to insert the enzyme into the interfacial film. Also, previous dissolution in the subphase, the other method where the enzyme is not processed in organic solvents, is the one that causes large expansion in the low-pressure regime. For the cases where HRP is dissolved in chloroform, we see significant changes compared to pure DPPG; what is more evident when performing co-dissolution. Even for co-spreading, where no dramatic changes are seen for the compression 10 min after film spreading, a clear change is seen 12 h after film formation. In contrast, the curve for injected HRP after 12 h is not significantly different from the curve obtained after 10 min, which indicates that the enzyme presents a fast adsorption kinetics over DPPG in this case. The authors discussed a few effects on film properties (static and dynamic elasticity) in each of the cases. As conclusion, they stated that injection is the most indicated method when improved enzyme incorporation is pursued.

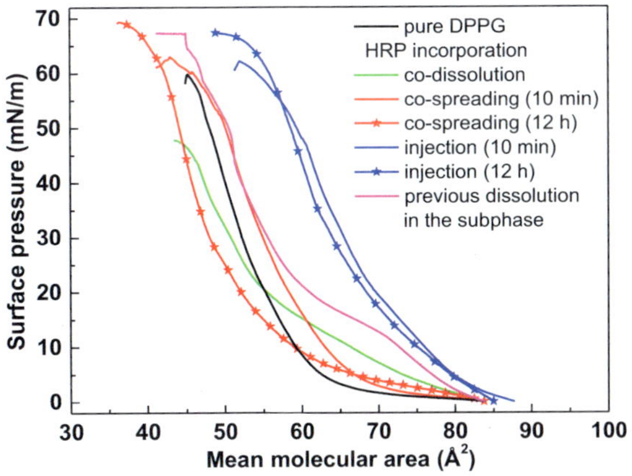

Fig. (7). Surface pressure-area isotherms for a DPPG monolayer to which the enzyme HRP has been incorporated in different ways (described in the inset). Adapted with permission from (10). Copyright 2008, Elsevier B. V. All rights reserved.

3. CHARACTERIZATION TECHNIQUES AND EXTRACTED INFORMATION

The techniques for characterizing Langmuir monolayers have been described in earlier reviews [11 - 13], to which readers are directed for further details. Here we intend to highlight the capabilities of the methods, and the sort of molecular-level information that can be obtained with them.

Surface pressure-area (π-A) isotherms are the oldest and most important way to characterize a Langmuir film. The surface pressure (π) is defined as the decrease in surface tension of the subphase ($\pi = \gamma_0 - \gamma$). Such isotherms are used to identify the so-called phases, corresponding to different packing arrangements of the film-forming molecules. Fig. (**8**) depicts a surface pressure isotherm for DPPC with the assigned gaseous, liquid (split into liquid-expanded - LE, and liquid-condensed - LC), and condensed phases, which are used to describe the states of packing of the film, from gas-like to crystalline-like. Also in Fig. (**8**), the molecular arrangement of the film is sketched for gas, LE and LC phases.

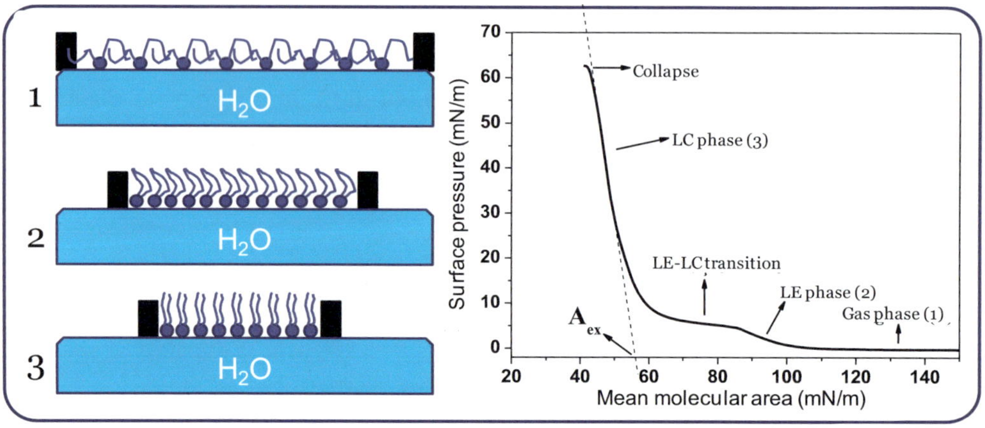

Fig. (8). Molecular packing of a Langmuir monolayer (molecules not to scale) for different stages of film compression on the left. Each stage is associated with the respective part of the surface pressure-area isotherm of DPPC on the right. Also indicated in the isotherm are the LE to LC transition region (plateau), film collapse pressure and extrapolated area (A_{ex}). For this material, a solid phase is never attained. [Adapted with permission from [2]. Copyright 2010, American Chemical Society].

A variety of other pieces of information can be derived from π measurements. For example, for systems where the biomolecules are adsorbed from the subphase, it is possible to determine the maximum insertion pressure (MIP - also known as exclusion pressure) - a limiting π for which higher packing states of the phospholipid film totally block the insertion of biomolecules. Also related to MIP is $\Delta\pi$, defined as π_f - π_i, where π_f is the pressure after biomolecules absorption and π_i the initial pressure of the lipidic film (MIP is the lowest π_i value for which $\Delta\pi$ is null) [14]. Several surface chemistry parameters can be estimated from π, including surface excess, surface coverage and critical surface micellization. Of high biological relevance is the in-plane elasticity (Cs^{-1}), obtained from π-A curves using Equation 1. As its name suggests, Cs^{-1} gives an estimative of how elastic a monolayer is. It also allows for evaluating effects from interaction of biomolecules. For instance, the phases of a monolayer are quantitatively determined using Cs^{-1} [15, 16].

$$Cs^{-1} = -A(\frac{d\pi}{dA}) \qquad \textbf{(1)}$$

Similar to Cs^{-1}, the dynamic elasticity also provides information about mechanical properties of floating monolayers, being commonly measured with the pendant drop technique [17]. Analogous experiments to measure surface rheology can be carried out in a Langmuir trough, simply by the oscillation of the barriers or the use of a dedicated experimental setup. When barrier oscillation is employed, the π value of the film is recorded alongside the movements, and its variation is compared with the imposed perturbation. In the dedicated surface rheometer, a floating needle at the interface (immersed in the film) is set to rotate with an external magnetic field, and its rotational movement is followed by a video camera. The movement observed is compared to the induced perturbation, as in barrier oscillation. In both cases, coherence, phase and periodicity of the movements are used to estimate the (visco)elasticity of the film.

Changes in the electrical character of model membranes can be informative of external agents interacting with real biomembranes. The electrical properties of floating monolayers can be evaluated using surface potential, conductance or Maxwell displacement current measurements. Surface potential is by far the most popular technique, with the vibrating capacitor method (a variation of the Kelvin probe method) being the most employed methodology to estimate it. Similar to the surface pressure, monolayer surface potential (ΔV) is determined relative to the uncovered water surface. With the most accepted theoretical treatment [11, 18], ΔV can be divided into contributions from three slabs in the film, *i.e.* (1) hydrophobic/tail region, (2) polar/headgroups region and (3) aqueous region underneath the film. Thus, ΔV can be described by the sum of V - in terms of dipole moment and the effective dielectric constant - for each of these layers (Equation 2). The term Ψ_0 in Equation 2 accounts for the contribution from the electrical double layer when the film is made of ionized materials. For charged phospholipids and charged interacting biomolecules, this is often the parameter with higher influence [19, 20]. Dynarowicz-Łatka *et al.* [11] bring a long list of examples where surface potential was used to characterize Langmuir monolayers, in addition to the less used methods with conductance and Maxwell displacement current measurements.

$$\Delta V = \frac{1}{A.\varepsilon_0}\left(\frac{\mu_1}{\varepsilon_1}+\frac{\mu_2}{\varepsilon_2}+\frac{\mu_3}{\varepsilon_3}\right)+\Psi_0 \qquad \textbf{(2)}$$

where A is the area per molecule, ε_0 is the vacuum permittivity, μ_1 is the normal component of the dipole moment arising from reorientation of water molecules at the interface owing to the presence of the Langmuir monolayer, μ_2 and μ_3 are the normal components of the dipole moments of the headgroups and tail groups, respectively, and ε_1, ε_2 and ε_3 are the effective dielectric constant for each layer.

In many cases where a biomolecule interacts with Langmuir films of phospholipids, morphology is altered which can be monitored by one of the following methods: fluorescence microscopy, second harmonic generation (SHG), Brewster angle microscopy (BAM) and imaging ellipsometry. The first usually requires a fluorescent label to be incorporated either to the film or to the interacting biomolecule. As the Langmuir film is ultrathin, optical sensitivity can be a problem when trying to visualize the films in this method. SHG also requires optical labeling, being again of limited use [21]. Microscopy at the Brewster angle is in turn label-free and, hence, more versatile for monolayers imaging [22]. BAM is based on the fact that *p*-polarized light is not reflected if impinged on a specific interface at its Brewster angle (BAM condition), which is defined by Equation 3 (θ_B is the Brewster angle, and n_1 and n_2 are the refraction index for the interfacing media - *e.g.* air and water). The measurement setup is zeroed under BAM conditions, and no light reaches the detector. When the interface is changed, BAM conditions are no longer valid and images are formed by contrast. For instance, regions of the surface coated or not with film, or regions with different thickness in a continuous film can be distinguished. The images in Fig.(**9A**) and (**9B**) illustrate, respectively, these two cases.

$$\theta_B = \arctan\left(\frac{n_2}{n_1}\right) \qquad \textbf{(3)}$$

Imaging ellipsometry is based on the change of the polarization state of the light upon reflection at the interface. Ellipsometry has been traditionally used for solid films to determine optical properties and thickness using polarizers and conventional optical detectors. However, in modern equipment, a CCD (charged-coupled device) camera is used as detector, and ellipsometric data can be collected with spatial resolution. When plotting such spatial data set, an image is formed based on the difference of optical properties among the many different detection spots (pixels) coming from the surface. Although images are similar to that obtained by BAM, state-of-the-art ellipsometers are advantageous in at least two aspects. Firstly, the angle of incidence of the incoming light can be varied in a wide range, outside the Brewster condition used in BAM, and contrast can be largely improved by finding brighter (ellipsometric) reflections at different angles. Secondly, white-light sources from lamps (not only laser, as typical in BAM experiments) can be used in the measurements. This is helpful to avoid undesirable photochemical and photophysical processes that might occur in the interfacial film. Moreover, this tuning of the incident wavelength opens the possibility of conducting spectroscopic ellipsometry, which also expands the range of possible application for ellipsometry as an optical characterization tool.

Surface spectroscopy techniques for characterizing Langmuir films were reviewed recently by Volpati *et al.* [23], and include spectroscopy in the ultraviolet and visible region (UV-Vis *in situ*), polarized-modulated infrared reflection-absorption spectroscopy (PM-IRRAS), and sum-frequency generation (SFG) spectroscopy. UV-Vis *in situ* is very useful to estimate the state of aggregation of materials at interfaces [24], but its use in membrane models is very restricted by the limited range of wavelengths that are both transmitted by water and absorbed by biomolecules. In contrast, PM-IRRAS and SFG are attractive for Langmuir films as biomembrane models.

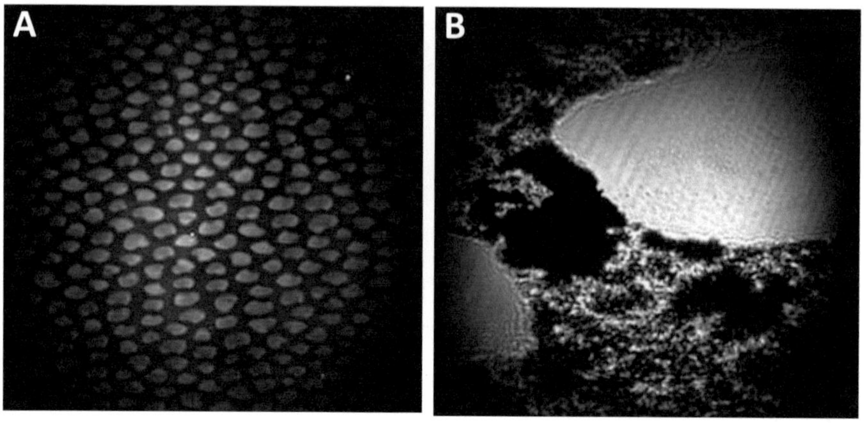

Fig. (9). (A) BAM image for a Langmuir film of DPPC in the transition region from liquid-expanded (LE) to liquid-condensed (LC) phases. Bright areas are LC domains immersed in the LE phase (dark matrix). (B) BAM image for a Langmuir monolayer of the azopolymer HPDR13 in the gaseous phase ($\pi = 0$ mN/m). In this case, bright domains are aggregates of the polymer, and dark regions are areas of the air-water interface not coated with the polymeric film.

While PM-IRRAS is surface specific due to the use of both polarizations of the light (*p* and *s*), and the calculation of a differential signal [25], SFG is intrinsically sensitive to interfaces due to the non-linear optical, second-order origin of the signal [26]. Both techniques are suitable for determining molecular orientation, since anisotropy is detected in SFG, and the introduction of polarization-modulation in the PM-IRRAS eliminates the contribution of randomly distributed groups (ex. water vapor and O_2) to the spectrum. They may be seen as complementary because molecular vibrations that appear in the film in pairs with specular distribution will vanish in the SFG spectra, but will be stronger in the PM-IRRAS signal. A clear example is given in Fig. (**10**), where the spectral region is shown for the vibration from alkyl chains of DPPG for both PM-IRRAS and SFG. While more abundant and mirrored distributed vibrations from CH_2 groups are dominant in the PM-IRRAS spectrum, unpaired and scarcer vibrations from CH_3 groups predominate in the SFG spectra. The signal from these spectroscopies can be used to infer about interacting groups in biomolecules and phospholipids, and to determine changes in molecular packing and alignment in the models, as illustrated in section 4.1 with the chitosan example.

In addition to the methods mentioned above, there are other less conventional, advanced methods for Langmuir monolayer characterization that are rarely used for biomembrane models. These methods include RAMAN spectroscopy, X-ray diffraction, scattering and reflectivity, and neutron reflectivity [5, 11, 15, 27]. The difficulty in acquiring good-quality RAMAN spectra for such an ultrathin and non-static system as a Langmuir film, and the lack of a well-defined crystalline structure for the monolayers might be the reason for the still-scarce use of these methods.

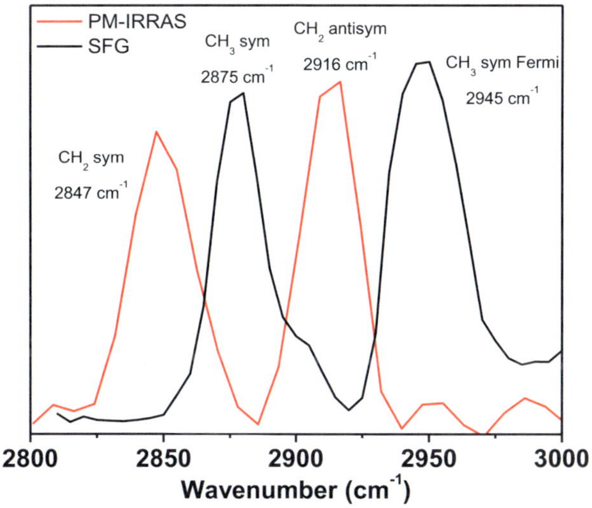

Fig. (10). PM-IRRAS (red curve) and SFG (black curve) vibrational spectra recorded for a Langmuir film of DPPG compressed to 30 mN/m. While vibrations from CH_2 groups dominate the PM-IRRAS spectrum because of the higher density of these groups at the interface, peaks from vibrations of the less abundant CH_3 groups are much more intense in the SFG spectrum due to their asymmetrical distribution in the film. The SFG spectrum is a courtesy of Paulo B. Miranda and Caio V. Rímoli.

4. BIOACTIVE BIOMOLECULES INTERACTING WITH LANGMUIR MONOLAYERS: EXAMPLES

In this section, a few selected examples are given to illustrate the kind of knowledge obtained using Langmuir films in biological systems, where interaction of biomolecules and membranes plays a central role. Text is distributed according to three main classes of biomolecules explored in such systems: polysaccharides, peptides and proteins. We again emphasize that a full coverage of the literature in the field was not aimed here. Our goal is rather to exemplify the potential of the modeling method described in this chapter. For a more comprehensive survey of a broader variety of systems, including the ones using the biomolecules below, the readers are referred to review papers [2, 28 - 30].

4.1. Polysaccharides

From the literature dealing with polysaccharides interfacing Langmuir monolayers we identify two distinct areas of research. First, there are several papers dealing with the interaction of chitosan and its effects over biomembrane models. Second, there is a much smaller number of works dealing with polysaccharides combined (often complexed) with proteins. Here, we start describing two noteworthy works from this second category, and finally comment on the main findings from studies involving chitosan.

Nobre *et al.* [31] combined SFG spectroscopy and Dynamic Surface Elasticity measurements with more conventional Langmuir characterization techniques to propose a model for the bridging between the protein alkaline phosphatase and phospholipids. This bridging is performed by the polysaccharide carragenan, for which divalent Zn^{2+} ions are crucial. These ions help the polymer to adopt a conformation in which sulfate groups along the chain preferentially face the membrane, and hydroxyl groups at the opposite side anchor the protein. These results have practical implications since the association of alkaline phosphatase with the membrane (mediated by carragenan) is relevant for regulation of "*in vivo*" physiological processes.

A second example of Langmuir monolayers to study association of polysaccharides with cell membranes involved mixed ODA-DOPC (ODA - octadecylamine, a positively charged lipid, DOPC - di-octadecyl-phosp-o-choline), and pure DOPC, interacting with anionic poly(sialic acid) (PSA) [19].

Mixed films expanded with increasing PSA concentration in the subphase, while pure films of DOPC showed less pronounced, concentration-independent expansion. Combined with results from liposomes, these data led to the conclusion that the electrical properties of the membrane are crucial for the association of polysaccharides. Specifically, positive surface potential and transmembrane potential are favorable for PSA binding. According to the authors, this favored interaction is relevant for PSA mediation of cell-cell bacterial interactions in neuroinvasive tumor development.

The number of studies covering chitosan-Langmuir monolayer interactions is larger, as reviewed recently [2, 30], and only major aspects are listed here as a way to illustrate the usefulness of the models. With only one exception [32], all papers in the literature report an expansion induced by chitosan on the Langmuir monolayers, regardless of the lipid composition [6, 20, 33 - 42]. Such an expansion is a consequence of chitosan penetration into the interfacial film, with concomitant decrease in membrane elasticity, since the "soft" polymeric chitosan turns the film more viscoelastic. Chitosan incorporation imposes some constraint to the area occupied by the phospholipids, causing them to be more oriented upon interaction. This increased lipid chain alignment has been proven by Pavinatto *et al.* [6, 34] using SFG and PM-IRRAS. Representative SFG spectra for DMPA in the presence and absence of chitosan are shown in Fig. (**11**). A so-called "order parameter" can be calculated from the ratio between the symmetric CH_3 and CH_2 vibrations (centered, respectively, at 2873 and 2843 cm^{-1}). As commented in section 3, oriented alkyl chains have the CH_2 SFG signal diminished due to the symmetric distribution of these groups. Thus, the higher this ratio the more aligned the lipid molecules. From Fig. (**11**), one observes that the CH_3/CH_2 ratio increased from 6.22 to 9.06 after interaction with chitosan. A similar increase occurred for mixed DMPA-cholesterol films [6], confirmed by PM-IRRAS results.

Another general trend is a saturation with concentration for the chitosan effects on membrane models, usually around 0.2 ± 0.1 mg/mL [33 - 36]. Above this concentration the disturbances imposed by chitosan (*e.g.* expansion and changes in elasticity) stop increasing monotonically with concentration, remaining constant. The existence of such saturation is of high importance for determining ideal chitosan doses in biological applications. Structural chemical parameters and media conditions also affect the action of chitosan over biomembranes, and this must be taken into account when predicting the concentration dependence. For instance, low-molecular weight chitosans are generally more powerful membrane disturbants, because they display faster diffusion and easier accommodation at the interface [20, 38]. Salts in solution can either compete with chitosan for interacting sites, hence decreasing its effectiveness [39], or make chitosan oligomers more soluble and less surface active, which suppresses their effects [33]. Regarding interaction forces acting in chitosan-phospholipids interactions, those of Coulombic origin are of major importance, as expected by the cationic nature of the polymer and the anionic character of phosphate groups in phospholipids. However, there is strong evidence that hydrophobic and polar interactions play an important role in specific actions of chitosan [37, 40, 41].

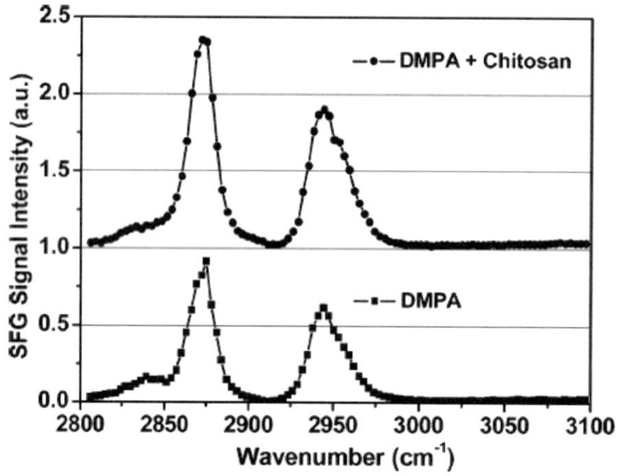

Fig. (11). SFG spectra for a Langmuir-Blodgett (LB) film of a phospholipid (DMPA - di-miristoyl-phosphatidic-acid) in the presence or absence of chitosan. Reprinted with permission from [34]. Copyright 2007, American Chemical Society.

4.2. Proteins

Proteins can adsorb at the air-water interface due to their amphiphilic nature, resulting in a surface excess. However, such adsorption may result in changes in the conformation of the proteins compared to that in solution, sometimes with unfolding of their polypeptide chains. It is convenient therefore to employ lipid Langmuir monolayers as matrices for adsorbing proteins coming from the aqueous subphase, which has been investigated in the last two decades [30]. The presence of a lipid Langmuir monolayer at the interface normally stimulates protein adsorption [43], providing a favorable environment to maintain the polypeptide chain conformation [44]. In these studies, the main aim is to obtain blends of proteins and lipids for applications in biotechnology [45 - 47], and the molecular interactions are not studied in the same level of detail as they are in investigations associated with cell membrane models. For the latter, lipid-protein synergistic effects are probed [48 - 50].

An interesting example of how Langmuir monolayers serve to identify molecular interactions was reported for the photoreceptor proteins *Retinitis Pigmentosa* 2 (RP2) and Recoverin [14]. They are believed to bind to cell membranes, affecting phospholipid oxidation related to diseases induced by toxic sub-products of the oxidation, such as macular degeneration. Protein binding was estimated with two parameters from surface pressure measurements: the maximum insertion pressure (MIP), and surface pressure increase ($\Delta\pi$). RP2 and Recoverin interacted with intact and oxidized 1,2-didocosahexaenoyl-sn-glycero-3-phosphocholine (DDPC) as shown in Fig. (**12**). Protein binding was not so evident for the oxidized phospholipid, which may lead to inadequate membrane processes and consequent cellular malfunctioning and diseases. For RP2, the physical state of the lipid monolayer influences protein binding [51]. The shape of liquid-condensed domains of phospholipid monolayers changed upon RP2 binding, suggesting RP2 preference for this physical state, with the β helix domain playing a major role.

Another remarkable collection of studies with Langmuir monolayers from polypeptides involve self-aggregation of proteins and peptides related to Alzheimer's disease (AD). Langmuir monolayers were shown as suitable to investigate the mechanism and origin of AD [52], providing information on the nature of short- and long-range interactions between lipids and the amyloid beta precursor proteins, the major constituents of hallmark senile plaques found in AD patients. Vibrational spectroscopic methods, in particular, provided insight into the structuring of the molecules at the monolayer, being useful to study the steps related to assembly of β-sheet structures and ensuing formation of amyloid fibrils. Progress in this area has been described by Thakur *et al.* [52]. For example, the full-length tau protein (hTau40) was reported to insert into anionic dimyristoyl phosphatidyl glycerol (DMPG) lipid monolayers [53]. The misfolding and aggregation of this intrinsically disordered protein into neurofibrillary tangles is associated to AD, but the molecular mechanisms are not known in detail. In spite of being highly charged and soluble, hTau40 was surface active, inserting selectively into DMPG monolayers. A combination of data obtained by X-ray reflectivity, grazing incidence X-ray diffraction, and neutron reflectivity indicated that hTAu40 is able to penetrate into DMPG lipid head and tail groups, disrupting the lipid packing, and adopting a conformation with density similar to that of a folded protein. Although this protein disrupts supported DMPG bilayers, it does not affect neutral DPPC bilayers. Overall, these results suggested that hTau40s has specificity to anionic lipids, disrupting the organization of lipid layers, which can be associated to its toxicity in neurodegenerative diseases.

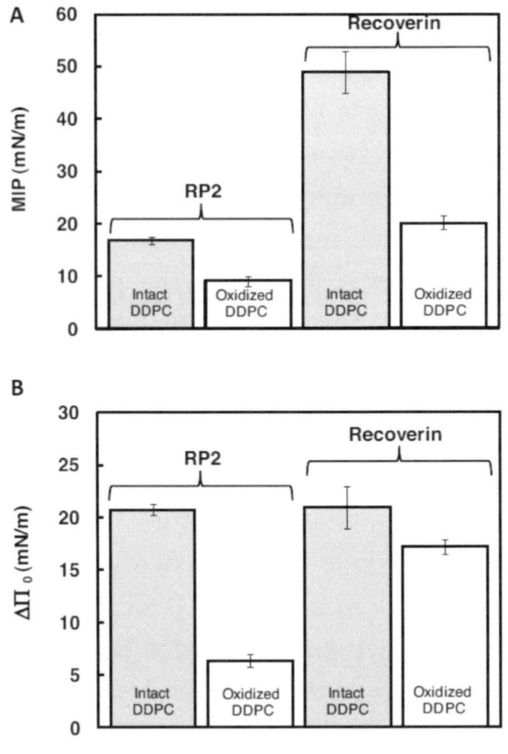

Fig. (12). Maximum Insertion Pressure (A), and surface pressure increase (B) for the proteins RP2 and Recoverin interacting with intact and oxidized DDPC phospholipid. [Reprinted with permission from (51). Copyright 2012, American Chemical Society].

For transmembrane proteins, we highlight a representative study by Bussières *et al.* [54], where a truncated version of lecithin (retinol acyltransferase -tLRAT), a membrane-associated protein that catalyzes the esterification of retinol into retynil ester, was investigated using Langmuir monolayers. tLRAT has shown a MIP value higher than the estimated lateral pressure of cell membranes (30-35 mN/m) [55], suggesting that tLRAT binds strongly in membranes despite the absence of its N- and C-terminal hydrophobic domains. Also, PM-IRRAS and circular dichroism showed that the protein adopts an α-helical secondary structure when bound to the membrane. LRAT is made of a core of four β-sheets surrounded by three α-helices, from which the authors proposed a schematic model where tLRAT is anchored in a biomembrane by its N- and C-terminal region (Fig. **13**). In this model, the active site of the protein (constituted by a triad of aminoacids - H60, Y154 and C161) is located underneath the head groups plane of the membrane, being in a favorable position to hydrolyze phospholipids.

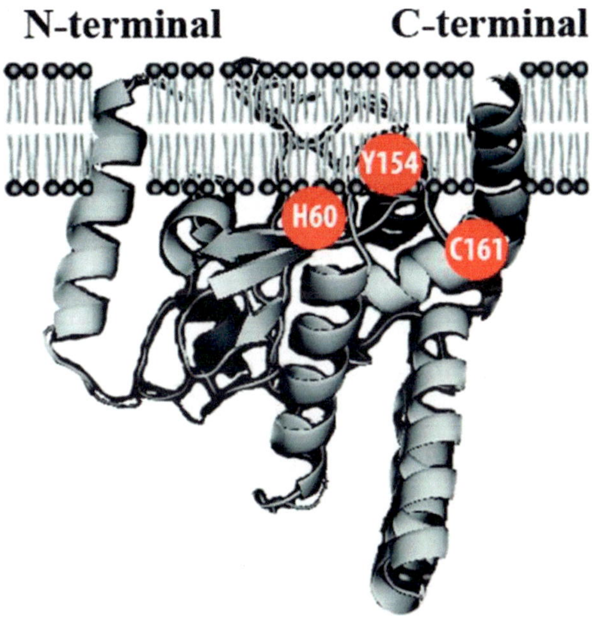

Fig. (13). Schematic representation for insertion of truncated protein LRAT through its N- and C-terminal regions into a biomembrane. Highlighted in red are the three aminoacids comprising the protein active site, which lie in a favorable configuration to perform enzymatic activity. [Reprinted with permission from [49]. Copyright 2012, American Chemical Society].

Lipid monolayers can be formed with more complex mixtures, for instance, using lipid extracts. By way of illustration, we mention Gallier *et al.* [56], who formed Langmuir monolayers with lipids extracted from milk fat globules. They found that β-casein added to the aqueous subphase induced clustering of sphingomyelin-cholesterol into micro-domains, which could be related to the natural molecular arrangement of lipids and proteins on the cell membrane outer layer into rafts.

As a final example of proteins investigated in Langmuir monolayers, we refer to a special property of chitosan which can remove the protein β-lactoglobulin (BLG) from negatively charged monolayers [57]. BLG inserted into the subphase was found to adsorb rapidly onto monolayers of negatively charged phospholipids, as shown by the increase in surface pressure and appearance of amide bands in the PM-IRRAS spectra. Chitosan, which has a low diffusion coefficient, was also adsorbed from the aqueous subphase, and then complexed with BLG at the lipid monolayer. These chitosan-BLG complexes were removed from the interface, thus causing a decrease in surface pressure and the amide bands disappeared from the PM-IRRAS spectra after a given time. In fact, chitosan not only complexed and removed BLG from the interface, but it also dragged lipid molecules from a stable monolayer. This only occurred for negatively charged lipids, but not for zwitterionic ones, being therefore highly specific.

4.3. Peptides

The class of materials with the largest variety of applications involving interactions with the cell membrane is surely the one of peptides. This is mainly due to the great flexibility and control in peptide synthesis, which allows for the design of precise architectures with a specific targeted activity over the membrane. Peptides are found in all living organisms, exhibiting a variety of functions and biological activities. In the next paragraphs we discuss systems in which peptides with specific action interact with cells. In every case, modeling of their interactions using Langmuir monolayers, and the biological relevance of the findings, are commented upon. In general, the adsorption of peptides at the air-water interface and their incorporation into membranes strongly depend on its amino acid sequence (primary structure). The ratio between polar and non-polar residues, in particular, is of primary importance, because the driving force for their adsorption/incorporation is dictated by the relative number of hydrophobic amino acids. Moreover, the location of the hydrophobic residues along the peptide chain is important, since it determines the secondary structure adopted by the molecule. This will also eventually dictate which region of the peptide will be exposed to air (or lipid) or face the aqueous subphase.

Antimicrobial peptides (AMP) are the most employed in Langmuir films, some of which possess antibacterial and antiviral properties. They may be considered a new class of antibiotics due to their high selectivity and broad spectrum of activity. Differently from conventional antibiotics that usually bind to a specific site, antimicrobial peptides are known as molecules that act by destabilizing the membrane. Despite their differences in amino acid sequence and secondary structure, most AMPs present between 12 and 50 amino acid residues and a positively net charge. Several properties of AMPs may be investigated with Langmuir films, especially their specificity toward certain cells. In this context, bacterial membranes are usually modeled by negatively charged phospholipids, and mammalian membranes by choline-containing phospholipids.

For the AMP Plantaricin 149 (Pln 149), a C-terminal amide analogue was synthesized (Pln 149a) and showed to form an amphipathic α-helix when bound to the negatively charged DPPG [58]. The incorporation into a DPPG monolayer increased with peptide concentration, and above 2 μM Pln 149a was able to disrupt the lipidic film, according to adsorption kinetics data. Fig. (**14**) shows the model proposed for the interaction, with accumulation of the peptide at the interface taking place with a concomitant removal of lipids, eventually leading to film disruption. These results were used to explain the mechanism of peptide antimicrobial activity in real biological systems.

Two other peptide families of high biological relevance studied with Langmuir models comprise β-amyloid and fusion peptides, which are involved in important illnesses such as Alzheimer's disease (AD) and acquired immune deficiency syndrome (AIDS), respectively. Specifically for β-amyloid, precursor proteins are widely studied, as mentioned in section 4.2. However, truncated peptides representing the key fragments of such precursor proteins (Aβ) are also explored in cell membrane models of neurons. Thakur *et al.* [52], for instance, observed aggregates in a Langmuir model membrane which supported the hypothesis that short Aβ segments from amyloid proteins are able to trigger formation of fibrils. Maltseva and co-workers [59] also showed that β-sheet structures are induced by adsorption of Aβ(1-40) peptide to interfacial phospholipid films, in a much similar fashion. Later, these and a few other authors expanded their studies, showing that the composition and electrical charge of the phospholipids head group at the interface [59 - 62], as well as the nature and amount of dissolved ions in the subphase [59, 63], strongly affect the kinetics and effectiveness of Aβ segments adsorption. Specifically, negative charges at the interface efficiently induce peptides surface activity. Lepère *et al.* [64] used a peptide named LSFD to model the Aβ(1-40) segment. Using data from π-A isotherms, IRRAS spectroscopy and X-ray diffraction experiments, they proposed that LSFD forms a well-ordered β-strand structure at the air-lipid interface, with anti-parallel β-sheet domains that lie flat on the water surface. This supramolecular structure, represented in Fig. (**1**) of reference [64], had dimensions of 4.75 and 9.7Å according to GIXD measurements. Such β-strand moiety is believed to be involved in the early stages of amyloid aggregation and development of AD.

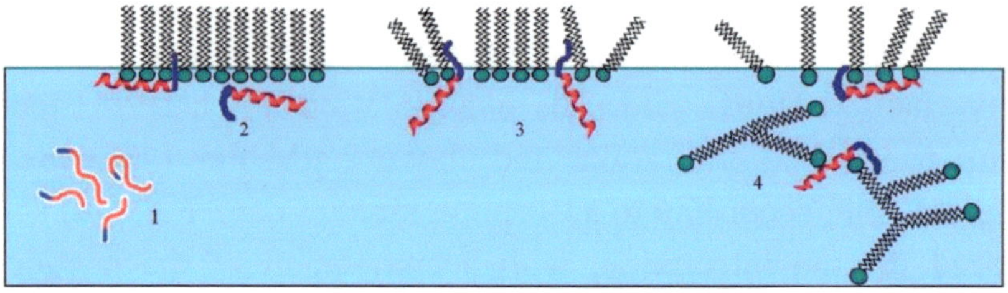

Fig. (14). Model for the accumulation of the peptide Pln 149a onto Langmuir monolayers of negatively charged phospholipids. In the sequence from 1 to 4 the peptide is believed to lose its original random-coil structure, develop induced surface activity and α-helix, and disturb the original packing state of the lipidic film, eventually causing its disruption. [Reprinted with permission from [58]. Copyright 2009, Elsevier B. V. All rights reserved].

Membrane fusion peptides in the protein gp41 are involved in the initial approach and contact between the enveloped human immunodeficiency virus (HIV) and cells. Schwarz *et al.* [65] published a seminal work in this field, where the sequence of the last 23 aminoacids in the N-terminus of gp41 was set to interact with Langmuir monolayers of the lipid POPC (palmitoyl-octadecyl-phosphatidl-choline). The molar fraction of the peptide incorporated to the film was determined, from which the peptide conformation could be estimated based on its occupied area at the interface. α-helices and α-hairpin lying perpendicular to the interface were the predominant structures at low packing ($\pi < 15$ mN/m), but for high film packing the conformation changed to β-strand units and then partitioning to the subphase and aggregation. Also observed was a strong dependence of aggregation state with the peptide-to-phospholipids ratio.

Similar findings were reported for the same N-terminal residue of gp41 by Castano e Desbat [66]. Besides a change in conformation from α-helix to β-sheet, the thickness of a Gibbs monolayer (interfacial film of the peptide formed by adsorption to a bare air-water interface) was determined as 13 Å through BAM images. This value is very close to that expected for a monolayer of anti-parallel β-sheets. In addition, by using PM-IRRAS, the authors inferred that the peptide is inserted in mixed monolayers in a tilted manner, with the main axis of the structure (either as α-helix to β-sheet) forming an angle of 30 to 60° with the water surface. This specific insertion may be key to the perturbation and fusion of the cell membrane in viral infection.

The same fusion peptide has been continually explored in subsequent years by different groups, and the use of Langmuir films as membrane models rendered extra important information on the system. For instance, oligomerization of gp41 and the persistence of the α-helix conformation with compression were observed with BAM and PM-IRRAS, respectively [67]. Also by using PM-IRRAS, it has been proven that the lipid composition of the membrane plays an important role in gp41 fusion activity [68].

Amide bands vanished from the spectra for DPPC and for a mixture of DPPC and cholesterol, but an alpha-helix structure was detected when the peptide interacted with DPPG (negatively charged). The axis of this helix is reoriented along film compression, from parallel to the water surface to an angle of 60° with the interface. Altogether, Langmuir modeling has contributed significantly to understanding the action of fusion peptides in the infection of cells by HIV.

SCOPE OF THE CHAPTER

“Bioactive molecules” is a generic term comprising a vast amount of molecules, including synthetic molecules, as most of the drugs available today. In order to reduce the chapter length and achieve text concision, we restrict our discussion to biological molecules; namely, polysaccharides, proteins and peptides. This is the reason why the term “Bioactive Biomolecules” is in the chapter's title, rather than “Bioactive Molecules” or “Bioactive Compounds”. Secondly, to be effective at having readers understanding the benefits from using Langmuir monolayers as membrane models, we opted not to discuss related systems and models. For instance, we do not cover Langmuir-Blodgett films, which are complementary to Langmuir monolayers; and vesicles and liposomes, which are alternative models. Finally, it is worth mentioning that although we do try to cover in section 2 all the characterization techniques available for Langmuir films, we do not aim at offering an exhaustive and detailed description of each of them. For this, the readers are referred to specialized literature.

CONCLUDING REMARKS

Simplified models of complex biological systems, such as Langmuir monolayers discussed in this chapter, are bound to be useful for many years for understanding fundamental aspects in biology, chemistry and physics. This is especially true for the interaction of bioactive biomolecules with cell membranes, since the features of both entities are hard to predict precisely. If on one hand the models sometimes neglect important characteristics of the whole system, on the other hand they allow for the precise characterization of individual phenomena that contribute with a high percentage to the global understanding of the interactions. For Langmuir films as cell membrane models, interaction forces and the effect of bioactive molecules on the phospholipid matrix are already widely explored.

From the experimental point of view, we have emphasized that planning and extreme care in the procedures are required to obtain reliable and biologically-meaningful results. The incorporation of proteins, peptides and other water soluble materials is particularly complicated in this respect. Surface activity, diffusion rate, bulk and interface partition are only a few features of the absorption of soluble biomolecules that must be characterized to avoid artifacts. In addition, the assembly and stability of mixed protein-phospholipids films require detailed investigation for establishing the foundations for more advanced studies with "closer-to-reality models". Nevertheless, efforts in this direction are worth pursuing, since we believe that a gradual refinement of the models is the best way to achieve results of ever increasing biological relevance.

CONFLICT OF INTEREST

The authors confirm that they have no conflict of interest to declare for this publication.

ACKNOWLEDGEMENTS

This work was supported by FAPESP (2013/14262-7) and CNPq (Brazil).

REFERENCES

[1] Langmuir I. The constitution and fundamental properties of solids and liquids. Part II: Liquids. J Am Chem Soc 1917; 39(9): 1848-906. [http://dx.doi.org/10.1021/ja02254a006]

[2] Pavinatto FJ, Caseli L, Oliveira ON. Chitosan in nanostructured thin films. Biomacromolecules 2010; 11(8): 1897-908. [http://dx.doi.org/10.1021/bm1004838] [PMID: 20590156]

[3] Brockman H. Lipid monolayers: why use half a membrane to characterize protein-membrane interactions? Curr Opin Struct Biol 1999; 9(4): 438-43. [http://dx.doi.org/10.1016/S0959-440X(99)80061-X] [PMID: 10449364]

[4] Tanford C. The hydrophobic effect: formation of micelles and biological membranes. 2nd ed., New York: Wiley-Interscience 1980.

[5] Giner-Casares JJ, Brezesinski G, Möhwald H. Langmuir monolayers as unique physical models. Curr Opin Colloid Interface Sci 2014; 19(3): 176-82. [http://dx.doi.org/10.1016/j.cocis.2013.07.006]

[6] Pavinatto FJ, Pacholatti CP, Montanha EA, *et al.* Cholesterol mediates chitosan activity on phospholipid monolayers and Langmuir-Blodgett films. Langmuir 2009; 25(17): 10051-61. [http://dx.doi.org/10.1021/la901019p] [PMID: 19705898]

[7] Pavinatto FJ. Interação entre quitosana e modelos de membrana celular: filmes de Langmuir e Langmuir-Blodgett (LB). PhD Thesis, São Carlos, Ciência e Engenharia de Materiais, University of São Paulo 2010;. Available from: http://www.teses.usp.br/teses/disponiveis/88/881.

[8] Pavinatto FJ, Fernandes EG, Alessio P, *et al.* Optimized architecture for Tyrosinase-containing Langmuir–Blodgett films to detect pyrogallol. J Mater Chem 2011; 21(13): 4995.

[http://dx.doi.org/10.1039/c0jm03864d]

[9] Hammes GG, Hammes-Schiffer S. Hydrodynamics of Macromolecules. In: Hammes GG, Hammes-Schiffer S, Eds. Physical Chemistry for the Biological Sciences. Hoboken: John Wiley & Sons, Inc 2015; pp. 415-40.

[10] Schmidt TF, Caseli L, Nobre TM, Zaniquelli MED, Oliveira ON. Interaction of horseradish peroxidase with Langmuir monolayers of phospholipids. Colloids Surf A Physiochem 2008; 321(1-3): 206-10. [http://dx.doi.org/10.1016/j.colsurfa.2008.01.039]

[11] Dynarowicz-Latka P, Dhanabalan A, Oliveira ON. Modern physicochemical research on Langmuir monolayers. Adv Colloid Interface Sci 2001; 91(2): 221-93. [http://dx.doi.org/10.1016/S0001-8686(99)00034-2] [PMID: 11392357]

[12] Barnes G, Gentle I. Interfacial science: An introduction. 2nd ed., Oxford: Oxford University Press 2011.

[13] Balogh DT, Ferreira M, Oliveira ON. Langmuir–blodgett–kuhn multilayer assemblies: Past, present, and future of the LB technology. In: Knoll W, Advincula RC, Eds. Functional Polymer Films. Weinheim: Wiley-VCH Verlag GmbH & Co. 2011; pp. 113-49.

[14] Boisselier E, Calvez P, Demers E, Cantin L, Salesse C. Effect of oxidation of polyunsaturated phospholipids on the binding of proteins in monolayers. Colloids Surf B Biointerfaces 2013; 109: 109-14. [http://dx.doi.org/10.1016/j.colsurfb.2013.03.021] [PMID: 23624278]

[15] Petty MC. Langmuir-blodgett films: An introduction. Cambridge: Cambridge University Press 1996. [http://dx.doi.org/10.1017/CBO9780511622519]

[16] Davies JT, Rideal EK. Interfacial phenomena. New York: Academic Press 1963.

[17] Rotenberg Y, Boruvka L, Neumann AW. Determination of surface tension and contact angle from the shapes of axisymmetric fluid interfaces. J Colloid Interface Sci 1983; 93(1): 169-83. [http://dx.doi.org/10.1016/0021-9797(83)90396-X]

[18] Oliveira ON Jr, Taylor DM, Morgan H. Modelling the surface potential-area dependence of a stearic acid monolayer. Thin Solid Films 1992; 210-211: 76-8. [http://dx.doi.org/10.1016/0040-6090(92)90172-8]

[19] Nowotarski K, Sapoń K, Kowalska M, Janas T, Janas T. Membrane potential-dependent binding of polysialic acid to lipid monolayers and bilayers. Cell Mol Biol Lett 2013; 18(4): 579-94. [http://dx.doi.org/10.2478/s11658-013-0108-x] [PMID: 24293107]

[20] Pavinatto A, Pavinatto FJ, Delezuk JA, *et al.* Low molecular-weight chitosans are stronger biomembrane model perturbants. Colloids Surf B Biointerfaces 2013; 104: 48-53. [http://dx.doi.org/10.1016/j.colsurfb.2012.11.047] [PMID: 23298587]

[21] Matsuoka Y, Taguchi D, Manaka T, Iwamoto M. Visualizing polarization structure of lipid Langmuir monolayer by surface second-harmonic generation technique. Thin Solid Films 2014; 2014(554): 8-12. [http://dx.doi.org/10.1016/j.tsf.2013.04.099]

[22] Vollhardt D. Brewster angle microscopy: A preferential method for mesoscopic characterization of monolayers at the air/water interface. Curr Opin Colloid Interface Sci 2014; 19(3): 183-97. [http://dx.doi.org/10.1016/j.cocis.2014.02.001]

[23] Volpati D, Aoki PH, Alessio P, *et al.* Vibrational spectroscopy for probing molecular-level interactions in organic films mimicking biointerfaces. Adv Colloid Interface Sci 2014; 207(1): 199-215. [http://dx.doi.org/10.1016/j.cis.2014.01.014] [PMID: 24530000]

[24] Dos Santos DS Jr, Pavinatto FJ, Balogh DT, Misoguti L, Oliveira ON Jr, Mendonça CR. *In situ* UV-vis absorbance measurements for Langmuir films of poly[4′-[[2-(methacryloyloxy)-ethyl]ethylami-o]-2-chloro-4-nitroazobenzene] (HPDR13) azopolymer. J Colloid Interface Sci 2004; 276(1): 138-42. [http://dx.doi.org/10.1016/j.jcis.2004.03.027] [PMID: 15219441]

[25] Mendelsohn R, Flach CR. Infrared reflection – absorption spectrometry of monolayer films at the air – water interface. In: Chalmers JM, Griffiths PR, Eds. Handbook of Vibrational Spectros. New Jersey: John Wiley & Sons Ltd 2001; Vol. 2: pp. 1028-41.

[26] Sung W, Kim D, Shen YR. Sum-frequency vibrational spectroscopic studies of Langmuir monolayers. Curr Appl Phys 2013; 13(4): 619-32. [http://dx.doi.org/10.1016/j.cap.2012.12.002]

[27] Stefaniu C, Brezesinski G, Möhwald H. Langmuir monolayers as models to study processes at membrane surfaces. Adv Colloid Interface Sci 2014; 208: 197-213. [http://dx.doi.org/10.1016/j.cis.2014.02.013] [PMID: 24612663]

[28] Leblanc RM. Molecular recognition at Langmuir monolayers. Curr Opin Chem Biol 2006; 10(6): 529-36. [http://dx.doi.org/10.1016/j.cbpa.2006.09.010] [PMID: 17015031]

[29] Maget-Dana R. The monolayer technique: A potent tool for studying the interfacial properties of antimicrobial and membrane-lytic peptides and their interactions with lipid membranes. Biochem Biophy Acta 1999; 1462(1-2): 109-40. [http://dx.doi.org/10.1016/S0005-2736(99)00203-5]

[30] Nobre TM, Pavinatto FJ, Caseli L, Barros-Timmons A, Dynarowicz-Łątka P, Oliveira ON Jr. Interactions of bioactive molecules & nanomaterials with Langmuir monolayers as cell membrane models. Thin Solid Films 2015; 593: 158-88. [http://dx.doi.org/10.1016/j.tsf.2015.09.047]

[31] Nobre TM, de Sousa e Silva H, Furriel RP, Leone FA, Miranda PB, Zaniquelli ME. Molecular view of the interaction between iota-carrageenan and a phospholipid film and its role in enzyme immobilization. J Phys Chem B 2009; 113(21): 7491-7. [http://dx.doi.org/10.1021/jp900841p] [PMID: 19415915]

[32] Fang N, Chan V. Chitosan-induced restructuration of a mica-supported phospholipid bilayer: an atomic force microscopy study. Biomacromolecules 2003; 4(6): 1596-604. [http://dx.doi.org/10.1021/bm034259w] [PMID: 14606885]

[33] Parra-Barraza H, Burboa MG, Sánchez-Vázquez M, Juárez J, Goycoolea FM, Valdez MA. Chitosan-cholesterol and chitosan-stearic acid interactions at the air-water interface. Biomacromolecules 2005; 6(5): 2416-26. [http://dx.doi.org/10.1021/bm050106s] [PMID: 16153076]

[34] Pavinatto FJ, Caseli L, Pavinatto A, *et al.* Probing chitosan and phospholipid interactions using Langmuir and Langmuir-Blodgett films as cell membrane models. Langmuir 2007; 23(14): 7666-71.

[http://dx.doi.org/10.1021/la700856a] [PMID: 17539668]

[35] Pavinatto FJ, Pavinatto A, Caseli L, *et al.* Interaction of chitosan with cell membrane models at the air-water interface. Biomacromolecules 2007; 8(5): 1633-40.
[http://dx.doi.org/10.1021/bm0701550] [PMID: 17419586]

[36] Wydro P, Krajewska B, Hac-Wydro K. Chitosan as a lipid binder: a langmuir monolayer study of chitosan-lipid interactions. Biomacromolecules 2007; 8(8): 2611-7.
[http://dx.doi.org/10.1021/bm700453x] [PMID: 17630796]

[37] Pavinatto A, Pavinatto FJ, Barros-Timmons A, Oliveira ON Jr. Electrostatic interactions are not sufficient to account for chitosan bioactivity. ACS Appl Mater Interfaces 2010; 2(1): 246-51.
[http://dx.doi.org/10.1021/am900665z] [PMID: 20356241]

[38] Krajewska B, Wydro P, Jańczyk A. Probing the modes of antibacterial activity of chitosan. Effects of pH and molecular weight on chitosan interactions with membrane lipids in Langmuir films. Biomacromolecules 2011; 12(11): 4144-52.
[http://dx.doi.org/10.1021/bm2012295] [PMID: 21936509]

[39] Cámara CI, Quiroga MV, Wilke N, Jimenez-Kairuz A, Yudi LM. Effect of chitosan on distearoylphosphatidylglycerol films at air/water and liquid/liquid interfaces. Electrochim Acta 2013; 94: 124-33.
[http://dx.doi.org/10.1016/j.electacta.2013.01.137]

[40] Krajewska B, Wydro P, Kyzioł A. Chitosan as a subphase disturbant of membrane lipid monolayers. The effect of temperature at varying pH: I. DPPG. Colloids Surfaces A Physicochem Eng Asp 2013; 434: 349-58.

[41] Krajewska B, Kyzioł A, Wydro P. Chitosan as a subphase disturbant of membrane lipid monolayers. The effect of temperature at varying pH: II. DPPC and cholesterol. Colloids Surfaces A Physicochem Eng Asp 2013; 434: 359-64.

[42] Pavinatto A, Souza AL, Delezuk JA, Pavinatto FJ, Campana-Filho SP, Oliveira ON Jr. Interaction of O-acylated chitosans with biomembrane models: probing the effects from hydrophobic interactions and hydrogen bonding. Colloids Surf B Biointerfaces 2014; 114: 53-9.
[http://dx.doi.org/10.1016/j.colsurfb.2013.09.034] [PMID: 24161506]

[43] Baszkin A. Molecular recognition on the supported and on the air/water interface-spread protein monolayers. Adv Colloid Interface Sci 2006; 128-130: 111-20.
[http://dx.doi.org/10.1016/j.cis.2006.11.008] [PMID: 17196538]

[44] Girard-Egrot AP, Godoy S, Blum LJ. Enzyme association with lipidic Langmuir-Blodgett films: interests and applications in nanobioscience. Adv Colloid Interface Sci 2005; 116(1-3): 205-25.
[http://dx.doi.org/10.1016/j.cis.2005.04.006] [PMID: 16181605]

[45] Toimil P, Prieto G, Miñones J Jr, Trillo JM, Sarmiento F. Monolayer and Brewster angle microscopy study of human serum albumin-dipalmitoyl phosphatidyl choline mixtures at the air-water interface. Colloids Surf B Biointerfaces 2012; 92: 64-73.
[http://dx.doi.org/10.1016/j.colsurfb.2011.11.022] [PMID: 22154096]

[46] Kundu S, Matsuoka H, Seto H. Zwitterionic lipid (DPPC)-protein (BSA) complexes at the air-water interface. Colloids Surf B Biointerfaces 2012; 93: 215-8.

[http://dx.doi.org/10.1016/j.colsurfb.2012.01.008] [PMID: 22300899]

[47] Fragneto G, Alexandre S, Valleton J-M, Rondelez F. Competition for space between a protein and lipid monolayers. Colloids Surf B Biointerfaces 2013; 103: 416-21.
[http://dx.doi.org/10.1016/j.colsurfb.2012.10.057] [PMID: 23261561]

[48] Lhor M, Bernier SC, Horchani H, *et al.* Comparison between the behavior of different hydrophobic peptides allowing membrane anchoring of proteins. Adv Colloid Interface Sci 2014; 207: 223-39.
[http://dx.doi.org/10.1016/j.cis.2014.01.015] [PMID: 24560216]

[49] Vicente EF, Nobre-Pavinatto TM, Pavinatto FJ, de Oliveira Junior ON, da Costa Filho AJ, Cilli EM. N-terminal microdomain peptide from human dihydroorotate dehydrogenase: structure and model membrane interactions. Protein Pept Lett 2015; 22(2): 119-29.
[http://dx.doi.org/10.2174/0929866521666140508125215] [PMID: 24810226]

[50] Talhari DT, Moraes ML, Castilho PV, Oliveira ON Jr, Beltramini LM, Araújo APU. Interaction of a C-terminal peptide of Bos taurus diacylglycerol acyltransferase 1 with model membranes. Biochim Biophys Acta - Biomembr 2009; 1788(10): 2320-5.

[51] Boisselier É, Calvez P, Demers É, Cantin L, Salesse C. Influence of the physical state of phospholipid monolayers on protein binding. Langmuir 2012; 28(25): 9680-8.
[http://dx.doi.org/10.1021/la301135z] [PMID: 22686284]

[52] Thakur G, Micic M, Leblanc RM. Surface chemistry of Alzheimer's disease: a Langmuir monolayer approach. Colloids Surf B Biointerfaces 2009; 74(2): 436-56.
[http://dx.doi.org/10.1016/j.colsurfb.2009.07.043] [PMID: 19726167]

[53] Damalio JC, Nobre TM, Lopes JL, Oliveira ON Jr, Araújo AP. Lipid interaction triggering Septin2 to assembly into β-sheet structures investigated by Langmuir monolayers and PM-IRRAS. Biochim Biophys Acta 2013; 1828(6): 1441-8.
[http://dx.doi.org/10.1016/j.bbamem.2013.02.003] [PMID: 23416254]

[54] Bussières S, Cantin L, Desbat B, Salesse C. Binding of a truncated form of lecithin:retinol acyltransferase and its N- and C-terminal peptides to lipid monolayers. Langmuir 2012; 28(7): 3516-23.
[http://dx.doi.org/10.1021/la203896n] [PMID: 22260449]

[55] Blume A. A comparative study of the phase transitions of phospholipid bilayers and monolayers. Biochim Biophys Acta - Biomembr 1979; 557(1): 32-44.
[http://dx.doi.org/10.1016/0005-2736(79)90087-7]

[56] Gallier S, Gragson D, Jiménez-Flores R, Everett DW. β-Casein–phospholipid monolayers as model systems to understand lipid–protein interactions in the milk fat globule membrane. Int Dairy J 2012; 22(1): 58-65.
[http://dx.doi.org/10.1016/j.idairyj.2011.08.007]

[57] Caseli L, Pavinatto FJ, Nobre TM, Zaniquelli ME, Viitala T, Oliveira ON Jr. Chitosan as a removing agent of beta-lactoglobulin from membrane models. Langmuir 2008; 24(8): 4150-6.
[http://dx.doi.org/10.1021/la7038762] [PMID: 18302443]

[58] Lopes JLS, Nobre TM, Siano Á, *et al.* Disruption of *Saccharomyces cerevisiae* by Plantaricin 149 and investigation of its mechanism of action with biomembrane model systems. Biochim Biophys Acta -

Biomembr 2009; 1788(10): 2252-8.
[http://dx.doi.org/10.1016/j.bbamem.2009.06.026]

[59] Maltseva E, Brezesinski G. Adsorption of amyloid beta (1-40) peptide to phosphatidylethanolamine monolayers. ChemPhysChem 2004; 5(8): 1185-90.
[http://dx.doi.org/10.1002/cphc.200400045] [PMID: 15446741]

[60] Ege C, Lee KY. Insertion of Alzheimer's A beta 40 peptide into lipid monolayers. Biophys J 2004; 87(3): 1732-40.
[http://dx.doi.org/10.1529/biophysj.104.043265] [PMID: 15345552]

[61] Brezesinski G, Maltseva E, Möhwald H. Adsorption of amyloid β (1-40) peptide at liquid interfaces. Z Phys Chem 2007; 221: 95-111.
[http://dx.doi.org/10.1524/zpch.2007.221.1.95]

[62] Thakur G, Pao C, Micic M, Johnson S, Leblanc RM. Surface chemistry of lipid raft and amyloid Aβ (1-40) Langmuir monolayer. Colloids Surf B Biointerfaces 2011; 87(2): 369-77.
[http://dx.doi.org/10.1016/j.colsurfb.2011.05.047] [PMID: 21708455]

[63] Lau T-L, Ambroggio EE, Tew DJ, *et al.* Amyloid-β peptide disruption of lipid membranes and the effect of metal ions. J Mol Biol 2006; 356(3): 759-70.
[http://dx.doi.org/10.1016/j.jmb.2005.11.091] [PMID: 16403524]

[64] Lepère M, Muenter AH, Chevallard C, Guenoun P, Brezesinski G. Comparative IR and X-ray studies of natural and model amyloid peptides at the air/water interface. Colloids Surfaces A Physicochem Eng Asp 2007; 303(1–2): 73-8.

[65] Schwarz G, Taylor SE. Polymorphism and interactions of a viral fusion peptide in a compressed lipid monolayer. Biophys J 1999; 76(6): 3167-75.
[http://dx.doi.org/10.1016/S0006-3495(99)77468-0] [PMID: 10354441]

[66] Castano S, Desbat B. Structure and orientation study of fusion peptide FP23 of gp41 from HIV-1 alone or inserted into various lipid membrane models (mono-, bi- and multibi-layers) by FT-IR spectroscopies and Brewster angle microscopy. Biochim Biophys Acta - Biomembr 2005; 1715(2): 81-95.

[67] Matar G, Nasir MN, Besson F. Interfacial properties and structure stability of the gp41 tryptophan-rich peptide from HIV-1. J Colloid Interface Sci 2010; 352(2): 520-5.
[http://dx.doi.org/10.1016/j.jcis.2010.08.075] [PMID: 20850757]

[68] Matar G, Besson F. Influence of the lipid composition of biomimetic monolayers on the structure and orientation of the gp41 tryptophan-rich peptide from HIV-1. Biochim Biophys Acta - Biomembr 2011; 1808(10): 2534-43.

CHAPTER 8

Protection of Bioactive Compounds

Stefan Dima[1,*] and **Cristian Dima**[2]

[1] *Department of Chemistry, Physics and Environment, Faculty of Sciences and Environment, "Dunarea de Jos" University of Galati, Romania*

[2] *Faculty of Food Science and Engineering, "Dunarea de Jos" University of Galati, Romania*

Abstract: The vegetal extracts are complex blends of chemical compounds from various classes: polyphenols, carotenoids, vitamins, essential oils, *etc.* It is known that these components are highly sensitive to the action of oxygen, light radiations and high temperature. These factors contribute to the degradation of bioactive compounds and implicitly to the decrease of their biological potential. Therefore, the encapsulation of bioactive compounds in different matrices is required in order to prevent the aforementioned shortcomings.

The purpose of this review is to synthesize the newest information about the biocomponents' encapsulation. The encapsulation of bioactive compounds, ensures the stability of the volatile compounds during thermic processing, the protection of bioactive components against environmental factors, the increase of the solubility of bioactive compounds, the controlled release of bioactive components, *etc.* Various techniques are used for the encapsulation of bioactive compounds, some of them being: emulsification, coacervation, gelation, extrusion, spray drying, spray cooling, *etc.*

Keywords: Biopolymers,Coacervation, Emulsions, Encapsulation, Gelation, Microparticles, Nanoparticles.

INTRODUCTION

Scientific history provides numerous examples of man's creativity manifesting itself by reproducing certain naturally occurring phenomena and processes.

* **Corresponding author Stefan Dima:** Department of Chemistry, Physics and Environment, Faculty of Sciences and Environment, "Dunarea de Jos" University of Galati, 47 Domneasca Street, 800008 Galati, Romania; E-mail: Stefan.Dima@ugal.ro

Encapsulation is a case in point. The encapsulation process is found in the primary stages of the universe, when the chemical elements and molecules got together into structural units wrapped in a semipermeable membrane able to protect them against external factors [1]. These structures, called coacervates, have evolved towards what we call today biological cell. The biological membrane and the membrane surrounding the microcapsule share the following characteristics:

- It delimitates the space within the cell or capsule from the outer environment;
- It protects the components of the cell or capsule against the aggression of the external factors;
- It communicates with the outer environment by transfer of mass, energy and information.

In order to better understand the encapsulation process, the present paper will attempt to briefly answer the following questions: What is encapsulation? Why encapsulate bioactive components? How to do the encapsulation of components? What are the advantages and disadvantages of encapsulation? The answers to these questions are found in the preoccupations of the scientists in the entire world, whose results are published in the thousands of papers issued annually in various specialised journals.

WHAT IS ENCAPSULATION?

Encapsulation is an innovative technology aimed at increasing the functionality of materials with applications in a wide range of areas: medicine, pharmacy, food industry, agriculture, *etc.*

This technology achieves the entrapping of a solid, liquid or gaseous substance, called encapsulate, active, core, payload, fill or internal phase into another substance, called shell material, wall material, carrier material, coating material, external phase or support phase, under the form of particles of various sizes [2]. These particles are a delivery system providing the embedding, transport, and release of component in various parts of the human body, such as skin, nose, lungs, mouth, stomach, the small intestine, or colon [3].

Certain particles obtained by encapsulation have a homogeneous internal morphology and are called core-shell particles or capsules. In these particles the active substance exists under the form of one or more nuclei surrounded by one or more layers of encapsulating material (Fig. **1a, b, c**). Other particles have a heterogeneous internal morphology and are called matrices or microspheres [4].

In these particles the active substance is diffused into a matrix, which may be a polymer hydrogel or a porous material (Fig. **1d**). Sometimes these matrices may be covered with one or several layers of material, thus improving the particle qualities (Fig. **1e**). A special group of particles in which bioactive compounds may be encapsulated are mixed micelles obtained by the self-assembling of amphiphile molecules, emulsions, nanoemulsions, microemulsions and liposomes.

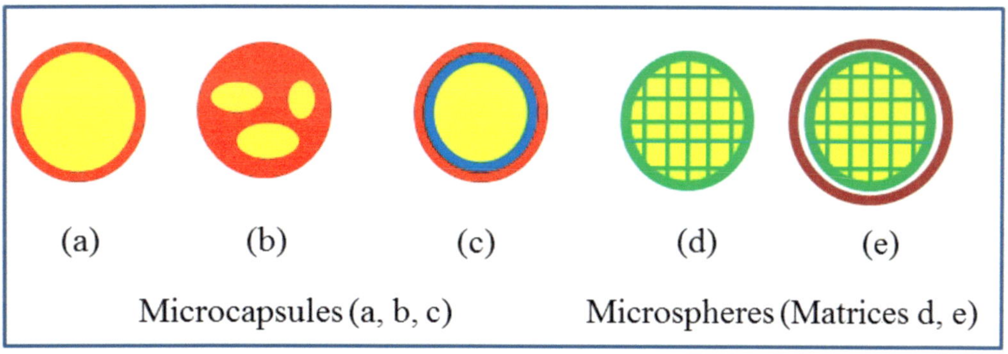

Fig. (1). Microcapsule (a, b, and c) and microspheres (d and e) morphology.

The classification of particles according to size is still a topic of debate among researchers [5 - 8]. Nevertheless, in most studies, the size of 100 nm is acknowledged as the upper limit for nanoparticles and the lower limit for microparticles. Thus, particles with sizes between 1nm and 100 nm are considered nanocapsules, and those with sizes between 100 nm and 1000 nm are considered microcapsules [4, 5, 7, 9, 10]. This threshold dimension (100 nm) is said to be the point where a sudden fall of the bulk material properties occurs. In fact, the threshold size of altering the bulk material properties depends on the nature of the material. That is why various scientific papers provide different size ranges for nano and microparticles.

Due to their small size and the high surface/volume ratio, nanoparticles have certain advantages, such as: high stability against the tendency of gravitational aggregation and separation, reduced light scattering, which allows for the improved appearance of foods, high solubility of lipophilic biocompounds, and better bioavailability [9 - 12].

WHY ENCAPSULATE BIOACTIVE COMPOUNDS?

The encapsulation of bioactive compounds has become a necessity, on the one hand due to their physical, chemical and physiological sensitivity, and on the other hand, due to technical and economic reasons, such as the improvement of transport and handling, the diversification of delivery methods, the increase of efficiency in their incorporation into food products. Thus, food manufacturers have noticed that there are biocomponents which cannot be directly introduced into the food matrices, such as the lipophilic biocomponents into hydrophilic matrices, and hydrophilic biocomponents into lipophilic matrices respectively. Also, bioavailability studies have shown that in food consumption certain bioactive substances may be chemically or enzymatically degraded, as a result of physical, chemical and physiological factors specific to the areas in the gastrointestinal tract [5, 13, 14].

Out of the many causes imposing the encapsulation of bioactive components, the most important refer to:

- *The protection of bioactive components* against environmental factors, such as: temperature, light, oxygen, enzymes, moisture, microorganisms, *etc.*;
- *The isolation of bioactive molecules* from the molecules in the food matrix they may react with. The possible reactions resulting in the degradation of the bioactive molecules in the food matrix are: oxidation, reduction, hydrolysis and isomerisation. A case in point consists of poly-unsaturated fatty acids (PUFA) susceptible to oxidation in the presence of oxygen, free radicals, oxido-reducing enzymes, transitional metals and other pro-oxidative components [15, 16]. Similarly, the minerals' encapsulation inhibits the pro-oxidative action of some minerals, preventing their precipitation in the presence of certain chemicals, such as oxalic acid, phosphoric acid, proteins, enzymes, *etc.*;

- *The increase of the solubility of bioactive compounds*. Thus, to the purpose of inserting lipophilic bioactive components (citrus oil, carotenoids, vitamins D, E, ω-3 fatty acids) into hydrophilic food matrices, such as beverages, they are encapsulated into water-soluble delivery systems, such as nano and microparticles based on hydrosoluble polymers [17], while in order to introduce hydrophilic components (polyphenols) into lipophilic matrices, they are encapsulated in lipid based micro and nano particles (emulsions, nanoemulsions, solid lipid nano and microparticles) [18];
- *The controlled release of bioactive components*. The control of the release from micro and nano particles refers to the release rate, as well as the centres where the molecules are to manifest their activity. In this respect, delivery systems are designed in order to provide a fast, slow, constant or pulsatory release, as well as a vectorised release, as is the case in releasing biocomponents in the stomach, small intestine, colon [19 - 21]. By the slow release of antioxidants (polyphenols, carotenoids) or antimicrobial substances (essential oils), the shelf life of food products may be extended;
- *The increase of the bioavailability of bioactive compounds*, due to the increase of bioaccessibility, as well as the absorption of biomolecules through biological barriers. Thus, in the absorption through the mucus layer and the epithelial cells of the small intestine, the delivery systems should provide a balance between the hydrophilic character, necessary for the solubilisation in the intestinal fluid, and the hydrophobic one, necessary for the passage of particles through the wall of the small intestine [22];
- *Masking the unpleasant taste and odour* of some biocomponents, such as peptides, fish oil, minerals, *etc.*;
- *The conversion of liquids into free flowing powders* whereby the biocomponent's mass and volume are reduced, and consequently the storage and transportation costs are cut down. Thus, by the spray drying method the solutions of biopolymers in which the bioactive components (vitamins, antioxidants, colourings, flavours, oils, *etc.*) are dispersed are atomised into very fine droplets, which dry into powder form;
- *The avoidance of decreasing the content of* bioactive substance in the food product by mass transfer (vaporisation, diffusion) of volatile components (flavours, essential oils).

DESIGN AND FABRICATION OF MICRO AND NANO PARTICLES FOR DELIVERY OF THE BIOACTIVE COMPOUNDS

Encapsulation in general and the encapsulation of the bioactive compounds in particular is a complex process in which imagination and artistic sense are intertwined with the precision of scientific principles. In designing the micro and nanoparticles used in food and medicine functionalization there is cooperation among researchers from various areas, such as: food science, pharmacy, medicine, materials science, chemistry, physics, biotechnology, microbiology, engineering, marketing, *etc.*

In food industry, encapsulation represents a new technique of manufacturing innovative foods, such as functional foods. Food functionalization has become necessary as a result of the increasing exigency of customers, who are requiring more and more foods able to provide nutritional value, as well as to preserve health and decrease the risk of disease [24]. These foods are obtained by introducing inside the food matrix the micronutrients and nutraceuticals which play an essential part in the human body (vitamins, minerals, polyunsaturated fatty acids) or have both a nutritional and a beneficial role in consumer health [21 - 25].

Designing bioactively-charged micro and nanoparticles is an integrated process, with interrelated stages, so that the manufactured micro and nanoparticles really contribute to developing the characteristics required from the functional food: safety and security, high nutritional value, health benefits, good sensorial qualities and acceptable cost.

"Anamnesis" of the Functional Food Product

According to the principles of reverse engineering, designing micro and nano particles should start by the "anamnesis" of the functional food product manufactured by means of encapsulated micronutrients and nutraceuticals. According to the characteristics of the functional food, the bioactive compounds are chosen and the characteristics of micro and nanoparticles are defined, which in turn depend on the nature of the encapsulating material and the encapsulating technique [26, 27].

This is the stage when the nutritional and health characteristics of the functional food are decided, as well as the target group of consumers: children, seniors, consumers with religious restrictions, special diet consumers, *etc.* It is also the stage of the analysis of the physico-chemical properties of the food matrix where the micro and nanocapsules are introduced (aggregation state, chemical composition, shape, appearance). Thus, it is important to know the aggregation state of the food matrix, as the characteristics of the micro and nano particles introduced into a liquid food (beverage, juice), are fundamentally different from the characteristics of the micro and nanoparticles introduced into a solid food.

The analysis of the chemical components present in the food matrix (lipids, proteins, carbohydrates, enzymes) is necessary to avoid their reactions with the bioactive components or even the wall material.

The knowledge of technological operations and the physico-chemical requirements used in the manufacture, packaging, and storage of the functional food is an important objective in designing micro and nanoparticles. Thus, it is possible to design micro and nanoparticles able to withstand high or low temperatures, pH or ionic strength variations, mechanical operations, light, moisture.

Choice of Core Materials

The core material is made up of the bioactive compounds entrapped into various classes of particles introduced in the functional food. In the manufacture of functional foods with micro and nanocapsules, the choice of the bioactive compounds is made according to the biological activity, and their physico-chemical properties influence the choice of the wall material and the encapsulation method. The biological activity of some biocomponents plays an important role in the product quality, as well as the health of the consumer. For example, antioxidants (polyphenols, carotenoids) extend the shelf life of the food product, but reduce the oxidative stress in human cells. Other biocomponents, such as vitamins, minerals bring a beneficial contribution to consumer health, having less influence on the sensorial qualities of foods.

The physico-chemical properties of bioactive compounds influence the encapsulation process.

Thus, certain micronutrients and nutraceuticals are hydrophilic compounds (B, C vitamins, certain polyphenols), and others are lipophilic compounds (PUFAs, carotenoids, essential oils). These characteristics are expressed by the oil/water partition coefficient ($K_{O/W}$) representing the ratio between the concentration of the bioactive component in the oil phase and the concentration of the bioactive component in the water phase. Most of the times, the standard oil phase is octanol [28, 29]. Also, to avoid the use of very small or very high values of the partition coefficient $K_{O/W}$, the logarithm of the oil/water partition coefficient (Log $K_{O/W}$) is used. Thus, lipophilic components have a high value of the partition coefficient $K_{O/W}$, respectively Log $K_{O/W} > 3$, while hydrophilic components have low values of the $K_{O/W}$ and respectively Log $K_{O/W} < 3$. For example, ascorbic acid has Log $K_{O/W}$ = -2.15 while quercetin-3 sulphate has Log $K_{O/W}$ = -1.11 in comparison with quercetin which has low water-solubility and Log $K_{O/W}$ = 1.82. The partition balance of the bioactive compounds between the water phase and the oil phase influences the encapsulation efficiency, the rate and mechanism of releasing biomolecules from particles and the bioavailability of the active components [28, 30, 31].

Table **1** shows the important physico-chemical properties of bioactive components influencing the encapsulation process.

Table 1. Physico-chemical properties of bioactive compounds.

Compound	Molar mass ($g \cdot mol^{-1}$)	Melting temperature T_m (°C)	Partition coefficient octanol/water Log $K_{O/W}$	Water solubility C_{SW} ($g \cdot L^{-1}$)	pK_a
Carotenoids					
β-carotene	536.9	180	14,76	1.5×10^{-9}	-
Licopene	536.9	175	14.5	3.7×10^{-9}	-
Lutein	568.9	180	11.5	5.6×10^{-5}	14.6
Zeaxanthin	568.9	205	10.9	8.0×10^{-5}	14.6
Astaxanthin	596.8	210	8.24	5.6×10^{-7}	12.3
Phytosterols					
β-sitosterol	414.7	138	10.5	8.7×10^{-7}	15.0
Stigmasterol	412.7	150	10.07	1.6×10^{-6}	15.0
Campesterol	400.7	155	9.97	2.0×10^{-6}	15.0
Vitamins					
Vitamin C	176.12	191	- 2.15	330	4.17/11.57

The solubility of biocomponents is another characteristic to be taken into account in their encapsulation. Thus, lipophilic components are lipid-soluble and are generally encapsulated in lipid-based micro and nanoparticles, such as: O/W emulsions, O/W nanoemulsions, micelles, microemulsions, liposomes, solid lipid nanoparticles (SLN), nanostructured lipid carriers (NLC), but also in biopolymer nanoparticles, filled hydrogel particles, molecular complexes (cyclodextrins), while hydrophilic compounds are encapsulated in other kinds of particles such as: reverse micelles, liposomes, W/O/W-multiple emulsions, biopolymer nanoparticles, hydrogel particles, polyelectrolyte complexes [30 - 34]. The molecular weight of bioactive compounds influence the aggregation state, solubility and the rate of their release from particles.

At room temperature, certain bioactive compounds are in a liquid state, such as: fatty acids (ω-3 fatty acids, conjugated linoleic acid), α-tocopherol, essential oils, while others are solid, such as polyphenols (curcumin, catechin, quercetin, resveratrol), phytosterols, carotenoids.

The encapsulation of crystalline lipophilic compounds is a real challenge as their presence in an emulsion-based delivery system may trigger: emulsion destruction, reduction of the biocomponent release rate, biocomponent precipitation in the gastrointestinal tract (GIT), reduction of their bioavailability [30].

To obtain delivery systems with a higher concentration of lipophilic bioactive component, the crystalline biocompound is first melted and then mixed with a transporter lipid heated to a temperature above its melting temperature. At body temperature, the transporter lipid is in a solid state. The hot liquid lipidic mixture is dispersed into the aqueous solution of a hydrophilic emulsifier, and through cooling solid lipid particles (SLP) of various sizes loaded with lipophilic bioactive compounds are obtained.

Choice of Wall Materials

The choice of the wall material is an essential stage in the strategy of encapsulating bioactive compounds and is performed in correlation with the other stages. The wall material influences the encapsulation efficiency, the microparticle characteristics, and the bioavailability of bioactive components.

The manufacture of micro and nanocapsules in food industry employs wall materials providing food safety (GRAS-Generally Regarded As Safe) with the following characteristics [2, 35, 36, 38]:

- They are not toxic and do not alter the food texture ;
- They are tasteless and odourless;
- Theyare easily dispersible in solvents (water, ethyl alcohol) and form solutions or dispersions with low viscosity at a relatively high concentration (η< 500 cps la c $\geq$ 45%);
- They are slightly or not at all hygroscopic;
- They have emulsifying properties and/or ability to form thin or elastic films;
- They have a good ability to eliminate the solvent;
- They show a good resistance to physico-chemical agents: light, temperature, pH, ionic strength, enzymes, oxygen, moisture;
- They have a crystalline, permeable structure, in order to ensure the controlled release of the bioactive components ;
- They are available in convenient quantities and at an accessible cost.

Table **2** shows the classes of biomolecules commonly used as wall materials in the food industry.

Table 2. Materials used for encapsulation of bioactives compounds [35].

Chemical classes (IUPAC)	Chemical compounds
Polysaccharides	Starch and derivatives: maltodextrins, syrups, dextrins, cyclodextrins Cellulose and derivatives: ethyl-, methyl-, hydroxipropil-,carboxymethyl-cellulose Plant exudates and extracts: gum arabic, gum karaya, gum tragacanth, mesquite gum, pectins, guar, locust bean, tara) Marine extracts: alginate -brown seaweed, carrageenan Microbial and animal polysaccharides: xanthan, gellan, dextran, chitosan
Proteins	Vegetable proteins: gluten, soy protein, zein, rice protein Animal proteins: gelatin, milk proteins, whey proteins
Lipids	Glycerides, fatty acids and fatty alcohols, phospholipids, waxes (beeswax, carnauba wax, candellia wax).
Other materials	Polyvinilpyrrolidone, polylactic acids and derivatives, shellac, paraffin, inorganic materials

Choice of Encapsulation Methods and Techniques for Bioactive Compounds

Nowadays there are many methods whereby bioactive components may be encapsulated. Most of them are only applied in research laboratories. Although the number of industrial scale encapsulation methods is lower, they are more efficient, in order to meet the ever increasing demands of the microcapsule market [37]. The choice of an encapsulation technique takes into account the properties of the bioactive compound, the properties of the wall material, the characteristics of the microcapsules, agreed upon in accordance with the attributes of the functional food.

Taking their size consideration, micro and nanoparticles may be obtained both by dividing the bulk material from a large size (mm, cm, m) to a micro and nanometric scale (top-down methods), and by associating molecules (10^{-9} m) into aggregates of nanometric sizes (bottom-up methods) (Fig. **2**). Some delivery systems are obtained by the successive application of the top-down and bottom-up methods. For example, the droplets of the emulsions prepared by the breaking and homogenisation of non-miscible liquids in the presence of emulsifiers (top-down method) may be coated in one or more layers of biopolymers by means of the layer-by-layer technique (bottom-up method).

Characteristics of Micro and Nanoparticles

The characteristics of micro and nanocapsules are required by the functional food manufacturer, in accordance with the food attributes, as they determine the stability in the particles in the food matrix and the GIT. Nowadays the characteristics of particles are analysed by modern means of investigation, such as: scanning electron microscopy (SEM), transmission electron microscopy (TEM), differential scanning calorimetry (DSC), X-ray diffraction analysis [34, 35].

The characteristics of micro and nanoparticles mainly depend on the preparation technique and the properties of the wall material. In the following we shall describe in short the main characteristics of micro and nanoparticles influencing the quality of the functional food, such as: particle size and morphology, electric charge, the ability to load and store biocomponents and the release mechanism.

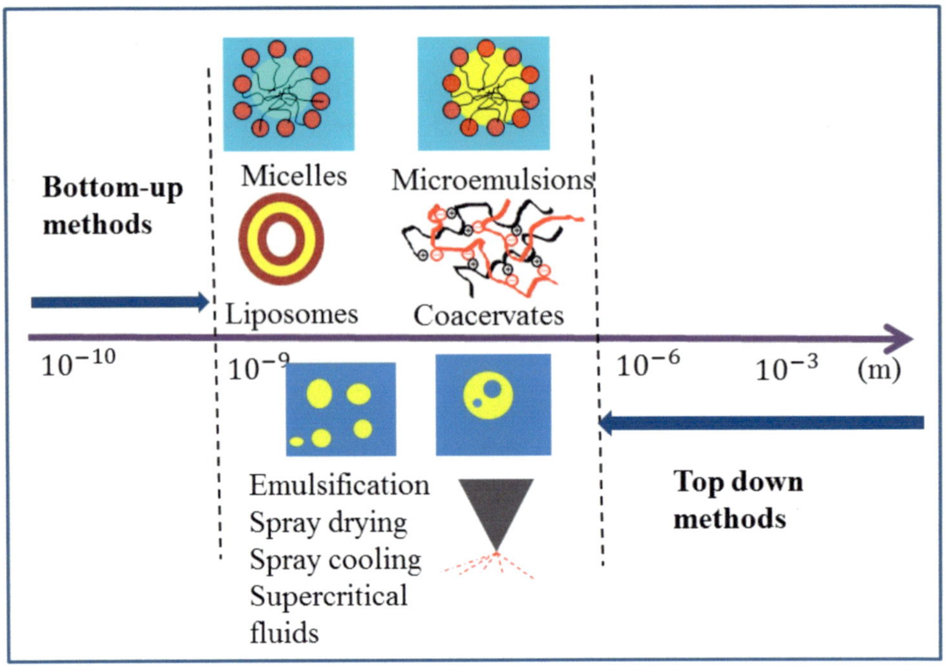

Fig. (2). Encapsulation methods of bioactive compounds.

- ***Size*** is one of the most important characteristics of food-grade particles as it affects the texture and optical clarity of food, the flow properties of powder, transport, storage and introduction into the food matrices, the loading capacity and the biocomponents' release rate [39 - 41]. Also, particle size has a determining role in their passage through the GIT wall, thus improving the bioavailability of the active components. Unfortunately, the small size of nanoparticles favours the occurrence of diseases [39]. Most colloidal systems coating the bioactive compounds (nanoemulsions, microemulsions, micelles, liposomes, polymer microspheres *etc.*) have a spherical shape and a different internal structure. These characteristics have a great impact on the mechanical and rheological properties and the release mechanism [39]. The size and shape of particles may be controlled by several factors such as: energy input, emulsifier type and concentration, properties of core and wall materials, temperature [41].
- ***The Particle Charge*** is mainly due to the molecules of the ionic surfactants or the polyelectrolytes adsorbed at the surface of colloidal particles. The electric

charge of the particles is measured by the electrokinetic potential zeta (ζ), whose value depends on the pH of the aqueous environment, ionic strength, and temperature. Changing the particles' charge may occur by varying the pH of the continuous phase or by coating them with layers of biopolymers with opposite electrical charges [42]. The particles' electrical charge influences the stability of colloidal systems, favours their interaction with other chemical compounds in the food matrix, and plays an important part in their adhesion to mechanic and biological surfaces [43].

- ***Loading Capacity, Encapsulation Efficiency and Retention Efficiency*** are characteristics which play a major role in selecting food-grade particles and deciding their cost.

Loading Capacity (LC) is the amount of biocomponent inside the particle (m_{cp}), compared to the amount of wall material (m_w). It is calculated by the relation **(1)**:

$$LC = \frac{m_{cp}}{m_w} \cdot 100 \quad \textbf{(1)}$$

Encapsulation Efficiency (EE) is the fraction of the total amount of bioactive component used for encapsulation (m_{cT}) which is inside the particle (m_{cp}). EE is calculated by the relation **(2)**:

$$EE = \frac{m_{cp}}{m_{cT}} \cdot 100 \quad \textbf{(2)}$$

Retention Efficiency (RE) is the fraction of the total amount of bioactive component inside the particle at the initial moment of measurement (m_{cpo}), remaining within the particle after a certain time t of storage or treatment (m_{cpt}). It is calculated by the relation **(3)**:

$$RE = \frac{m_{cpt}}{m_{cpo}} \cdot 100 \quad \textbf{(3)}$$

The higher the values of LC, EE and RE, the better the particles and the better their response to the demands of the functional food manufacturers.

• ***Release of Bioactive Compounds***

An important objective of microencapsulation is the controlled release of the encapsulates. In food industry, the controlled released of biocomponents ensures the maintenance of the nutritional and sensorial characteristics of foods and the extension of their shelf-life. In this respect, it is imperative to study the biocomponent release mechanisms in various types of microcapsules. The release studies are generally carried out "*in vitro*", as microcapsules are introduced in model systems, simulating the physiological conditions of the gastrointestinal tract, as well as the food processing conditions [40, 47, 52].

Research has shown that the release of a biocomponent out of micro and nanoparticles is based on four main phenomena: swelling, diffusion, erosion/dissolution and fragmentation [48 - 50].

The release mechanisms depend on several factors, among which: the nature of the wall material, the nature of the core, the particle geometry and morphology, the release conditions (solvent, pH, ionic strength, temperature) and the method of particle preparation [44, 45].

Most of the times, the release of the bioactive substances from micro and nanoparticles occurs according to zero-order **(4)**, first-order **(5)**, Higuchi **(6)**, and Korsmeyer-Peppas **(7)** equations [46].

$$\frac{Q_t}{Q_\infty} = 100(1 - k_o \cdot t) \quad \textbf{(4)}$$

$$\frac{Q_t}{Q_\infty} = 100\left(1 - e^{k_1 t}\right) \quad \textbf{(5)}$$

$$\frac{Q_t}{Q_\infty} = k_H \cdot t^{0.5} \quad \textbf{(6)}$$

$$\frac{Q_t}{Q_\infty} = k \cdot t^n \quad \textbf{(7)}$$

where: Q_t is the cumulated amount of substance released at time t

Q_∞ is the cumulated amount of substance released at the equilibrium state

n is release exponent, indicative of the transport mechanism;

k_o, k_1, k_H and k are constants incorporating geometrical and structural

characteristics of the wall material and compounds bioactive.

Exponent n for polymeric controlled delivery systems of spherical geometry has values of n=0.43 for Fickian diffusion, 0.43<n<1.0 for anomalous (non-Fickian) transport, and n>1.0 for super case-II transport [40, 49, 51].

PREPARATION OF MICRO AND NANO DELIVERY SYSTEMS OF BIOACTIVE COMPOUNDS

Emulsion-Based Micro and Nanoparticles

Conventional Emulsions and Nanoemulsions

Emulsions and nanoemulsions are the most common delivery systems of bioactive components. Most encapsulation techniques of lipophilic bioactive compounds have emulsion preparation as an important stage.

That is why this paragraph will detail the main aspects related to emulsion stability and preparation.

Emulsions are colloidal systems made up of a discontinuous phase, consisting of the droplets of a liquid, dispersed into another liquid, called continuous phase. Depending on the properties of the two liquids, emulsions may be O/W, where the discontinuous phase is made up of the droplets of an unpolar (oily) liquid, dispersed into a polar liquid, and W/O, where the discontinuous phase is made up of the droplets of a polar (aqueous) liquid, dispersed into an unpolar liquid [53 - 57]. A special kind of emulsions is multiple emulsions, which may be water-i--oil-in-water emulsions ($W_1/O/W_2$) and oil-in-water-in-oil emulsions ($O_1/W/O_2$). In these emulsions, the discontinuous phase is the primary W_1/O or O_1/W

emulsion dispersed into an aqueous or oily continuous phase [58, 59] (Fig. **3**).

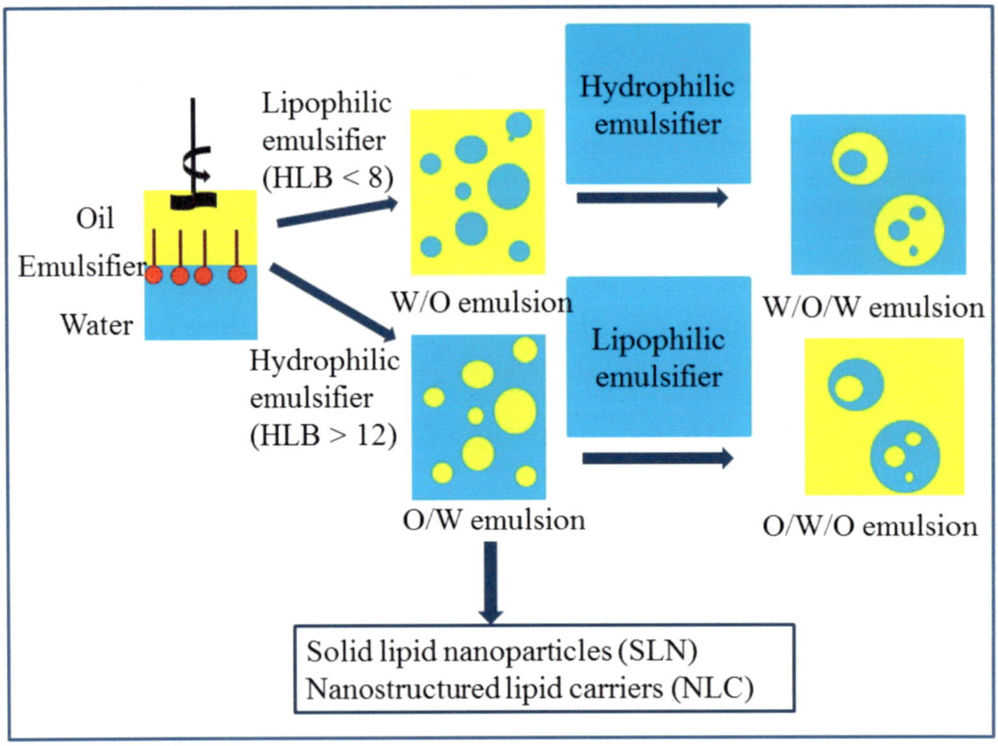

Fig. (3). Emulsion-based micro and nanoparticles.

When manufacturing emulsions, an important role is played by surfactants, which, together with other system characteristics (oil type, temperature, composition, ionic strength, *etc.*) may decide the type of emulsion formed.

The surfactant molecules contain both a polar (hydrophilic) group, and an un-polar (lipophilic) group. The ratio between the hydrophilic and lipophilic character defines an empirical dimension called HLB (Hydrophile-Lipophile Balance), specific to each surfactant. For practical uses another parameter is used to describe the nature of the oily phase, called recurred HLB (HLB_{req}). The oil HLB_{req} is the value of the HLB of the surfactant forming the most stable emulsion with the same composition. It was experimentally proved that in general the surfactants which disperse in water (HLB >12) act as emulsifiers for O/W emulsions, while the surfactants which do not disperse in water (HLB < 8) act as emulsifiers for W/O emulsions. The emulsifier molecules are adsorbed at the O/W interface, reduce the interfacial tension, and favour the breaking of the film between the two non-miscible liquids with the formation of discontinuous phase droplets. Similarly, the emulsifier molecules are adsorbed on the surface of the droplets forming a layer able to ensure the electrostatic, kinetic or steric stability of the emulsion [60, 61].

According to the size of the discontinuous phase droplets, the emulsions may be: conventional emulsions, whose droplet diameter range is 100-1000 nm and miniemulsions or nanoemulsions, whose droplet diameter range is 10-100 nm [55, 56]. Both types of emulsions are unstable thermodynamic systems, as in storage they decay and break, releasing free energy. Nevertheless, when analysed over long periods of time, emulsions preserve their important properties, especially their kinetic stability. The emulsions' kinetic stability or metastability is due on the one hand to the balance between the droplets' guided displacement and their Brownian movement, and on the other hand, the balance between the attraction and the rejection forces that occur among droplets. Thus, the speed of the droplets' movement through the continuous phase under the action of gravity is calculated by the relation **(8)**

$$v_{creaming/sedimentation} = \frac{2(\rho_2 - \rho_1)r^2}{9 \cdot \eta_1} g \qquad \textbf{(8)}$$

where: ρ_2 și ρ_1 are the densities of the dispersed phase, and the continuous phase respectively, r is the radius of the droplets which are considered spherical in shape, η_1 is the shear viscosity of the continuous phase and g is the acceleration due to gravity.

In the case of the O/W emulsions used as food products, where the density of the edible oil is lower than the density of the water phase ($\rho_2 > \rho_1$) the oil droplets have the tendency to move opposite the gravitational force, gathering in the upper part under the form of a cream. This phenomenon is termed creaming. On the other hand, in the W/O emulsions used as food products, the water droplets, whose density is higher than the density of the oil in the continuous phase ($\rho_2<\rho_1$), gather in the lower part of the emulsion, in a phenomenon called sedimentation.

Although nanoemulsions are more thermodynamically unstable than macroemulsions, due to the larger interphase surface, they are kinetically stable due to the smaller size of the droplets determining the decrease of the creaming/sedimentation velocity.

An easily controllable factor of the creaming/sedimentation velocity is the shear viscosity of the continuous phase. Thus, the kinetic stability of the food emulsions (O/W) may be improved by adding some thickening agents, such as: proteins, carbohydrates, gums, *etc.*, in the aqueous continuous phase [61].

Emulsion stability is also influenced by the interaction potential among the droplets in an emulsion. According to the DLVO theory (proposed by Derjaguin, Landeau, Vervew and Overbeek), the droplets of an emulsion are simultaneously affected by van der Waals attractive forces (**9**), and electrostatic repulsive forces) **(10)**:

$$U_{VDW} = -\frac{A_{212}r}{12h} \qquad \textbf{(9)}$$

where A_{212} is the Hamaker constant, which depends on the chemical composition of the two phases, r is the droplet radius, and h is the distance among droplets.

The strength of the van der Waals interactions decreases with the droplets' separation and increases with the droplet size and with the physical properties of the droplets and continuous phase.

$$U_{electrostatic} = 2\pi\varepsilon_0\varepsilon_R r\psi_0^2 \ln\left(1 + exp(-\kappa h)\right) \qquad \textbf{(10)}$$

where ε_0 and ε_R are the dielectric constant of a vacuum, and the relative dielectric constant of the solution respectively, ψ_0^2 is surface potential and κ^{-1} is Debye length and defines the distance where the counter ions in solution may influence the electrical properties of the interphase O/W.

$$\kappa^{-1} = \sqrt{\frac{\varepsilon_0 \varepsilon_R k_B T}{e^2 \sum n_{0i} Z_i^2}} \qquad \textbf{(11)}$$

where k_B is the Boltzmann constant, T is temperature, n_{0i} is the number of ions per cubic meter bulk electrolyte solution, Z_i is valence of ionic species of type i from the bulk electrolyte solution and e is the elementary charge (1.602×10^{-19}C).

In accordance with relations **(10)** and **(11)** repulsive interactions are influenced by the droplet size, superficial charges, and the ionic strength of the water phase.

The ionic species in the aqueous phase modify parameter κ^{-1} and may lead to the electrostatic stabilisation of emulsions or their destruction by coagulation of the polymeric chains of the polyelectrolytes in the aqueous phase (proteins, polysaccharides, polymer complexes) which are adsorbed on the surface of the droplets, resulting in either the steric stabilisation of emulsions, at higher polymer concentrations, or their destruction by bridging flocculation, at low polymer concentrations [62 - 67]. Studies have shown that branched polysaccharides, such as pectins, stabilise emulsions better than linear polysaccharides, like alginates [66]. The protein-polysaccharide complexes provide emulsions with higher stability than proteins alone, as the chains of the polyelectrolyte-complex undergo conformation alterations, being adsorbed at the interface in a more readily manner [65].

The dominant interactions decide on the stability or instability of emulsions. Thus, in the presence of polyvalent ions, at a higher concentration than the critical flocculation concentration, the repulsive potential falls in favour of the van der Waals interactions triggering the destabilisation of emulsion.

As compared to conventional emulsions, nanoemulsions are stable against droplet aggregation, as the small size of the droplets essentially reduces the intensity of the van der Waals attractive forces. Also, droplet size influences the appearance of emulsions. Thus, conventional emulsions are muddy or opaque, due to the size of the droplets of the same order as the wavelength of light, which results in strong light scattering. On the contrary, nanoemulsions have a transparent appearance, as their droplet size is much smaller than the wavelength of light and so the light scattering is weak [63].

Emulsions and nanoemulsions are prepared by high energy methods and low energy methods [64]. In the case of high energy methods, the devices that are used produce strong mechanic forces, able to break the two non-miscible liquids, or the larger droplets into smaller ones. The most commonly used in emulsion manufacture are mechanic homogenisers, ultrasound homogenisers, and high pressure homogenisers [53 - 55].

Obtaining emulsions with reduced energy consumption presupposes their spontaneous formation when the interface tension is reduced by various methods: temperature variation (phase inversion temperature-PIT method), the variation of chemical composition (phase inversion composition or PIC method) or the change of solvent quality [56].

Double emulsions $W_1/O/W_2$ are prepared in two stages. The first stage means the preparation of the primary emulsion W_1/O by the strong homogenisation of a mixture made up of the aqueous solution of the hydrophilic bioactive component, and an oil containing a lipophilic emulsifier. In the second stage, the primary emulsion W_1/O is added in an aqueous solution of a hydrophilic emulsifier, being then homogenised in gentler conditions [68, 69].

Emulsions and nanoemulsions are colloidal systems able to ensure the delivery of both lipophilic bioactive components, and hydrophilic bioactive components.

When encapsulating lipophilic bioactive compounds, an important role is played by the oil in which they are dissolved. On the one hand, the carrier oils should ensure a good solubility of the biocomponents, even the crystalline ones, and on the other hand they should ensure a high bioavailability of biocomponents.

Thus, it was proved that the bioavailability of lipophilic components in emulsions and nanoemulsions depends on the oil composition, droplet size and the nature of the emulsifier [10]. Studies on "*in vitro*" digestion have shown that the bioaccessibility of emulsified β-carotene is higher when it is dissolved in long chain triglycerides (LCT) than in medium chain triglycerides (MCT). This effect is due to the long chain fatty acids produced by the LCT digestion, which is able to solubilise lipophilic components more readily than medium chain fatty acids [72].

Other researchers have encapsulated β-carotene by means of the homogenisation-evaporation method [70, 71]. They dissolved β-carotene in ethyl-acetate, and then emulsified it in an aqueous phase where various proteins (sodium caseinate, whey protein isolate and soybean protein isolate) had been solubilised.

After the homogenisation, ethyl acetate was removed from the emulsions by means of a rotary evaporator. The authors proved that the β-carotene in emulsion has a superior antioxidant activity than free β-carotene.

Among the hydrophilic components, ascorbic acid is largely used in food preparation due to its antioxidant properties. Since the L-ascorbic acid is easily degradable in aqueous solution and at high temperature, it is necessary to protect it in various colloidal systems, among which the W/O and W/O/W emulsions are the most common [73]. Thus, Khalid *et al.* [74] prepared W/O and W/O/W emulsions with L-ascorbic acid, using soybean oil as external phase. The same encapsulation in multiple emulsions was also applied to other hydrophilic components, such as: procyanidin [68], vitamin B_2 [75], vitamin A [76, 77], resveratrol [25, 76]; anthocyanidin extract [78].

A special practical importance is to be given to emulsions prepared by the layer-by-layer technique [42]. Multilayer O/W emulsions are prepared in several stages by using successively aqueous solutions of polyelectrolytes of different electrical charges (*e.g.* alginate (-) / chitosan(+)/k-carrageenan (-)). The polymeric chains are set one upon another around the oil droplets forming a multilayer of variable thickness. These emulsions have several advantages, like the fact that they are stable, their droplets have a controlled size and surface charge, and are capable of encapsulating both lipophilic components, in the lyophilise core, and hydrophilic components in the wall material. Unfortunately, this technique cannot be used industrially as it is too intricate and requires a strict control of the experimental environment (pH, ionic strength, temperature, *etc.*)

Solid Lipid Particles (SLP) and Nanostructured Lipid Carriers (NLC)

Solid lipid particles (SLP) are innovative colloidal delivery systems raising a special interest from researchers and producers of medicines and functional foods [79 - 82].

SLP are solid lipid matrices similar in size to the droplets in conventional emulsions or nanoemulsions. In preparing SLP high energy methods are applied, by means of which the oil phase and the aqueous phase are homogenised together in energetic conditions (sonication, high pressure), in the presence of a hydrophilic emulsifier, at a temperature above the melting point of the lipid phase. The O/W emulsion formed is subjected to controlled cooling, with the occurrence of total or partial crystallisation of the lipid in the droplets of the discontinuous phase. The lipophilic bioactive compounds in liquid or crystalline state are first solubilised in the melted lipid phase.

The SLP size and morphology is influenced by numerous factors, like: the lipid's properties, the type of surfactant, the composition surfactant-oil-water, the properties of the bioactive component, the preparation technique.

The lipids used in preparing food-grade SLPs should meet the following demands:

- Melting points above 40° C;
- Ability to ensure a good solubilisation of bioactive compounds;
- Stability and resistance to chemical and physico-chemical factors ;
- Biodegradability and ability not to turn into harmful compounds.

The characteristics of the encapsulated bioactive compounds influences the internal structure of SLP. Thus, the bioactive components which are liquid at room temperature, with melting points under the melting point of the lipid phase, form particles of the core-shell type in which one or several droplets of a liquid bioactive component are surrounded by a solid lipid shell. Bioactive compounds which are solid at room temperature, with high melting points, are found inside or at the surface of SLPs under the form of crystals dispersed into a solid lipid matrix.

For example, Matos-Jr. *et al.* [80] prepared SLM loaded with ascorbic acid by means of hydrogenate palm oil and vegetable glycerol monostearate as lipidic phase. The process yield was over 81% and the encapsulation efficiency ranged from 74% to 84%. The SEM images evinced the presence of ascorbic acid microcrystals on the surface of microparticles.

Pandita *et al.* [81] encapsulated resveratrol in SLN prepared by the solvent diffusion-solvent evaporation method, using phosphatidyl choline from soybean lecithin (PHOSPHONLIPON®90) as lipid carrier, and Poloxamer 188 as emulsifier. The encapsulation efficiency of resveratrol in SLNs was over 88%. Pharmacokinetic studies showed that resveratrol-loaded SLNs have superior bioavailability, after oral administration, as compared to resveratrol suspension.

In comparison to other delivery systems of bioactive compounds, solid lipid micro and nanoparticles have the following advantages:

- High encapsulation efficiency;
- Prolonged and controlled release of biocomponents;
- The preparation techniques ensure sterilisation and industrial scale manufacture;
- Avoidance of organic solvents during preparation.

Nevertheless, SLPs also have a series of disadvantages requiring their cautious use in medicine and food manufacture. The most obvious disadvantages are the low loading capacity, and the dispersion of biomolecules on the particle surface followed by their expulsion due to the lipid polymorphic transition.

In order to eliminate these disadvantages [83], used a lipid phase made up of a mixture of liquid and solid lipids, resulting in lipid particles with high loading capacity. The particles so obtained are termed nanostructured lipid carriers (NLCs) and have known a considerable development due to their superior characteristics as compared to SLPs. The most common liquid lipids used to produce NLCs are MCT and oleic acid. MCT are saturated lipids, food-grade, faster digestible than LCT and have a high stability against oxidation.

The presence of an oil together with a solid lipid improves the qualities of nanoparticles and modifies their internal structure. Thus, the lipophilic bioactive component is solubilised in the liquid lipid found into the core of lipid particles, thus increasing their loading capacity. Also, the crystallinity degree in the mixture of oil and solid lipid, thus avoiding the expulsion of biomolecules out of the particles. Therefore, NLCs are more stable than SLPs, have a higher loading capacity and encapsulation efficiency than SLPs, provide better protection of bioactive components and their controlled release, are easier to digest, improving the bioavailability of bioactive components [82]. Aditya *et al.* [84] encapsulated quercetin in nanoemulsions, SLN and NLC, and underlined important differences between the colloidal delivery systems under study. So, the sizes of lipid particles were 82-83nm, 103-127nm and 34-47 in nanoemulsions, SLN and NLC respectively. The maximum value of bioaccessibility was seen in NLC and nanoemulsions (60%), as compared to SLN (35%) and the free quercetin solution (7%). The efficiency of encapsulating quercetin in NLC was higher than 90%.

Self Assembling-Based Nanoparticles

Micelles and Microemulsions

Self-assembling is the capacity of certain chemical species (molecules, macromolecules) to self-organise due exclusively to internal, structural and energetic causes. One of the main driving forces of self-assembling is amphiphilicity. Amphiphilicity is a structural characteristic referring to the chemical species (molecules, polymers) having a polar (hydrophilic) portion and an unpolar (hydrophobic) portion. In contact with water, amphiphilic molecules have the tendency to assemble in such a manner that the hydrophobic portions should enter as little as possible in contact with water, and the hydrophilic portions should interact with the water molecules. The result is the formation of nanostructures with different shapes and properties, such as: micelles, reverse micelles, linear micelles, vesicles, liposomes, hexagonal and bicontinuous phases, *etc.* [55, 57, 133].

Thermodynamic studies have shown that miccelisation is a spontaneous process, occurring when the free Gibbs energy decreases. The negative value of the free Gibbs energy of micellisation is due to the "hydrophobic effect" and the hydrophobic interactions which trigger a considerable increase of entropy [85].

It is worth mentioning that not all amphiphilic molecules or polymers have the ability to self-assemble.

Low-molecular-weight amphiphiles (surfactants) are molecules with a so-called "head-tail" structure where the hydrophilic 'head' may be a non-ionic polar group, or an ionic group (anionic, cationic, or zwitterionic) and hydrophobic 'tail(s)'are one or two hydrocarbon chains, saturated or unsaturated, with 8–20 carbons per chain.

The self-assembly of low-molecular-weight amphiphiles depends on several factors, such as: the molecule structure (head group area, volume and chain length of hydrophobic tail), the "head-head", and "tail-tail" interactions (van der Waals interactions, hydrophobic interactions, steric interactions and electrostatic interactions), interactions between the micelle surface and the outer environment (electrostatic interactions, van der Waals interactions) as well as the characteristics of the outer environment (polarity, pH, ionic strength, temperature) [85].

In polar solvents, amphiphile molecules (especially ionised molecules) self-assemble in such a manner that the hydrophobic 'tail(s)' form a hydrophobic core inside the micelle, and the hydrophilic heads are outside the micelle, in contact with the water molecules and the counter ions in the aqueous solution (Fig. **4a**). These aqueous micelles allow the encapsulation of lipophilic bioactive compounds by entrapping them inside the hydrophobic core, while the hydrophilic biocomponents can be encapsulated in the core of reverse micelles (Fig. **4b**).

The presence of an amount of lipophilic bioactive compounds inside micelles determines the latter's swelling and the formation of systems similar to nanoemulsions [30, 31]. That is why these systems made up of water, oil and surfactant/cosurfactant are called microemulsions (Fig. **4c**), although they essentially differ from nanoemulsions by their being thermodynamically stable in certain conditions of composition and temperature [55, 135]. Nanoemulsions are transparent systems, but they are destroyed during storage, while microemulsions remain intact for a long time. Due to their special properties (thermodynamic stability, optical clarity and low viscosity), microemulsions are used as delivery systems for lipophilic bioactive components, such flavours, nutraceuticals, oil-soluble vitamins and oil-soluble antimicrobials, especially in soft drinks [31, 86, 87, 100].

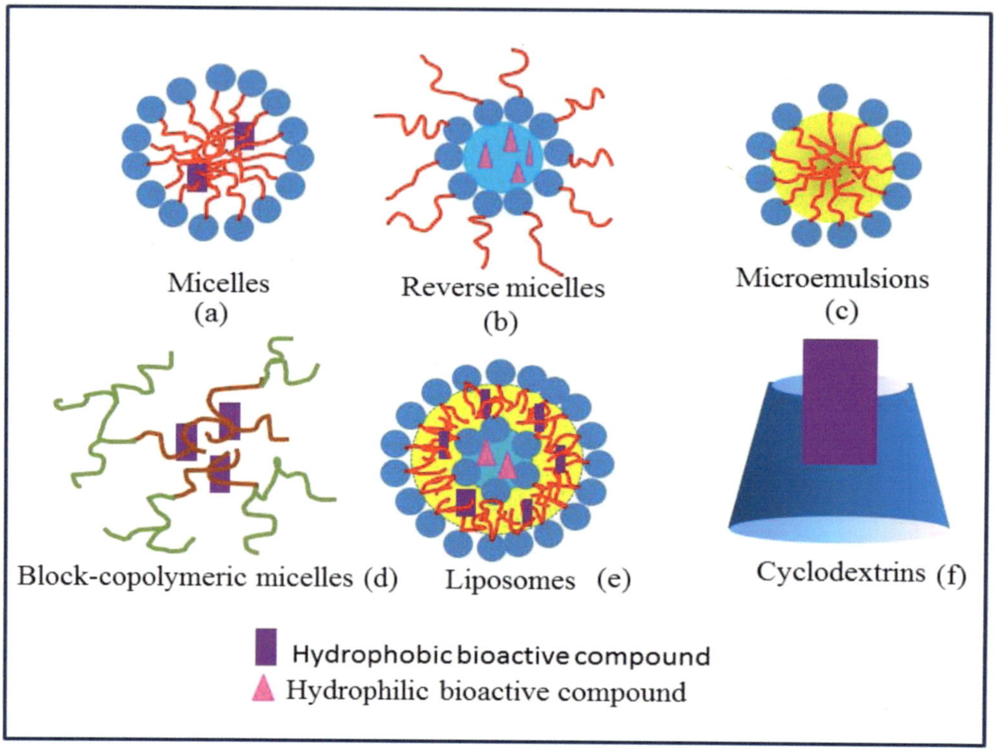

Fig. (4). Self assembling-based nanoparticles (not drawn to scale).

In the past few years, special attention has been granted to polymers with amphiphile structure [88]. Polymeric amphiphiles are macromolecular compounds made up of identical monomers (photopolymers) or different monomers successively arranged in a certain order (heteropolymers or co-polymers). Co-polymers arranged successively in groups of identical polymers (*e.g.* alginate: ---M-M-G-G-G-) are called block-copolymers. According to their chemical composition, these monomer blocks may be hydrophilic or hydrophobic. The capacity of block-co-polymers to self-assemble into in micellar aggregates is influenced by numerous factors, among which: polymer architecture (linear, grafted, branched, dendritic), size-distribution of the blocks, hydrophobicity/hydrophilicity ratio, stereochemical structure (configuration and conformation), polymer-solvent interactions (bad solvent, good solvent, teta solvent), pH and ionic strength of solvent (Fig. **4d**).

The most common example of amphiphilic polymer is proteins. Out of the 20 amino acids participating in protein construction, some are hydrophobic (alanine, valine, leucine, isoleucine, phenylalanine, tyrosine and tryptophan), and others are hydrophilic, either with a neutral polar structure (asparagine, cysteine, glutamine, methionine, serine and threonine), or with ionisable groups (aspartic acid, glutamic acid, arginine, lysine, histidine).

The hydrophobic or hydrophilic character of amino acids affects the secondary, tertiary, and even quaternary structure of proteins. Thus, certain proteins (whey proteins) adopt a globular structure in order to minimise the exposure of hydrophobic segments to water molecules. Other proteins, like gelatine and casein, containing proline, have an open tertiary structure, which is why, in order to avoid the contact between the hydrophobic areas and water molecules, their polymeric chains self-assemble and form micellar aggregates with sizes in the range 50-500 nm [89].

Like in the case of low-molecular-weight surfactant aqueous micelles, the hydrophobic segments of polymeric amphiphiles form a hydrophobic microdomain inside the block-co-polymer micelles where the lipophilic compound molecules may be entrapped [90].

As compared to low-molecular-weight surfactant aqueous micelles, the use of block-co-polymer micelles as delivery systems for lipophilic bioactive compounds has the following advantages:

- Block-copolymers micelles are stable, as the micellisation velocity is higher than the transfer velocity of the polymeric amphiphiles from the micelle into the solution;
- The critical micellisation concentration (CMC) of block-copolymers is much

lower than CMC of low-molecular-weight surfactants;

- Block-copolymers micelles provide a better protection of the encapsulated lipophilic bioactive compounds due to the more compact structure of the external layer, made up of polymeric chains with hydrophilic segments;
- Block-copolymers micelles have higher loading capacity and encapsulation efficiency due to the large volume of the micelle's hydrophobic core on the one hand, and the biopolymers' capacity to link the molecules of some bioactive components, on the other hand.

Specialised literature includes many papers on the applications of the polymeric amphiphiles to the encapsulation of bioactive components [90 - 96]. Some researchers used milk and whey proteins to protect the bioactive compounds against oxidation, such as β-carotene [71], vitamin D [97], folic acid [98, 99], phytosterols [99], and others used plant proteins, such as zein, which was used to encapsulate curcumin and quercetin [67, 93, 101 - 103].

Liposomes

Liposomes are colloidal systems made up by the self-assembly of the amphiphilic lipids into closed two-layer structures, similar to biological membranes (Fig. **4e**). According to the number of spherical double layers successively distributed (one on top of the other), liposomes are: unilamellar liposomes, and multilamellar liposomes, while according to size, liposomes are called small unilamellar vesicles-SUVs, with sizes ranging from 20 to 50 nm, and large unilamellar vesicles-LUVs, with sizes ranging from 50 to 100 nm. An important class of liposomes is multilamellar vesicles (MLVs), whose size varies very widely, from 0.05 to 10 μm. Due to the concentrical double layers, MLVs are stable than unilamellar liposomes and may entrap a larger amount of hydrophobic compounds than hydrophilic compounds. Liposome stability is influenced by several factors, such as: the structure and concentration of phospholipids, the charge of phospholipids, the aqueous phase's pH and ionic strength, temperature, the presence of other ingredients. Liposomes are delivery systems both for lipophilic components and hydrophilic components (Fig. **4e**)). Thus, lipophilic components are entrapped in the portion formed of the hydrocarbon lipid chains, and hydrophilic components are entrapped in the hydrophilic core of the liposome) or the hydrophilic spaces between the double layers. There is a considerable number of preparation techniques for liposomes with improved characteristics, described in numerous articles [104 - 106, 112, 113, 114]. The most common liposome preparation techniques are thin layer hydration (TLH), reverse phase evaporation (REV) and dehydration-rehydration vesicles (DRV), each having its advantages and disadvantages [102]. There are also newer methods, like microfluids, supercritical fluids and electroformation methods [101]. Like microemulsions, liposomes are less used in food industry because of the presence of ingredients with a toxicity potential (surfactants, organic solvents), and the difficulties in preparing them on an industrial scale. Nevertheless, there are many papers researching the bioactive compounds' encapsulation in liposomes [107 - 112].

Cyclodextrins

A special type of oligomer amphiphiles are cyclodextrins.

Cyclodextrins (CDs) are cyclic oligosaccharides obtained by the enzymatic degradation of starch. The most common cyclodextrins contain 6, 7 or 8 glucose units linked by α (1-4) glucoside bonds and are named α-, β-, and γ-CD respectively. Cyclodextrins are considered natural compounds, with a high level of biocompatibility, approved by the FDA [115, 118]. The conformation of glucose residues makes cyclodextrins adopt a tridimensional truncated cone shape with an inner hydrophobic cavity and an external hydrophilic surface [116]. Due to this structure, CDs behave like a "host", receiving a "guest" with special attributes (hydrophobic molecules of various sizes according to the cavity volume) (Fig. **4f**). The inclusion complex formed by the" host-guest" mechanism increases the water solubility of lipophilic molecules, protecting them against external factors (temperature, light, oxygen, pH). In order to improve the physico-chemical properties and the inclusion capacity, native CDs turned into CD-derivatives. Thus, by the reaction of the reactive hydroxyl groups with different chemical agents, native β-CD turned into hydrophilic β-CD derivatives, such as 2-hydroxypropyl-β-CD or hydrophobic β-CD derivatives, such as 2,6-diethyl-β-CD [116].

In the past few years, more and more poorly water-soluble bioactive compounds are encapsulated in CDs and CD-derivatives: curcumin [119], essential oils [120 - 122], vitamins [115], resveratrol [121, 122], antioxidants [116, 117, 123 - 125].

Biopolymer-Based Nanoparticles and Microparticles

Food-grade biopolymer-based nanoparticles and microparticles are prepared by means of the main classes of natural biopolymers, generally recognised as safe (GRAS), such as proteins and polysaccharides. The choice of biopolymers depends on several factors, like: particle characteristics (size, charge, loading capacity, digestibility, bioavailability, release mechanism), biopolymer properties (linear or branched structure, polar or ionic groups, digestibles or nondigestibles), and the properties of bioactive compounds (polarity, stability, liquid or solid crystalline).

Both polysaccharides and proteins provide a good protection for the bioactive compounds sensitive to physico-chemical factors and allow their controlled release into the various GIT areas. For example, pectin may be used to manufacture polymer particles to release bioactive compounds in the colon, as it is not digested by the enzymes in the stomach, nor those in the small intestine, but it is easily degraded by the microflora in the colon [126].

Individual Polymers

Biopolymer nanoparticles or microparticles may be formed of individual polymers, or mixtures of biopolymers. The formation of polymer particles out of a single polymer is based on the assembly of polymer chains into a tridimensional network (gel) in whose pores various colloidal systems (conventional emulsions, nanoemulsions, microemulsions, solid lipid particles, liposomes) containing bioactive compounds may be entrapped. Both proteins and polysaccharides are capable of gelation by specific methods. The theoretical bases of the behaviour of the proteins and polysaccharides used to encapsulate bioactive compounds are developed in excellent articles published by certain researchers [88, 89, 95, 127 - 129].

Globular proteins are jellified by heating at a temperature above the thermal denaturation temperature, when the self-assembly of polymeric chains occurs by hydrophobic interactions or disulphide bonds. Other proteins jollify by the variation of pH, ionic strength, or in the presence of chemical compounds (multivalent ions, glutaric aldehyde, genipin) or enzymes, such as transglutaminase. Casein jellifies by adjusting the pH in the proximity of the isoelectric point (pH 4.6), by adding Ca^{2+} ions or by adding rennet. The most commonly used proteins in manufacturing polymer particles are: gelatine, whey protein (β-lactoglobulin, α-lactalbumin), bovine serum albumin, casein, soy proteins, zein, gliadin [88 - 90].

Polysaccharides jellify by three main mechanisms: ionotropic, cold-set and heat-set gelation [92, 130, 131].

Ionotropic gelation is specific to polysaccharides whose structure consists of repeating different types of sugar units on which ionic groups ($-COO^-$, $-OSO_3^-$; $-NH_3^+$) are grafted.

Thus, both pectin, which is a linear polysaccharide mainly formed of α (1-4)-D-galacturonic acid residues, and alginate, which is a block-copolymer formed of β-D-mannuronic acid and α-L-guluronic acid residues form gels in the presence of multivalent cations (Ca^{2+}, Mg^{2+}). Alginate hydrogel particles have a pH dependent behaviour, allowing the release of bioactive compounds in certain GIT sectors. So, at a low pH (*e.g.* stomach) alginate beads shrink because of the protonation of the carboxyl groups and do not allow the release of biocomponents, while at a high pH (*e.g.* small intestine), the polyanionic chains electrostatically reject each other because of the large number of ionised carboxyl groups, while the alginate beads swell and favour the release of biocomponents [126].

Carrageenan is another type of linear anionic polysaccharide in whose structure there are α-1,4 and β-1,3 linked anhydrogalactose residues with a different content of sulphate groups. According to the number of sulphate groups, there are many types of carrageenan, among which the most important are kappa (κ-), iota (ι-) and lambda (λ-) carrageenan. Due to the different number of negative charges, carrageenans have different structures, and different gelation mechanisms respectively. Thus, for κ-carrageenan gelation occurs in the presence of potassium ions, while for γ-carrageenan gelation occurs in the presence of the calcium ions.

By heating, κ-carrageenan forms strong and rigid gels, and γ-carrageenan forms soft and elastic gels. λ-carrageenan does not form gels [130, 132].

Chitosan is a linear polysaccharide made up of units of glucosamine and N-acetylated glucosamine. It is a non-toxic, biocompatible and biodegradable polymer, with bioadhesive and antibacterial properties, obtained by the alkaline deacetylation of chitin. Its properties depend on the deacetylation degree. It is soluble at low pH, when the amino groups are protonated and may react with polyanions (alginate, carrageenan, tripolyphosphate) forming pH-sensitive hydrogels. The chitosan hydrogel swells at low pH due to the electrostatic repulsion among the protonated amino groups, and at high pH chitosan precipitates. This behaviour allows the use of chitosan in the manufacture of gastric delivery systems [65].

In addition to the polysaccharides described above, the manufacture of polymer particles may also use other polysaccharides, such as: starch and derivatives, cellulose and derivatives, arabic gum, locust gum, dextran, inulin, agar *etc.*

Mixture Polymers

When manufacturing food-grade polymer nano and microparticles, one may use mixtures of polyelectrolites with an opposite electrical charge, such as proteins and polysaccharides.

Due to their zwitterionic structure, proteins ale polyelectrolites whose electrical charge is pH dependent. The pH value where the proteins' net electrical charge is zero is called isolelectric point (pI). At pH values which are much higher than the isoelectric point, proteins possess a high net negative charge. Under these circumstances, electrostatic repulsions occur between the protein chains and the anionic polysaccharide chains, resulting in a diluted solution of the two biopolymers. At high concentrations two phases rich in one of the two polymers become separated. At pH values close to the isoelectric point, (pH~pI), proteins have no net electrical charge, but on the polymeric chain there are both positive and negative ionic groups. Under these circumstances, weak attractive interactions occur between the positive and negative ionic groups of the polysaccharide, resulting in the formation of soluble complex products. With the pH decrease, the number of positive groups on the protein chain decreases, which leads to the increase of the attractive interactions with anionic polysaccharides, forming proteins-polysaccharides complexes. This process occurs with phase separation and is termed coacervation. At pH $<<$ pI, proteins have positive net charge and form precipitates with anionic polysaccharides. Such phenomena also occur when anionic polysaccharides are mixed with cationic polysaccharides, such as: alginate-chitosan; carrageenan-chitosan *etc.*

The formation of non-covalent protein-polysaccharides complexes is a method to diversify the types of particles applied to functional food production [13, 26, 95, 134].

Biopolymer-Based Particles Fabrication Methods

The manufacture of polymeric particles employs a large variety of techniques, minutely described in many studies. The following will briefly describe the main methods of entrapping bioactive compounds into polymeric particles, applied to functional food production.

Complex Coacervation consists in entrapping bioactive compounds into a protein-polysaccharide complex or anionic – cationic polysaccharides complex. Biopolymer complex coacervates are formed by the electrostatic attractions between the electrical charges of the two biopolymers in specific pH and ionic strength circumstances. The encapsulation of lipophilic bioactive compounds by complex coacervation takes place in 4 stages. In short, first the lipophilic component is emulsified in a protein solution (*e.g.*. gelatine) in the presence of a hydrophilic emulsifier. In the O/W emulsion, where the oil droplets are covered in a protein layer, a solution of anionic polysaccharide (*e.g.* arabic gum) is added. When the pH is adjusted below the pI, electrostatic attractions occur between the protein chains with a positive net electrical charge and the anionic polysaccharide chains, with formation of coacervates within which the oil droplets are entrapped. Gelation of the protein-polysaccharide complex occurs either by cooling or by the thermal denaturation of protein. To solidify the particles one may use enzymatic reticulation agents (transglutaminase) or chemical reticulation agents (glutaraldehyde, adipic acid derivatives, malonic acid derivatives, epichlorohydrin).

Because of religious or toxicity restrictions, numerous studies published in the past few years replace mammalian gelatines with fish gelatine or proteins from milk or plants (whey protein isolate, β-lactoglobulin, soybean protein isolate, α-gliadin, zein) which form coacervates with a wide range of anionic polysaccharides (arabic gum, pectin, carrageenan, xanthan, alginate).

Our laboratory has recently encapsulated pimenta dioica essential oil in a chitosan-κ carrageenan complex, studying the influence of pH and temperature on the swelling and release processes [8].

The Extrusion Method is a convenient method of manufacturing cell-loaded polymer microparticles, hydrophilic or lipophilic compounds. The encapsulation of bioactive components by the extrusion method takes place by covering the following stages: preparing biopolymer solutions (alginate, pectin, κ-carrageenan, chitosan) in which bioactive products (cells, polyphenols, essential oils, ω-3 fatty acids, carotenoids, *etc.*) are included; dropping droplets of biopolymer solution and bioactive into a gelation bath, by means of various devices: pipette, syringe, vibrating nozzle, spraying nozzle, jet cutter, atomizing disc, electric field, coaxial air flow and concentric nozzle; gelation of polymer droplets which may occur in the presence of multivalent ions (Ca^{2+}, Mg^{2+}, K^{+}, sodium tripolyphosphate); separation, drying and storage of polymer microspheres. This method yields microspheres with sizes ranging within 10-800μm. The characteristics of the microspheres obtained by extrusion are influenced by the viscosity of the biopolymer solution, the concentration of the gelation agents, the dropping method. To the purpose of improving the characteristics of the pimenta dioica essential oil microspheres, we first prepared chitosan microspheres, then they were immersed into a κ-carrageenan solution, resulting in chitosan microspheres covered in a k-carrageenan layer. The pimenta dioica essential oil chitosan/κ-carrageenan microspheres manifested better stability and release profile, different from the one shown by the pimenta dioica essential oil chitosan microspheres [8].

Emulsion-internal Gelation Method, is applied to the encapsulation of cells and hydrophilic biocomponents. The method is based on preparing a W/O emulsion, whose discontinuous phase is made up of droplets containing the biopolymer aqueous solution, the bioactive component and a source of multivalent ions producing the gelation of biopolymers (*e.g.* $CaCO_3$). They are released from their salts by adding an acid (acetic acid) in the W/O emulsion. The external phase of the W/O emulsion may be a vegetable oil where a lipophilic surfactant (Span 80) was solubilised. The most difficult stage of this method is the separation of the microspheres in the system. Some authors produce the breaking of emulsion by adding a $CaCl_2$ solution and the separation of the two phases. The microspheres are separated in the aqueous phase by filtration. The microspheres obtained have sizes between 0.5 and 100μm [136].

Spray Drying is one of the earliest and most common methods to encapsulate bioactive compounds. A suspension or solution of biopolymers in water containing bioactive component is atomized and sprayed in a drying chamber, where the water in the fine droplets is suddenly vaporized, and thus produces microparticles whose size is in the range 10-100μm. The characteristics of microparticles are influenced by the physico-chemical properties of the materials (biopolymer concentration, the water content, the wall material-core material ratio, viscosity, *etc.*), and the technological parameters (atomizing gas flow rate, drying gas flow rate, inlet temperature, outlet temperature, *etc.*). The published articles mention a wide variety of wall materials (whey proteins, plant proteins, animal proteins and polysaccharides) used to encapsulate various classes of bioactive compounds (polyphenols, essential oils, polyunsaturated fatty acids, vitamins, probiotic bacteria, *etc.*)

The disadvantages of this method consist in the restrictive use of only the hydrophilic wall material and the application of high temperatures which may degrade thermosensitive components.

FUTURE TRENDS

Today's consumers also consider food as a source of health. They require that food bring an important contribution to preventing disease and improving health. That is why food manufacturers are adopting new strategies regarding the use of traditional technologies on the one hand, and the use of natural components in the special delivery systems, on the other hand, able to provide protection against environmental factors. The encapsulation of bioactive components has led to the diversification of functional food and the improvement of their quality. The interest in developing new encapsulation technologies and new and functional GRAS (Generally Regarded as Safe) encapsulant materials ensuring the ever need for food-grade micro and nanoparticles has been on the rise lately. So, according to "Food Encapsulation: A Global strategic Business Report" by Global Industry Analysts, Inc. (GIA), showing the analysis of the market trends in encapsulated food systems in the US, Canada, Japan, Europe, Asia-Pacific, Latin America, the global market of encapsulated food products will increase to $39 billion in 2015.

The research in the field of encapsulation of bioactive compounds used in food manufacture focuses on improving the characteristics of micro and nanoparticles, so that they may withstand the processing operations, and the food texture, colour, appearance, taste and odour should remain unaffected. Also, more attention is granted to the manufacture of new delivery systems for the bioactive compounds able to increase bioactivity by the controlled release in certain TIG areas. In this respect, the researchers showed greater interest in obtaining wall materials with low allergenic risk, and in materials that may be used in preparing personalised foods (nutragenomics) or foods aimed at special consumer groups (children, seniors, athletes, consumers with religious constraints).

It is obvious that food nanotechnologies have lately known an unprecedented development as a result of the cooperation between the researchers in academic fields (chemistry, biochemistry, medicine, pharmacy) and the ones in the technological, industrial and marketing-related fields.

CONFLICT OF INTEREST

The author confirms that author has no conflict of interest to declare for this publication.

ACKNOWLEDGEMENTS

Declared none.

REFERENCES

[1] Poncelet D, Dreffier C, Subra-Paternault P, Vandamme TF. Mecroencapsulation des sciences aux technologies. Paris: Lavoisier 2007; pp. 3-8.

[2] Sobel R, Versic R, Gaonkar AG. Microencapsulation in the food industry a practical implementation guide. London: Elsevier Inc 2014; pp. 3-12.

[3] Wilson CG. Fundamentals and applications of controlled release drug delivery. New York: Controlled Release Society, Springer 2012; pp. 3-19.
[http://dx.doi.org/10.1007/978-1-4614-0881-9_1]

[4] McClements DJ. Encapsulation technologies and delivery systems for food ingredients and nutraceuticals. Cambridge: Woodhead Publishing Limited 2012; pp. 3-18.
[http://dx.doi.org/10.1533/9780857095909.1.3]

[5] Joye IJ, Davidov-Pardo G, McClements DJ. Nanotechnology for increased micronutrient bioavailability. Trends Food Sci Technol 2014; 40: 168-82.
[http://dx.doi.org/10.1016/j.tifs.2014.08.006]

[6] Cushen M, Kerry MJ, Morris M, Cruz-Romero M, Cummins E. Nanotechnologies in the food industry - Recent developments, risks and regulation. Trends Food Sci Technol 2012; 24: 30-46.
[http://dx.doi.org/10.1016/j.tifs.2011.10.006]

[7] Fathi M, Mozafari MR, Mohebbi M. Nanoencapsulation of food ingredients using lipid based delivery systems. Trends Food Sci Technol 2012; 23: 13-27.
[http://dx.doi.org/10.1016/j.tifs.2011.08.003]

[8] Dima C, Cotarlet M, Alexe P, Dima S. Microencapsulation of essential oil of pimento *Pimenta dioica*(L) Merr. by chitosan/k-carrageenan complex coacervation method. Innov Food Sci Emerg Technol 2014; 20: 203-11.

[9] Joye IJ, McClements DJ. Production of nanoparticles by antisolvent precipitation for use in food systems. Trends Food Sci Technol 2013; 34: 109-23.
[http://dx.doi.org/10.1016/j.tifs.2013.10.002]

[10] Joye IJ, McClements DJ. Biopolymer-based nanoparticles and microparticles: Fabrication, characterization, and application. Curr Opin Colloid Interface Sci 2014; 19: 417-27.
[http://dx.doi.org/10.1016/j.cocis.2014.07.002]

[11] Li Z, Jiang H, Xu C, Gu L. A review: Using nanoparticles to enhance absorption and bioavailability of phenolic phytochemicals. Food Hydrocoll 2015; 43: 153-64.
[http://dx.doi.org/10.1016/j.foodhyd.2014.05.010]

[12] McClements DJ. Enhancing nutraceutical bioavailability through food matrix design. Curr Opin Food Sci 2015; 4: 1-6.
[http://dx.doi.org/10.1016/j.cofs.2014.12.008]

[13] Champagne CP, Fustier P. Microencapsulation for the improved delivery of bioactive compounds into foods. Curr Opin Biotechnol 2007; 18(2): 184-90.
[http://dx.doi.org/10.1016/j.copbio.2007.03.001] [PMID: 17368017]

[14] De Voss P, Faas MM, Spasojevic M, Sikkema J. Encapsulation for preservation of functionality and targeted delivery of bioactive food components. Int Dairy J 2010; 20: 292-302. [http://dx.doi.org/10.1016/j.idairyj.2009.11.008]

[15] Gadeyne F, Van Ranst G, Vlaeminck B, Vossen E, Van der Meeren P, Fievez V. Protection of polyunsaturated oils against ruminal biohydrogenation and oxidation during storage using a polyphenol oxidase containing extract from red clover. Food Chem 2015; 171(171): 241-50. [http://dx.doi.org/10.1016/j.foodchem.2014.08.109] [PMID: 25308665]

[16] Perez AA, Andermatten RB, Rubiolo AC, Santiago LG, Rubiolo AC, Santiago LG. β-Lactoglobulin heat-induced aggregates as carriers of polyunsaturated fatty acids. Food Chem 2014; 158: 66-72. [http://dx.doi.org/10.1016/j.foodchem.2014.02.073] [PMID: 24731315]

[17] Sagalowicz L, Leser ME. Delivery systems for liquid food products. Curr Opin Colloid Interface Sci 2010; 15: 61-72. [http://dx.doi.org/10.1016/j.cocis.2009.12.003]

[18] Gadkari VP, Balaraman M. Catechins: Sources, extraction and encapsulation: A review. Food Bioprod Process 2015; 93: 122-38. [http://dx.doi.org/10.1016/j.fbp.2013.12.004]

[19] Mackie A. Encapsulation technologies and delivery systems for food ingredients and nutraceuticals. Cambridge: Woodhead Publishing Limited 2012; pp. 49-70. [http://dx.doi.org/10.1533/9780857095909.1.49]

[20] McClements DJ. Edible lipid nanoparticles: digestion, absorption, and potential toxicity. Prog Lipid Res 2013; 52(4): 409-23. [http://dx.doi.org/10.1016/j.plipres.2013.04.008] [PMID: 23664907]

[21] McClements DJ, Li Y. Structured emulsion-based delivery systems: controlling the digestion and release of lipophilic food components. Adv Colloid Interface Sci 2010; 159(2): 213-28. [http://dx.doi.org/10.1016/j.cis.2010.06.010] [PMID: 20638649]

[22] Dima S, Dima C, Iordachescu G. Encapsulation of functional lipophilic food and drug biocomponents. Food Eng Rev 2015; 7(4): 417-38. [http://dx.doi.org/10.1007/s12393-015-9115-1]

[23] Bigliardi B, Galati F. Innovation trends in the food industry: the case of functional foods. Trends Food Sci Technol 2013; 31: 118-29. [http://dx.doi.org/10.1016/j.tifs.2013.03.006]

[24] Augustin MA, Sanguansri L. Encapsulation technologies and delivery systems for food ingredients and nutraceuticals. Cambridge: Woodhead Publishing Limited 2012; pp. 19-48. [http://dx.doi.org/10.1533/9780857095909.1.19]

[25] Davidov-Pardo G, McClements DJ. Resveratrol encapsulation: designing delivery systems to overcome solubility, stability and bioavailability issues. Trends Food Sci Technol 2014; 38(2): 88-103. [http://dx.doi.org/10.1016/j.tifs.2014.05.003]

[26] Cerqueira MA, Pinheiro AC, Silva HD, *et al.* Design of bio-nanosystems for oral delivery of functional compounds. Food Eng Rev 2014; 6: 1-19. [http://dx.doi.org/10.1007/s12393-013-9074-3]

[27] Van Boekel MA. Innovation in agri-food systems: product quality and consumer acceptance. Wageningen: Wageningen Academic Publishers 2005; pp. 147-72.

[28] Rothwell JA, Day AJ, Morgan MR. Experimental determination of octanol-water partition coefficients of quercetin and related flavonoids. J Agric Food Chem 2005; 53(11): 4355-60. [http://dx.doi.org/10.1021/jf0483669] [PMID: 15913295]

[29] O'Driscoll CM, Griffin BT. Biopharmaceutical challenges associated with drugs with low aqueous solubility--the potential impact of lipid-based formulations. Adv Drug Deliv Rev 2008; 60(6): 617-24. [http://dx.doi.org/10.1016/j.addr.2007.10.012] [PMID: 18155800]

[30] McClements DJ. Crystals and crystallization in oil-in-water emulsions: implications for emulsion-based delivery systems. Adv Colloid Interface Sci 2012; 174: 1-30. [http://dx.doi.org/10.1016/j.cis.2012.03.002] [PMID: 22475330]

[31] McClements DJ. Nanoemulsions *versus* microemulsions: terminology, differences, and similarities. Soft Matter 2012; 8: 1719-29. [http://dx.doi.org/10.1039/C2SM06903B]

[32] McClements DJ. Edible nanoemulsions: fabrication, properties, and functional performance. Soft Matter 2011; 7: 2297-316. [http://dx.doi.org/10.1039/C0SM00549E]

[33] McClements DJ, Rao J. Food-grade nanoemulsions: formulation, fabrication, properties, performance, biological fate, and potential toxicity. Crit Rev Food Sci Nutr 2011; 51(4): 285-330. [http://dx.doi.org/10.1080/10408398.2011.559558] [PMID: 21432697]

[34] Silva HD, Cerqueira MA, Vicente AA. Nanoemulsions for food applications: development and characterization. Food Bioprocess Technol 2012; 5: 854-67. [http://dx.doi.org/10.1007/s11947-011-0683-7]

[35] Wandrey C, Bartkowiak A, Harding SE. In: Zuidan NJ, Nedouic VA, Eds. Encapsulation Technologies for Active Food Ingredients and Food Processing. Zuidam, N.J.; Nedović, V.A., Eds., Springer Science+Business Media, LLC, New York, 2010; pp. 31-100.

[36] Dima S. Microîncapsularea în Sisteme alimentare. Galati: Academica 2009; pp. 5-105.

[37] Perez R, Gaonkar AG. Microencapsulation in the Food Industry A Practical Implementation Guide. London: Elsevier Inc 2014; pp. 543-9.

[38] Sanguansri P, Augustin MA. Nanoscale materials development - a food industry perspective. Trends Food Sci Technol 2006; 17: 547-56. [http://dx.doi.org/10.1016/j.tifs.2006.04.010]

[39] Coles D, Frewer LJ. Nanotechnology applied to European food production – A review of ethical and regulatory issues. Trends Food Sci Technol 2013; 34: 32-43. [http://dx.doi.org/10.1016/j.tifs.2013.08.006]

[40] Mastromatteo M, Mastromatteo M, Conte A, Del Nobile MA. Advances in controlled release devices for food packaging applications. Trends Food Sci Technol 2010; 21: 591-8. [http://dx.doi.org/10.1016/j.tifs.2010.07.010]

[41] Piorkowski DT, McClements DJ. Beverage emulsions: Recent developments in formulation,

production, and applications. Food Hydrocoll 2014; 42: 5-41.
[http://dx.doi.org/10.1016/j.foodhyd.2013.07.009]

[42] Shchukina EM, Shchukin DG. LbL coated microcapsules for delivering lipid-based drugs. Adv Drug Deliv Rev 2011; 63(9): 837-46.
[http://dx.doi.org/10.1016/j.addr.2011.03.009] [PMID: 21447362]

[43] Hombach J, Bernkop-Schnürch A. In: Schäfer-Korting M, Ed. Drug Delivery; Springer- Verlag Berlin, Heidelberg; 2010; pp. 251-266.

[44] Anal AK, Stevens WF. Chitosan-alginate multilayer beads for controlled release of ampicillin. Int J Pharm 2005; 290(1-2): 45-54.
[http://dx.doi.org/10.1016/j.ijpharm.2004.11.015] [PMID: 15664129]

[45] Oztop MH, McCarthy KL. Mathematical modeling of swelling in high moisture whey protein gels. J Food Eng 2011; 106: 53-9.
[http://dx.doi.org/10.1016/j.jfoodeng.2011.04.007]

[46] Sun W, Zhang N, Li X. Release mechanism studies on TFu nanoparticles-in-microparticles system. Colloids Surf B Biointerfaces 2012; 95: 115-20.
[http://dx.doi.org/10.1016/j.colsurfb.2012.02.019] [PMID: 22424827]

[47] Čalija B, Cekić N, Savić S, Rolf Daniels R, Marković B, Milić J. pH-sensitive microparticles for oral drug delivery based onalginate/oligochitosan/Eudragit®L100-55 "sandwich" polyelectrolyte complex. Colloid Surface B Biointerfaces 2013; 110: 395-402.
[http://dx.doi.org/10.1016/j.colsurfb.2013.05.016] [PMID: 23751419]

[48] Zou X, Zhao X, Ye L, Wang Q, Li H. Preparation and drug release behavior of pH- responsive bovine serum albumin-loaded chitosan microspheres. J Ind Eng Chem 2015; 21: 1389-97.
[http://dx.doi.org/10.1016/j.jiec.2014.06.012]

[49] Hadjitheodorou A, Kalosakas G. Analytical and numerical study of diffusion-controlled drug release from composite spherical matrices. Mate Sci Eng C Meta Biol Appl 2014; 42: 681-90.
[http://dx.doi.org/10.1016/j.msec.2014.06.009]

[50] Li L, Wang L, Shao Y, Ni R, Zhang T, Mao S. Drug release characteristics from chitosan-alginate matrix tablets based on the theory of self-assembled film. Int J Pharm 2013; 450(1-2): 197-207.
[http://dx.doi.org/10.1016/j.ijpharm.2013.04.052] [PMID: 23624081]

[51] Siepmann J, Siepmann F. Mathematical modeling of drug release from lipid dosage forms. Int J Pharm 2011; 418(1): 42-53.
[http://dx.doi.org/10.1016/j.ijpharm.2011.07.015] [PMID: 21802501]

[52] Dima C, Pătraşcu L, Cantaragiu A, Alexe P, Dima Ş. The kinetics of the swelling process and the release mechanisms of *Coriandrum sativum* L. essential oil from chitosan/alginate/inulin microcapsules. Food Chem 2016; 195: 39-48.
[http://dx.doi.org/10.1016/j.foodchem.2015.05.044] [PMID: 26575710]

[53] Chang Y, McLandsborough L, McClements DJ. Fabrication, stability and efficacy of dual-component antimicrobial nanoemulsions: essential oil (thyme oil) and cationic surfactant (lauric arginate). Food Chem 2015; 172: 298-304.
[http://dx.doi.org/10.1016/j.foodchem.2014.09.081] [PMID: 25442557]

[54] Hashtjin AM, Abbasi S. Nano-emulsification of orange peel essential oil using sonication and native gums. Food Hydrocoll 2015; 44: 40-8. [http://dx.doi.org/10.1016/j.foodhyd.2014.08.017]

[55] Rao J, McClements DJ. Impact of lemon oil composition on formation and stability of model food and beverage emulsions. Food Chem 2012; 134(2): 749-57. [http://dx.doi.org/10.1016/j.foodchem.2012.02.174] [PMID: 23107687]

[56] Rao J, McClements DJ. Food-grade microemulsions and nanoemulsions: Role of oil phase composition on formation and stability. Food Hydrocoll 2012; 29: 326-34. [http://dx.doi.org/10.1016/j.foodhyd.2012.04.008]

[57] Lawrence MJ, Rees GD. Microemulsion-based media as novel drug delivery systems. Adv Drug Deliv Rev 2012; 64: 175-93. [http://dx.doi.org/10.1016/j.addr.2012.09.018] [PMID: 11104900]

[58] Muschiolik G. Multiple emulsions for food use. Curr Opin Colloid Interface Sci 2007; 12: 213-20. [http://dx.doi.org/10.1016/j.cocis.2007.07.006]

[59] Jiménez-Colmenero F. Potential applications of multiple emulsions in the development of healthy and functional foods. Food Res Int 2013; 52: 64-74. [http://dx.doi.org/10.1016/j.foodres.2013.02.040]

[60] Su J, Flanagan J, Singh H. Improving encapsulation efficiency and stability of water-in-oil-in- water emulsions using a modified gum arabic (*Acacia (sen)* SUPER GUMTM). Food Hydrocoll 2008; 22: 112-20. [http://dx.doi.org/10.1016/j.foodhyd.2007.03.005]

[61] Salager JL. Pharmaceutical emulsions and suspensions. ; New York: CRC Press 2000.

[62] Salminen H, Weiss J. Electrostatic adsorption and stability of whey protein–pectin complexes on emulsion interfaces. Food Hydrocoll 2014; 35: 410-9. [http://dx.doi.org/10.1016/j.foodhyd.2013.06.020]

[63] Komaiko J, McClements DJ. Food-grade nanoemulsion filled hydrogels formed by spontaneous emulsification and gelation: Optical properties, rheology, and stability. Food Hydrocoll 2015; 46: 67-75. [http://dx.doi.org/10.1016/j.foodhyd.2014.12.031]

[64] Komaiko J, McClements DJ. Low-energy formation of edible nanoemulsions by spontaneous emulsification: Factors influencing particle size. J Food Eng 2015; 146: 122-8. [http://dx.doi.org/10.1016/j.jfoodeng.2014.09.003]

[65] Scholten E, Moschakis T, Biliaderis CG. Review- Biopolymer composites for engineering food structures to control product functionality. Food Struct 2014; I: 39-54. [http://dx.doi.org/10.1016/j.foostr.2013.11.001]

[66] Tokle T, Lesmes U, Decker EA, McClements DJ. Impact of dietary fiber coatings on behavior of protein-stabilized lipid droplets under simulated gastrointestinal conditions. Food Funct 2012; 3(1): 58-66. [http://dx.doi.org/10.1039/C1FO10129C] [PMID: 22086435]

[67] Pan Y, Tikekar RV, Wang MS, Roberto J, Avena-Bustillos RJ, Nitin N. Effect of barrier properties of zein colloidal particles and oil-in-water emulsions on oxidative stability of encapsulated bioactive compounds. Food Hydrocoll 2015; 43: 82-90.
[http://dx.doi.org/10.1016/j.foodhyd.2014.05.002]

[68] Berendsen R, Güell C, Ferrando M. A procyanidin-rich extract encapsulated in water-in-oil-in- water emulsions produced by premix membrane emulsification. Food Hydrocoll 2015; 43: 636-48.
[http://dx.doi.org/10.1016/j.foodhyd.2014.07.023]

[69] Schmidt US, Bernewitz R, Guthausen G, Schuchmann HP. Investigation and application of measurement techniques for the determination of the encapsulation efficiency of O/W/O multiple emulsions stabilized by hydrocolloid gelation. Colloid Surface A. 2015. (Abstract)

[70] Aaron SL, Griffin C, Roos YH. Stability and loss kinetics of lutein and β-carotene encapsulated in freeze-dried emulsions with layered interface and trehalose as glass former. Food Res Int 2014; 62: 403-9.
[http://dx.doi.org/10.1016/j.foodres.2014.03.059]

[71] Yi J, Lam TI, Yokoyama W, Luisa W, Cheng LW, Zhong F. Beta-carotene encapsulated in food protein nanoparticles reduces peroxyl radical oxidation in Caco-2 cells. Food Hydrocoll 2015; 43: 31-40.
[http://dx.doi.org/10.1016/j.foodhyd.2014.04.028]

[72] McClements DJ, Li Y. Structured emulsion-based delivery systems: controlling the digestion and release of lipophilic food components. Adv Colloid Interface Sci 2010; 159(2): 213-28.
[http://dx.doi.org/10.1016/j.cis.2010.06.010] [PMID: 20638649]

[73] Carrillo-Navas H, Cruz-Olivares J, Varela-Guerrero V, Alamilla-Beltrán L, Vernon-Carte EJ, Pérez-Alonso C. Rheological properties of a double emulsion nutraceutical system incorporating chia essential oil and ascorbic acid stabilized by carbohydrate polymer-protein blends. Carbohydr Polym 2012; 87: 1231-5.
[http://dx.doi.org/10.1016/j.carbpol.2011.09.005]

[74] Khalid N, Kobayashi I, Marcos A, Neves MA, Uemura K, Nakajima M. Preparation and characterization of water-in-oil emulsions loaded with high concentration of L – ascorbic acid. LWT -. Food Sci Technol (Campinas) 2013; 51: 448-54.

[75] Li B, Jiang Y, Liu F, Chai Z, Li Y, Li Y. Synergistic effects of whey protein– polysaccharide complexes on the controlled release of lipid-soluble and water-soluble vitamins inW1/O/W2 double emulsion systems. Int J Food Sci Technol 2012; 47: 248-54.
[http://dx.doi.org/10.1111/j.1365-2621.2011.02832.x]

[76] Hemar Y, Cheng LJ, Oliver CM, Sanguansri L, Agustin M. Encapsulation of resveratrol using water-in-oil-in-water double emulsions. Food Biophys 2010; 5: 120-7.
[http://dx.doi.org/10.1007/s11483-010-9152-5]

[77] Sauvant P, Cansell M, Sassi AH, Atgié C. Vitamin A enrichment: Caution with encapsulation strategies used for food applications. Food Res Int 2012; 46: 469-79.
[http://dx.doi.org/10.1016/j.foodres.2011.09.025]

[78] Frank K, Walz E, Gräf V, Greiner R, Köhler K, Schuchmann HP. Stability of anthocyanin-rich w/o/w-

emulsions designed for intestinal release in gastrointestinal environment. J Food Sci 2012; 77(12): N50-7.
[http://dx.doi.org/10.1111/j.1750-3841.2012.02982.x] [PMID: 23240975]

[79] Mehnert W, Mäder K. Solid lipid nanoparticles.Production, characterization and applications. Adv Drug Deliv Rev 2012; 64: 83-101.
[http://dx.doi.org/10.1016/j.addr.2012.09.021] [PMID: 11311991]

[80] Matos-Jr FE, Di Sabatino M, Passerini N, Favaro-Trindade CS, Albertini B. Development and characterization of solid lipid microparticles loaded with ascorbic acid and produced by spray congealing. Food Res Int 2015; 67: 52-9.
[http://dx.doi.org/10.1016/j.foodres.2014.11.002]

[81] Pandita D, Kumar S, Poonia N, Lather V. Solid lipid nanoparticles enhance oral bioavailability of resveratrol, a natural polyphenol. Food Res Int 2014; 62: 1165-74.
[http://dx.doi.org/10.1016/j.foodres.2014.05.059]

[82] Tamjidi F, Shahedi M, Varshosaz J, Nasirpour A. Nanostructured lipid carriers (NLC): A potential delivery system for bioactive food molecules. Innov Food Sci Emerg Technol 2013; 19: 29-43.

[83] Müller RH, Petersen RD, Hommoss A, Pardeike J. Nanostructured lipid carriers (NLC) in cosmetic dermal products. Adv Drug Deliv Rev 2007; 59(6): 522-30.
[http://dx.doi.org/10.1016/j.addr.2007.04.012] [PMID: 17602783]

[84] Aditya NP, Macedo AS, Doktorovova S, *et al.* Development and evaluation of lipid nanocarriers for quercetin delivery: A comparative study of solid lipid nanoparticles (SLN), nanostructured lipid carriers (NLC), and lipid nanoemulsions (LNE). Food Sci Technol 2014; 59: 115-21.

[85] Hiemenz PC, Rajagopalan R. Principles of Colloid and Surface Chemistry. 3rd ed. New York: Marcel Dekker, Inc. 1997; pp. 367-78.

[86] Lawrence MJ, Rees GD. Microemulsion-based media as novel drug delivery systems. Adv Drug Deliv Rev 2012; 64: 175-93.
[http://dx.doi.org/10.1016/j.addr.2012.09.018] [PMID: 11104900]

[87] Garti NA, Aserin A. Encapsulation technologies and delivery systems for food ingredients and nutraceuticals. Cambridge: Woodhead Publishing Limited 2012; pp. 211-51.
[http://dx.doi.org/10.1533/9780857095909.3.211]

[88] Livney YD. Encapsulation technologies and delivery systems for food ingredients and nutraceuticals. Cambridge: Woodhead Publishing Limited 2012; pp. 251-6.

[89] Livney ZD. Milk proteins as vehicles for bioactives. Curr Opin Colloid Interface Sci 2010; 15: 73-83.
[http://dx.doi.org/10.1016/j.cocis.2009.11.002]

[90] Tavares GM, Croguennec T, Antonio F, Carvalho AF, Bouhallab S. Milk proteins as encapsulation devices and delivery vehicles: Applications and trends. Trends Food Sci Technol 2014; 37: 5-20.
[http://dx.doi.org/10.1016/j.tifs.2014.02.008]

[91] Wichchukit S, Oztop MH, McCarthy MJ, McCarthy KL. Whey protein/alginate beads as carriers of a bioactive component. Food Hydrocoll 2013; 33: 66-73.
[http://dx.doi.org/10.1016/j.foodhyd.2013.02.013]

[92] Fathi M, Martín Á, McClements DJ. Nanoencapsulation of food ingredients using carbohydrate based delivery systems. Trends Food Sci Technol 2014; 39: 18-39. [http://dx.doi.org/10.1016/j.tifs.2014.06.007]

[93] Patel AR, Heussen PC, Hazekamp J, Drost E, Velikov KP. Quercetin loaded biopolymeric colloidal particles prepared by simultaneous precipitation of quercetin with hydrophobic protein in aqueous medium. Food Chem 2012; 133(2): 423-9. [http://dx.doi.org/10.1016/j.foodchem.2012.01.054] [PMID: 25683415]

[94] Patel AR, Velikov KP. Zein as a source of functional colloidal nano- and microstructures. Curr Opin Colloid Interface Sci 2014; 19: 450-8. [http://dx.doi.org/10.1016/j.cocis.2014.08.001]

[95] Schmitt C, Turgeon SL. Protein/polysaccharide complexes and coacervates in food systems. Adv Colloid Interface Sci 2011; 167(1-2): 63-70. [http://dx.doi.org/10.1016/j.cis.2010.10.001] [PMID: 21056401]

[96] Elzoghby AO, El-Fotoh WS, Elgindy NA. Casein-based formulations as promising controlled release drug delivery systems. J Control Release 2011; 153(3): 206-16. [http://dx.doi.org/10.1016/j.jconrel.2011.02.010] [PMID: 21338636]

[97] Haham M, Ish-Shalom S, Nodelman M, *et al.* Stability and bioavailability of vitamin D nanoencapsulated in casein micelles. Food Funct 2012; 3(7): 737-44. [http://dx.doi.org/10.1039/c2fo10249h] [PMID: 22569895]

[98] Penalva R, Esparza I, Agüeros M, Gonzalez-Navarro CJ, Gonzalez-Ferrero C, Irache JM. Casein nanoparticles as carriers for the oral delivery of folic acid. Food Hydrocoll 2015; 44: 399-406. [http://dx.doi.org/10.1016/j.foodhyd.2014.10.004]

[99] Chen Q, McGillivray D, Wen J, Zhong F, Quek SY. Co-encapsulation of fish oil with phytosterol esters and limonene by milk proteins. J Food Eng 2013; 117: 505-12. [http://dx.doi.org/10.1016/j.jfoodeng.2013.01.011]

[100] Lin C-C, Lin H-Y, Chen H-C, Yu M-W, Lee M-H. Stability and characterisation of phospholipid-based curcumin-encapsulated microemulsions. Food Chem 2009; 116: 923-8. [http://dx.doi.org/10.1016/j.foodchem.2009.03.052]

[101] Patil YP, Jadhav S. Novel methods for liposome preparation. Chem Phys Lipids 2014; 177: 8-18. [http://dx.doi.org/10.1016/j.chemphyslip.2013.10.011] [PMID: 24220497]

[102] Eloy JO, Claro de Souza M, Petrilli R, Barcellos JP, Lee RJ, Marchetti JM. Liposomes as carriers of hydrophilic small molecule drugs: strategies to enhance encapsulation and delivery. Colloids Surf B Biointerfaces 2014; 123: 345-63. [http://dx.doi.org/10.1016/j.colsurfb.2014.09.029] [PMID: 25280609]

[103] Gibis M, Thellmann K, Thongkaew C, Weiss J. Interaction of polyphenols and multilayered liposomal-encapsulated grape seed extract with native and heat-treated proteins. Food Hydrocoll 2014; 41: 119-31. [http://dx.doi.org/10.1016/j.foodhyd.2014.03.024]

[104] Malheiros PS, Daroit DJ, Brandelli A. Food applications of liposome- encapsulated antimicrobial Peptides. Trends Food Sci Technol 2010; 21: 284-92.

[http://dx.doi.org/10.1016/j.tifs.2010.03.003]

[105] Kumarn GP, Rajeshwarrao P. Nonionic surfactant vesicular systems for effective drug delivery—an overview. Acta Pharm Sin B 2011; 1: 208-19. [http://dx.doi.org/10.1016/j.apsb.2011.09.002]

[106] Mozafari MR, Johnson C, Hatziantoniou S, Demetzos C. Nanoliposomes and their applications in food nanotechnology. J Liposome Res 2008; 18(4): 309-27. [http://dx.doi.org/10.1080/08982100802465941] [PMID: 18951288]

[107] Pando D, Beltrán M, Gerone I, Matos M, Pazos C. Resveratrol entrapped niosomes as yoghurt additive. Food Chem 2015; 170: 281-7. [http://dx.doi.org/10.1016/j.foodchem.2014.08.082] [PMID: 25306346]

[108] Marsanasco M, Márquez AL, Wagner JR. Silvia del V. Alonso, S.V.; Chiaramoni, N.S., Liposomes as vehicles for vitamins E and C: An alternative to fortify orange juice and offer vitamin C protection after heat treatment. Food Res Int 2011; 44: 3039-46. [http://dx.doi.org/10.1016/j.foodres.2011.07.025]

[109] Toniazzo T, Berbel IF, Cho S, Favaro-Trindade CS, Moraes IC, Pinho SC. β-carotene-loaded liposome dispersions stabilized with xanthan and guar gums: Physico-chemical stability and feasibility of application in yogurt. LWT -. Food Sci Technol (Campinas) 2014; 59: 1265-73.

[110] Tan C, Zhang Y, Abbas S, Feng B, Zhang X, Xia S. Modulation of the carotenoid bioaccessibility through liposomal encapsulation. Colloids Surf B Biointerfaces 2014; 123: 692-700. [http://dx.doi.org/10.1016/j.colsurfb.2014.10.011] [PMID: 25456993]

[111] Sebaaly C, Jraij A, Fessi H, Charcosset C, Greige-Gerges H. Preparation and characterization of clove essential oil-loaded liposomes. Food Chem 2015; 178: 52-62. [http://dx.doi.org/10.1016/j.foodchem.2015.01.067] [PMID: 25704683]

[112] Rashidinejad A, Birch EJ, Sun-Waterhouse D, Everett DW. Delivery of green tea catechin and epigallocatechin gallate in liposomes incorporated into low-fat hard cheese. Food Chem 2014; 156: 176-83. [http://dx.doi.org/10.1016/j.foodchem.2014.01.115] [PMID: 24629955]

[113] Singh A, Thompson A, Corredig M. Encapsulation technologies and delivery systems for food ingredients and nutraceuticals. Cambridge: Woodhead Publishing Limited 2012; pp. 287-318. [http://dx.doi.org/10.1533/9780857095909.3.287]

[114] Luciani P, Berti D, Baglioni P. Colloids in Drug Delivery. London: Taylor & Francis Group, LLC 2010; pp. 395-408. [http://dx.doi.org/10.1201/9781439818268-c17]

[115] Vilanova N, Solans C, Vitamin A. Vitamin A Palmitate-β-cyclodextrin inclusion complexes: characterization, protection and emulsification properties. Food Chem 2015; 175: 529-35. [http://dx.doi.org/10.1016/j.foodchem.2014.12.015] [PMID: 25577116]

[116] Pinho E, Grootveld M, Soares G, Henriques M. Cyclodextrins as encapsulation agents for plant bioactive compounds. Carbohydr Polym 2014; 101: 121-35. [http://dx.doi.org/10.1016/j.carbpol.2013.08.078] [PMID: 24299757]

[117] López-Nicolás JM, Rodríguez-Bonilla P, García-Carmona F. Cyclodextrins and antioxidants. Crit Rev

Food Sci Nutr 2014; 54(2): 251-76.
[http://dx.doi.org/10.1080/10408398.2011.582544] [PMID: 24188271]

[118] Astray G, Gonzales-Barreiro C, Mejuto JC, Rial-Otero R, Simal-Gadara J. A review on use of cyclodextrins in foods. Food Hydrocoll 2009; 23: 1631-40.
[http://dx.doi.org/10.1016/j.foodhyd.2009.01.001]

[119] López-Tobar E, Blanch GP, Ruiz del Castillo ML, Sanchez-Cortes S. Encapsulation and isomerization of curcumin with cyclodextrins characterized by electronic and vibrational spectroscopy. Vib Spectrosc 2012; 62: 292-8.
[http://dx.doi.org/10.1016/j.vibspec.2012.06.008]

[120] Cabral HM. A review on cyclodextrin encapsulation of essential oils and volatiles. Flavour Fragrance J 2010; 25: 313-26.
[http://dx.doi.org/10.1002/ffj.2019]

[121] Summerlin N, Soo E, Thakur S, Qu Z, Jambhrunkar S, Popat A. Resveratrol nanoformulations: challenges and opportunities. Int J Pharm 2015; 479(2): 282-90.
[http://dx.doi.org/10.1016/j.ijpharm.2015.01.003] [PMID: 25572692]

[122] Cevallos PA, Buera MP, Elizalde BE. Encapsulation of cinnamon and thyme essential oils components (cinnamaldehyde and thymol) in b-cyclodextrin: Effect of interactions with water on complex stability. J Food Eng 2010; 99: 70-5.
[http://dx.doi.org/10.1016/j.jfoodeng.2010.01.039]

[123] Wang J, Cao Y, Sun B, Wang C. C., Physicochemical and release characterisation of garlic oil-β-cyclodextrin inclusion complexes. Food Chem 2011; 127: 1680-5.
[http://dx.doi.org/10.1016/j.foodchem.2011.02.036]

[124] Hill LE, Gomes C, Taylor TM. Characterization of beta-cyclodextrin inclusion complexes containing essential oils (trans-cinnamaldehyde, eugenol, cinnamon bark, and clove bud extracts) for antimicrobial delivery applications. Food Sci Technol (Campinas) 2013; 51: 86-93.

[125] Martínez-Alonso A, Losada-Barreiro S, Bravo-Díaz C. Encapsulation and solubilization of the antioxidants gallic acid and ethyl, propyl and butyl gallate with β-cyclodextrin. J Mol Liq 2015; 210: 143-50.
[http://dx.doi.org/10.1016/j.molliq.2014.12.016]

[126] Zhang Z, Zhang R, Chen L, Tong Q, McClements DJ. Designing hydrogel particles for controlled or targeted release of lipophilic bioactive agents in the gastrointestinal tract. Eur Polym J 2015; 72: 698-716.
[http://dx.doi.org/10.1016/j.eurpolymj.2015.01.013]

[127] Donati I, Paoletti S. Alginates: biology and applications. Berlin, Heidelberg: Springer- Verlag 2009; pp. 1-54.
[http://dx.doi.org/10.1007/978-3-540-92679-5_1]

[128] Yin Y, Li J. Chitosan-based hydrogels Functions and applications, kangde yao, k. Boca Raton, London: Taylor & Francis Group, LLC 2012; pp. 179-234.

[129] Rinaudo M. Chitosan-Based Systems for Biopharmaceuticals Delivery, Targeting and Polymer Therapeutics, Saramento, B. Chichester, UK: John Wiley & Sons, Ltd 2012; pp. 23-44.

[http://dx.doi.org/10.1002/9781119962977.ch2]

[130] Hezaveh H, Muhamad II. Modification and swelling kinetic study of kappa-carrageenan-based hydrogel for controlled release study. J Taiwan Inst Chem Eng 2013; 44: 182-91. [http://dx.doi.org/10.1016/j.jtice.2012.10.011]

[131] Jones OG, McClements DJ. Recent progress in biopolymer nanoparticle and microparticle formation by heat-treating electrostatic protein-polysaccharide complexes. Adv Colloid Interface Sci 2011; 167(1-2): 49-62.
[http://dx.doi.org/10.1016/j.cis.2010.10.006] [PMID: 21094486]

[132] Campo VL, Kawano DF, Braz da Silva D Jr, Carvalho I. Carrageenans: Biological properties, chemical modifications and structural analysis – A review. Carbohydr Polym 2009; 77: 167-80. [http://dx.doi.org/10.1016/j.carbpol.2009.01.020]

[133] Ziani K, Fang Y, McClements DJ. Encapsulation of functional lipophilic components in surfactant-based colloidal delivery systems: vitamin E, vitamin D, and lemon oil. Food Chem 2012; 134(2): 1106-12.
[http://dx.doi.org/10.1016/j.foodchem.2012.03.027] [PMID: 23107734]

[134] Hackley VH, Clogston JD. In: Characterization of Nanoparticles Intended for Drug Delivery, Scott E. McNeil, S.E., Ed.; Springer Science+Business Media, LLC: Berlin, 2011; pp. 35-52.

[135] Yaghmur A, de Campo L, Glatter O. Microemulsions Properties and Applications. Boca Raton, London, New York: Taylor & Francis Group 2009; pp. 185-202.

[136] Cotirlet M. Dima, Ş.; Bahrim, G., Psychrotrophic *Streptomyces* spp. cells immobilization in alginate microspheres produced by emulsification-internal gelation. J Microencapsul 2014; 31(1): 93-. [http://dx.doi.org/10.319/02652048.2013.808279]

SUBJECT INDEX

A

C

D

E

F

G

H

I

K

L

M

N

Q

R

S

Made in the USA
Las Vegas, NV
04 November 2026

57425981R00193